Fundamentals of Thermal and Nuclear Power Generation

JSME Series in Thermal and Nuclear Power Generation

Fundamentals of Thermal and Nuclear Power Generation

Edited by

Yasuo Koizumi

Graduate School of Information and Engineering, The University of Electro-Communications, Chofu, Tokyo, Japan

Tomio Okawa

The University of Electro-Communications, Chofu, Tokyo, Japan

Shoji Mori

Kyushu University, Fukuoka, Fukuoka, Japan

ELSEVIER

Elsevier
Radarweg 29, PO Box 211, 1000 AE Amsterdam, Netherlands
The Boulevard, Langford Lane, Kidlington, Oxford OX5 1GB, United Kingdom
50 Hampshire Street, 5th Floor, Cambridge, MA 02139, United States

British Library Cataloguing-in-Publication Data
A catalogue record for this book is available from the British Library

Library of Congress Cataloging-in-Publication Data
A catalog record for this book is available from the Library of Congress

ISBN: 978-0-12-820733-8

Publisher: Candice Janco
Acquisitions Editor: Maria Convey
Editorial Project Manager: Sara Valentino
Production Project Manager: Kamesh Ramajogi
Cover Designer: Alan Studholme

Typeset by MPS Limited, Chennai, India

Contents

List of contributors

Hitoshi Asano Kobe University, Kobe, Hyogo, Japan

Motonari Haraguchi Mitsubishi Power, Ltd., Yokohama, Japan

Daisuke Ito Kyoto University, Kumatori, Osaka, Japan

Kei Ito Kyoto University, Kumatori, Osaka, Japan

Yasuo Koizumi Graduate School of Information and Engineering, The University of Electro-Communications, Chofu, Tokyo, Japan

Ryosuke Matsumoto Kansai University, Suita, Osaka, Japan

Hiroshi Miyano Hosei University, Chiyoda, Tokyo, Japan

Shoji Mori Kyushu University, Fukuoka, Fukuoka, Japan

Takao Nakagaki Waseda University, Shinjuku, Tokyo, Japan

Takao Nakamura Osaka University, Suita, Osaka, Japan

Tomio Okawa The University of Electro-Communications, Chofu, Tokyo, Japan

Mamoru Ozawa Kansai University, Takatsuki, Osaka, Japan

Cheol Ho Pyeon Kyoto University, Kumatori, Osaka, Japan

Yasushi Saito Kyoto University, Kumatori, Osaka, Japan

Hisashi Umekawa Kansai University, Suita, Osaka, Japan

About the authors

Hitoshi Asano is a professor of Department of Mechanical Engineering, Kobe University, Japan. He graduated from Kobe University in 1990. He started research on two-phase flow dynamics as a research associate of Kobe University after 3.5 years working for Daikin Industries, Ltd. He received his doctoral degree from Kobe University in 2000, and was promoted to associate professor in 2001. From 2001 to 2002 he visited Stuttgart University, Germany, as a research fellow of the Alexander von Humboldt foundation. His interests is focused on thermofluid dynamics in compact heat exchangers for HVAC systems, boiling heat transfer enhancement, gas-liquid two-phase flows in power systems, especially the effect of surface tension and gravity on gas-liquid two-phase flows in small diameter tubes or under microgravity.

Motonari Haraguchi earned a bachelor's degree in mechanical engineering from University of Tokyo and started his career in Hitachi Ltd. in 1976 and was transferred to Mitsubishi Hitachi Power Systems (MHPS) in 2014. Altogether, he has been engaged in steam turbine design work for about 45 years with various fields such as turbine stage thermodynamic design, blade design, and new technology development. And he earned a Master of Science in mechanical engineering from University of California Berkeley in 1984. He is the cowriter of *Steam Turbine* in Japanese and *Advances in Steam Turbines for Modern Power Plants*.

Daisuke Ito received a BE (2005) degree in mechanical engineering and ME (2006) and DE (2009) degrees in nuclear engineering from Tokyo Institute of Technology. He joined Nuclear Energy Laboratory at ETH Zurich as a postdoctoral researcher in 2010 and researched nuclear thermal hydraulics and advanced two-phase flow measurement technique. He moved to Institute for Integrated Radiation and Nuclear Science, Kyoto University as an Assistant Professor in 2012. His current interests include multiphase flow dynamics, flow measurement instrumentations, and neutron transmission imaging.

Kei Ito is an associate professor of Institute for Integrated Radiation and Nuclear Science, Kyoto University. He received BE (1999) and ME (2001) degrees in mechanical engineering from Waseda University, and DE (2009) degree in nuclear engineering from Kyoto University. He belonged Japan Atomic Energy Agency from 2001 to 2015 as a research engineer for the study on various multiphase flows in a sodium-cooled fast reactor. His current research interests also include the numerical simulations of various multiphase flows, especially high-precision simulations of gas-liquid two-phase flows with an interface-tracking model.

Yasuo Koizumi is a research promotor at the University of Electro-Communications. He had been an invited researcher of the Japan Atomic Energy Agency for 5 years before now. He received his PhD degree from the University of Tokyo in 1977. He started his research career at the Japan Atomic Energy Research Institute in 1977 as a research engineer for nuclear reactor safety. He stayed at the Idaho National Engineering Laboratory from 1981 through 1983. He moved to the Department of Mechanical Engineering of Kogakuin University in 1989. Then, he moved to the Department of Functional Machinery and Mechanics of Shinshu University in 2008. He retired as professor in 2014 and he had been in the Japan Atomic Energy Agency since then. His research is focused in the areas of pool and flow boiling, critical heat flux, condensation heat transfer, and two-phase flow. He is also interested in heat transfer and fluid flow on the microscale. Since his research field is closely related to energy systems, he has great interest in thermal and nuclear power stations and energy supply in society.

Ryosuke Matsumoto has been a professor at Kansai University since 2016. He received BE (1994) and ME (1994) degrees and his doctoral degree from Doshisha University in 2001. He was a research fellow of the Alexander von Humboldt Foundation in 2002. His research interests include heat and mass transfer in a microscale, combustion system, and frost formation in refrigerating and heat pump systems. He received the Hartnett-Irvine Award from the International Center for Heat and Mass Transfer (ICHMT) in 2016.

Hiroshi Miyano started his career in Toshiba in 1971. He was engaged in R&D of BWR systems in the laboratory, and subsequently in primary system design and maintenance work as a leader of the system design department, after which he was appointed as Chief Nuclear Engineer of Toshiba. After that, he became a director of a group company. Throughout his career at Toshiba, he has contributed to the society of scientists and engineers. At the Japan Society of Mechanical Engineers (JSME) and the Japan Atomic Energy Society (AESJ), he was the member of committee on the development of codes and standards, and at the Japan Atomic Energy Society, he was chairman of the standards committee. Currently, he is chairman of the Fukushima Decommissioning Committee. Meanwhile, he was awarded the "Fellow" title by JSME and AESJ. In 2000 he joined the Harvard Business School, Advanced Management Program AMP-159, and completed his career. While working, he graduated from the doctoral program at Hokkaido University and received his doctorate in 2012. He taught at Hosei University and retired in 2019.

Shoji Mori is a Professor of Mechanical Engineering at Kyushu University, Japan. He received his PhD degree in Engineering from Kyushu University in 2003. He joined the Department of Chemical Engineering at Yokohama National University as an assistant professor in 2004, and he became an associate professor in 2007.

He has been a faculty member at Kyushu University since 2019. From 2009 to 2011, he worked on cryopreservation and thermal therapies at Bioheat and Mass Transfer Laboratory, Department of Mechanical Engineering, University of Minnesota, Minneapolis, as a visiting professor. His research interests are currently focusing on the enhancement of boiling critical heat flux and quenching using porous materials and two-phase annular flow.

Takao Nakagaki is a professor of Department of Modern Mechanical Engineering in Waseda University. He joined Toshiba Corporation in 1992 and worked for developing advanced power generation systems such as Molten Carbonate Fuel Cell and Chemically Recuperated Gas Turbine for 15 years. He received his doctoral degree from Waseda University in 2004. He has been involved in many CO2 capture, utilization, and storage (CCUS) R&D programs for 20 years. He is a member of technical advisory committee of Tomakomai CCS project operated by Japan CCS Co., Ltd., and advisory committee on science and technology of Research Institute of Innovative Technology for the Earth (RITE).

Takao Nakamura joined the Kansai Electric Power company in 1972 and transferred to the Department of Sustainable Energy and Environment at Osaka University in 2010 after over thirty years' job experience of construction and management work of PWR-type nuclear power plant in Japan. He also joined the code committee of American Society of Mechanical Engineers as committee member since 1997 to 2010. He received his PhD degree from the University of Tokyo in 2008. He retired Osaka University in 2015 and joined the Center for Open Innovation on Research and Education at Osaka University and works as research professor. He has also worked as Technical Advisor at Japan Nuclear Safety Institute since 2015.

Tomio Okawa started his research career at the Central Research Institute of the Electric Power Industry in 1990 and earned a doctor of engineering degree from the Tokyo Institute of Technology in 1995 after receiving bachelor's and master's degrees from Tokyo Institute of Technology. Then, he moved to the Department of Mechanical Engineering at Osaka University in 1999, and to his present position (The University of Electro-Communications, Tokyo, Japan) in 2011. His main research areas are gas-liquid two-phase flow and heat transfer with phase change. At present, he is doing research on subcooled flow boiling, departure from nucleate boiling, boiling heat transfer of nanofluid, application of nanoparticle layer to cooling of electronic devices, splashing during drop and liquid jet impingements on a liquid film, and thermal hydraulics encountered in freeze plug used in molten salt reactors.

Mamoru Ozawa is a professor emeritus and a senior researcher of Research Center for Societal Safety Sciences, Kansai University. He received his doctoral degree from Osaka University in 1977. His research interests include boiling heat transfer, two-phase flow dynamics, combustion system, boiler, and safety in industrial systems. He has authored and coauthored more than 150 refereed journals, more than 150 conference papers, and several handbooks and many books related to thermal hydraulics in steam power. His edited book, *Science of Societal Safety*, has been downloaded more than 50,000 times. He served Board of Directors and President of Heat Transfer Society of Japan and Chairman of Power and Energy System Division of Japan Society of Mechanical Engineers (JSME). He was awarded several times from Visualization Society of Japan, Heat Transfer Society of Japan, and JSME.

Cheol Ho Pyeon received a BE (1995) and ME (1997) in Nuclear Engineering, Nagoya University, and a PhD (2000) degree in Energy Science, Kyoto University. He joined Division of Neutron Standard at National Institute of Advanced Industry of Science and Technology in Japan as a postdoctoral researcher, focusing on neutron standard calibration of neutron detectors and pulsed-neutron generators. He moved to Institute for Integrated Radiation and Nuclear Science, Kyoto University as Assistant Professor in 2002, and Associate Professor now. His current interests include experimental analyses of statics and kinetics parameters in reactor physics, and nuclear experiment education for young generation, with the use of experimental facility, such as Kyoto University Critical Assembly.

Yasushi Saito has been a professor of Institute for Integrated Radiation and Nuclear Science, Kyoto University since 2012. He received his doctoral degree from the Department of Chemical Engineering at Kyoto University in 1998. Then he joined present institute, where he has been studying mainly multiphase flows in relation to nuclear thermal hydraulics, including severe accident, and also development of measurement technique for multiphase flow research. He stayed at Universitaet Karlsruhe (TH) as a research fellow of the Alexander von Humboldt Foundation from 2006 to 2008. His main research interests are nuclear thermal hydraulics, refrigeration system, and neutron imaging by using research reactors and accelerator-based neutron sources.

Hisashi Umekawa has been a professor of Kansai University since 2009. He received his PhD degree from Kansai University in 1998, and also received a master's degree from Kansai University in 1988. He started his career at Daikin industries, Ltd. in 1988, and then moved to the Department of Mechanical Engineering of Kansai University in 1991. He stayed at the Institute for Nuclear and Energy Technologies (IKET) in Research Center of Karlsruhe (FZK) from 1999 through 2000, supported by the Alexander von Humboldt Research Fellowship. His research is focused in the areas of flow boiling, critical heat flux, and two-phase flow. Since his research field is closely related to energy systems, he has great interest in thermal and nuclear power stations and energy supply in society. He served Chairman of Power and Energy System Division of Japan Society of Mechanical Engineers (JSME).

Preface of JSME Series in Thermal and Nuclear Power Generation

Electric power supply is a fundamental and principal infrastructure for modern society. Modern society uses power generation through heat. This series of books consists of eight volumes describing thermal and nuclear power generation, taking Japan as the example, and referring the other countries. The Volume 1 discusses how power supply is attained historically, focusing on the thermal and the nuclear power generation along with minimum-required scientific and technological fundamentals to understand this series of books. Then, the present status of the thermal and the nuclear power generation technique is displayed in detail in Volumes 2 through 8.

The rehabilitation and reconstruction of Japan after World War II was initiated through the utilization of a large amount of coal for boilers of the thermal power plants. Meanwhile, environmental pollution caused by coal combustion became serious, and then oil was introduced to the boilers. Due to two worldwide oil crises and because of carbon dioxide issues, natural gas has also begun to be used for boilers. Current thermal power generation in Japan is based on coal and gas utilization. As a result of enough power supply, Japan has become one of the leading countries economically and technologically in the world. The thermal power technology that started from introducing technology from abroad has been transfigured Japan into one of the most advanced in the world through the research and development of Japanese industry, government, and academia during this process.

Global warming related to excess carbon dioxide emissions has become a worldwide issue in recent years. Reducing carbon dioxide emission in thermal power generation is important to help cope with this issue. One direction is to change the fuel of a boiler from coal to gas that exhausts less carbon dioxide. Another important direction is to endeavor to enhance the thermal efficiency of coal thermal power plants as well as oil and gas. Many developing countries in the world need more thermal power plants in future. Although oil and/or gas thermal power plants may be introduced into these countries, it is supposed that coal thermal power plants will still be used due to economical reasons. Considering these situations, the publication of this series of books that displays and explains the developing history and the present status of the most advanced thermal power plants in Japan and other advanced countries is a timely planning for engineers and researchers in the advanced countries to pursue the further advancement and for engineers and researchers in developing countries to learn and acquire this knowledge.

Nuclear power generation technology in Japan started after being introduced from abroad approximately 60 years ago. Then, it reached the matured nuclear power technology through untiring endeavors for research and development. However, nuclear power plants at the Fukushima Daiichi nuclear power station were heavily damaged by huge tsunamis caused by the Great East Japan Earthquake in 2011 to result in contamination in the large area around the power station. Taking measures to improve nuclear power reactors to be more robustly is currently underway by analyzing the factors that caused this serious situation. Technical vulnerability can be solved by technology. Nuclear power generation technology is one of the definite promising technologies that should be used in the future. The nuclear power generation is still expected as one of the main ways to supply electricity in the framework of the basic energy plan of Japan as well as thermal power generation. It implies that the construction of new power reactors will be required to replace the nuclear reactors that will reach their useful lifetime. Looking overseas, many developing countries are introducing nuclear power generation technology as a safe and economically excellent way to obtain electricity. Transfer of the nuclear power generation technology developed and matured in Japan to those countries is naturally the obligation of Japan. In these situations, the necessity of human resource development in the field of nuclear power generation technology in the developing countries, as well as in Japan, is beyond dispute. Thus, it is an urgent task to summarize the nuclear power generation technology acquired by Japan to provide it.

The Power and Energy Systems Division (PESD) of the Japan Society of Mechanical Engineers is celebrating its 30th anniversary from establishment in 1990. This department is entrusted with handling power supply technology in mechanical engineering. Responding to the earlier-mentioned is truly requested. This task cannot be done by others but the PESD that is composed of leading engineers and researchers in this field in Japan. In view of these circumstances, summarizing Japan's and other countries' power generation technology and disseminating it not only in Japan but also overseas seems significantly important. So, it has been decided to execute this book series, publishing as one of the 30th anniversary events of the PESD.

Authors of this book series are those who have engaged in the most advanced research and development for the thermal power and nuclear power generation in Japan and Canada. Their experience and knowledge is reflected in their writing. It is not an introduction of what others did, but living knowledge based on their own experiences and thoughts are described. We hope that this series of books becomes learning material that is not yet in existence in this field. We hope that readers acquire a way of thinking as well as whole and detailed knowledge by having this book series in hand.

This series is the joint effort of many individuals, generously sharing and writing from their expertise. Their efforts are deeply appreciated. We are very thankful for the unbiased and heartful comments given from many reviewers to make this series superb. Special thanks should be given to Maria Convey and Sara Valentino of the editorial staff at Elsevier.

Editors in Chief

Yasuo Koizumi
Graduate School of Information and Engineering,
The University of Electro-Communications, Chofu, Tokyo, Japan

Mamoru Ozawa
Kansai University, Takatsuki, Osaka, Japan

March 16, 2020

Preface to Volume 1: Fundamentals of Thermal and Nuclear Power Generation

The importance of supplying power to society and maintaining human civilization is first described. Then, the history of when human beings acquired power that originates from heat is described. Next, an outline of the modern thermal and nuclear power generation are described, focusing on electric power generation. Following it, the scientific-technological fundamentals required to at least understand what will be presented in Volumes 2 through 8, which provide details of those power generation technologies; the fundamentals of thermodynamics, hydrodynamics, thermal engineering, combustion engineering, and nuclear physics. Finally, the present situation of power supply and current issues, such as carbon dioxide emissions that cause global warming, various accident risks, and others, are discussed.

The targets of this volume are postgraduate course students, researchers, and engineers in the field of mechanical engineering, nuclear engineering, and chemical engineering. The areas are not only in Japan but also the United States, Canada, China, and other countries.

We wish that readers acquire a way of thinking as well as whole and detailed knowledge in this volume. It is our great pleasure if readers enhance their talent and expand their knowledge.

Editors

Yasuo Koizumi[1], Tomio Okawa[1] and Shoji Mori[2]
[1]Graduate School of Information and Engineering, The University of Electro-Communications, Chofu, Tokyo, Japan,
[2]Kyushu University, Fukuoka, Fukuoka, Japan
March 16, 2020

Dawn of power for human beings/power from steam

1

Yasuo Koizumi
Graduate School of Information and Engineering, The University of Electro-Communications, Chofu, Tokyo, Japan

1.1 Civilization progress and energy

No one can argue against the fact that energy is essential for human beings. However, it is not noticed in daily life since the energy is easily acquired. Even so, we are forced to one day notice the importance of energy. Water service stops; an air conditioner does not work; or a candle is needed for light at night if electricity service stops because of some accident at an electric power station. Electricity is one form of energy. If the oil that produces the energy to make a car work is lost, the car does not work and we are forced to walk. Energy supply to the society is fundamental in our life.

It is written in the *Epic of Gilgamesh* that a big forest existed in the Mesopotamian in ancient times. The forest there has disappeared. This change might be caused by human activity. It is pointed out that a similar thing happened in Japan [1]. The following is what is written in Reference [1]. Although it is one view of history, it is convincing in considering the role of energy for a human being.

It is widely understood that ancient Japanese civilization began in the Nara Basin around the 3rd century although there are various opinions. The ancient civilized Japanese nation was founded there, and the capital of the nation was built. The area was surrounded by mountains that served as defense barriers for the attack of invaders. The land was blessed with clean water and flat fields and surrounded by wooded forests in the mountains. The wood from the forest was utilized as building materials of houses and firewood for heating, cooking, and lighting. Trees were cut down continuously for a long time, and then the wooded forest had disappeared. The activity resulted in the extinction of the forest. People there lost the energy source of wood from the forests. The extinction also led to repeat the natural disasters such as flood and the spread of plague. Finally, people at that time decided to leave the Nara Basin and relocate their capital to Kyoto at the end of the 8th century. Kyoto was surrounded by mountains and also by the wooded forests such as the former Nara Basin. Considering this historical event from the standpoint of an engineer, the capital relocation has one side that it was motivated by searching for wooded forests—enough woods as a source of energy. It implies that the capital relocation was caused by the forest energy collapse. The forest energy collapse seems similar to what happened in ancient Mesopotamia.

The history of the capital relocation in Japan was repeated again. Trees around Kyoto and then near Kyoto were cut down for construction and for firewood to get energy. Deep forests disappeared in the Kyoto area, which resulted in the forest

Fundamentals of Thermal and Nuclear Power Generation. DOI: https://doi.org/10.1016/B978-0-12-820733-8.00001-7

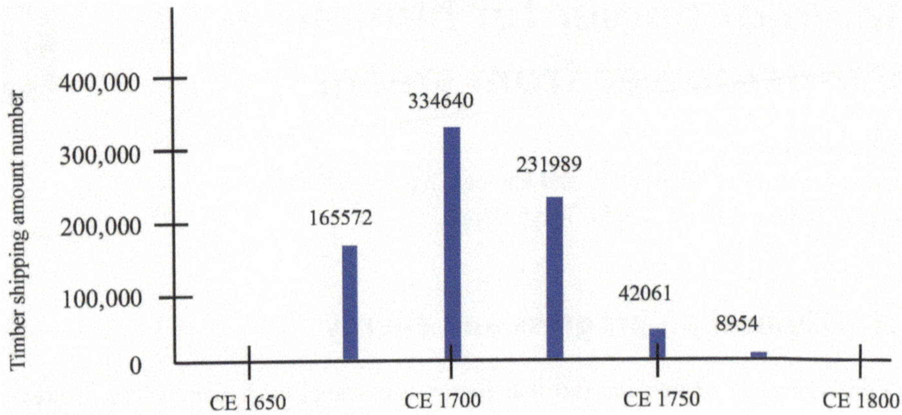

Figure 1.1 Timber Shipping Transition from Tenryu river region in Japan.
Source: Reproduced from K. Takemura, Japanese culture and energy, J. Jpn. Soc. Energy Resour. 40 (3) (2019) 221–224 [1].

energy collapse again. Then, the capital of Japan was relocated to a new frontier, Edo, which is now Tokyo, at the beginning of the 17th century.

Edo became a big city having about 1.5 million residents. Enough wood for construction and energy were provided from surrounding area to Edo. The construction of the capital Edo began around the early 17th century. At the early stage of the Edo construction, the wood was supplied from a near area, and then shifted slowly to farther area with deforestation. Fig. 1.1 expresses the timber shipping amount variation in the Tenryu river region at about 250 km east from Edo. After a while from the Edo construction beginning early in the beginning of the 17th century, the timber logging there increased drastically; however, it began to decrease. It suggests that the deforestation expanded from the area near Edo to a farther area. Edo almost faced the forest energy collapse again. Fortunately enough, Japan succeeded in the renovation of the government system via the civil war in 1867. The new Japanese government positively introduced modern technology such as energy production using steam. It was a blessing of God that plenty of coal was available in Japan. The energy source was changed from wood to coal, which resulted in breaking away from wood energy. This change led to Japan's drastic growth. This story is the proof of the importance of the energy for civilization.

1.2 Historical significance of getting force from fire for human beings

Many moments and/or events in history have led to the foundation of modern civilization. If it is asked to say what are the most important among them, one answer is that humankind

1. came to walk bipedally,
2. came to use fire,
3. acquired the technology to create and use great force, and
4. had technology to get force from fire in hand.

Hands were released from walking to be used for handling utensils. A head was raised to lead to the evolution of brains. Those worked together, and many useful techniques for life had been developed. Fire protected humankind from animal attack and relieved from cold. It was used for cooking, which led to improve the circumstances of food and health. Fire greatly contributed to improving the foundation of life, however, it was only limited in those. Human beings acquired evolved intelligence, and enhanced health and life support came to form social groups and get technical skill to move heavy things such as huge logs and rocks. Stout houses were built using the heavy logs and rocks. Huge tombs and sanctuaries made of huge rocks were constructed using the technology. It is the proof that many ancient civilizations such as ancient Egyptian, Mesopotamian, and Mexican civilizations had left archeological megalithic monuments in common. The skill to move heavy things and intelligence accumulation came to develop large machines such as sailing ships, spinning machines, and so on. However, the source to work the machines was limited to humankind itself, domestic animals, and even wind and water flow. Quite suddenly, humankind noticed in the 18th century that fire creates great force and succeeded in getting force from fire. The force originated from the fire opened the door to the modern industrial civilization. Humankind had come to get huge force easily and to operate large machines. For example, people do not need to walk because a car conveys them; they do not need to wash manually using only their hands; and they can travel to the moon by using newly developed machines. This suggests that getting the skill to get force from fire is quite significant in history.

1.3 Power

What is power? A simple answer to the question is that power is the source that makes a machine or a tool that provides a substitute for human hard labor. If power is not provided for an electric washing machine, it is only a box. If power is provided to the washing machine, it relieves people from hard manual washing. In this case, power is the electric motor. If a horse is connected to a wagon, people are relived from carrying heavy things on their bags. In this case, the power is the horse. The power source is activated by the energy such as electricity to the motor and food for the horse, respectively. Thus, the power has the source in energy.

Let us consider the situation as shown in Fig. 1.2. Some load is on the ground. The gravity force is acting on the load. The body does not move, thus no work is done physically. If the load is connected to a horse with a rope and the horse pulls the rope as shown in Fig. 1.3, the load is carried by the horse. Sharing force comes out between the bottom of the load and the ground. The horse pulls the load against the sharing horse with pulling force that is equal to the sharing force. The horse is doing work physically as a result of the pulling force and moving. The hard work of a human to pull the load on the ground is substituted by the work of the horse that the horse pulls the body and carries the load. Thus, the power is close to the work in the physical meaning. The work in the washing machine is that the motor rotates and stirs water in the washing machine.

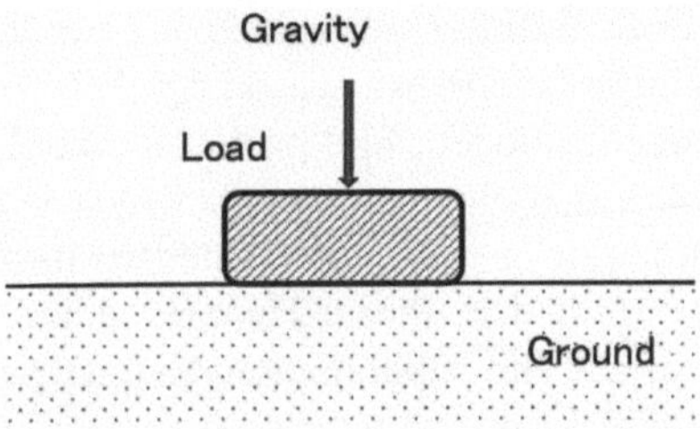

Figure 1.2 Just force acting.

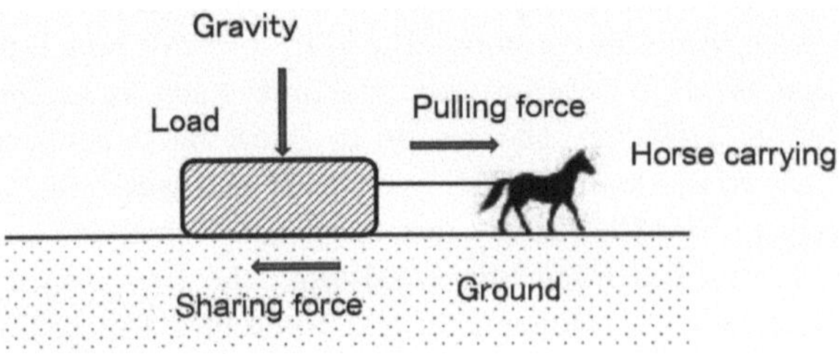

Figure 1.3 Force resulting in work.

Time that the work is performed for is important. How long the motor stirs water in the washing machine or how long it takes to move the load by the horse are important in the aforementioned examples and have physical meaning. It makes sense to see how efficiently the work is done. If the work is efficient, the force is used efficiently. It may be expressed by a work rate. If the work rate is high, the force is used efficiently. This is the definition of the power in the physical meaning. If the heavy item is carried efficiently, the power is high. Thus, the power is equal to the work rate and the unit of the power is Nm/s = J/s; W. The power that a human being is enjoying in the present civilization is generated from heat energy originally and used in other forms such as stirring water in the washing machine, running trains, and so on. The power generation systems originated from heat are the subjects of this book series.

1.4 History of getting power

Civilization has been seeking for power since humankind wanted ease. Power is the base for providing the ease. Fig. 1.4 illustrates the history of energy usage by humankind. In the figure, the energy consumption of oil equivalents per day in the world and the energy consumption by a person are included.

A person's energy was just for their own activities such as walking, running, and so on in the old days millions of years ago. The energy generated from food. Then, humankind started to use fire for heating and cooking hundreds of thousands of years ago and the energy usage for those was added to the food energy.

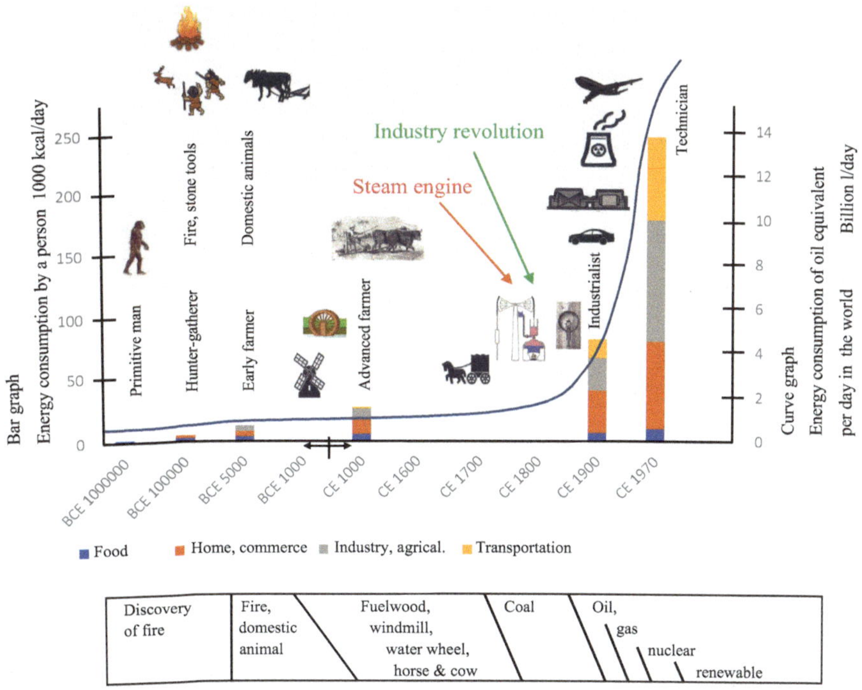

Figure 1.4 History of Power Production by Mankind.
Source: Reproduced from Japan Atomic Energy Relation Organization, 1-1-1 Relation of human being and energy. <https://www.ene100.jp/zumen/1-1-1>, 2017 [2].

As time proceeded, agriculture began around the 10th century BCE. Mankind cultivated itself in the early days, and then started to use domestic animals for farming to avoid hard work and to increase productivity around 5000 years ago. The energy from domestic animals was added to human energy usage. The energy usage by mankind increased gradually as time went on. Ancient societies and civilizations such as Mesopotamian and Egyptian civilizations came out in those days.

Then, humans learned to catch energy from natural wind and water flow as civilization progressed, and had developed mechanical devices and equipments such as sailing ships, windmills, water wheels, and so on. The daily life of humans became further convenient and easy by using those forces. Large sailing ships carried large cargoes between Asia and Europe and water wheels were used to operate spinning machines in the 18th century. The natural energy usage was added to the energy usage of humans. Firewood and even coal were used for heating and cooking. The energy used by humans further increased gradually, however, the increasing was still slow.

The increasing rate of human energy usage changed apparently and drastically around 1800. What happened in those days? People suddenly noticed that fire creates force and that force produces energy. Then, humans acquired ways to turn energy into power.

Let's consider the situation shown in Fig. 1.5. Steam produced by burning firewood or coal is filled in a can such as A. Then, water is injected in the can as B, resulting in steam condensation. The pressure P_i in the can becomes considerably lower than the atmospheric pressure P_a as shown in C. Thus, force resulting from the pressure difference between P_a and P_i acts on the surface of the can and the can may be crushed in some case. Now, suppose that this series of steam condensation in a container is performed in a cylinder-piston system as in D. When the steam in the cylinder is condensed and the pressure in the cylinder becomes P_i, the piston is pushed with the force.

$$F = (P_a - P_i) \times A. \tag{1.1}$$

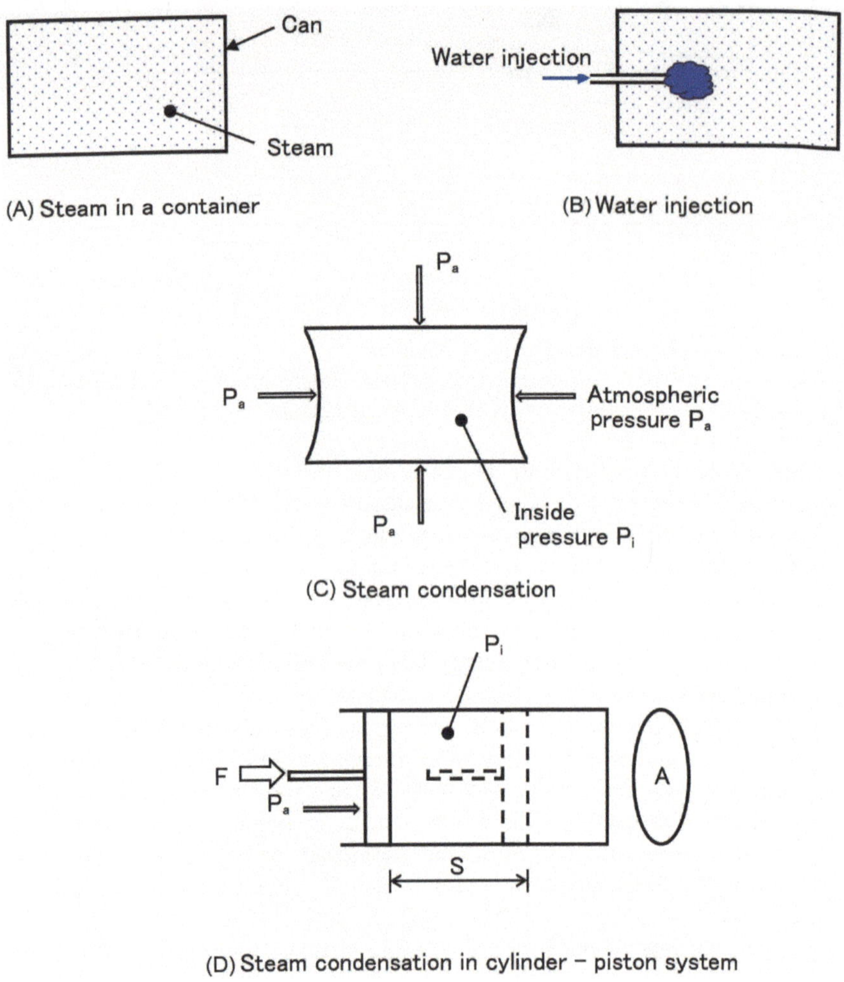

Figure 1.5 Powering from steam condensation. (A) Steam in a container (B) Water injection (C) Steam condensation (D) Steam condensation in cylinder-piston system.

Then, the piston moves by distance S. The force F on the piston and the movement S results in work

$$L = F \times S \tag{1.2}$$

in the physical meaning. If the work is taken out and turned into another form such as pulling the rope that is connected to a heavy item, the movement of the piston serves to carry the heavy item: work. If this work is changed to a work rate, that is how efficiently the work is performed, it is the definition of the power in the physical meaning.

Newcomen condensed steam in the cylinder with the piston as shown in Fig. 1.5D and developed a steam engine that provides power in 1712. In those days, coal was used to get heat. Coal lied deep underground. It was necessary to efficiently pump well water out to a coal mining site. To bring coal up from the deep underground to the ground was also hard work. This work was done by a human and a horse in those days. The developed steam engine began to be used for pumping out water from the coal mine. As long as coal burning was continued to produce steam, and the water supply was also continued to the steam boiler and to the cylinder to condense the steam periodically, the steam engine kept on carrying out the work of pumping out water without a break. The work corresponded to the work of several horses. As a result, coal was efficiently mined up and the coal was further used for getting more coal. Originating from it, the power used by mankind has increased sharply since the late 18th century.

It was said in Section 1.2 that a human being never used fire to get power before the 18th century. However, there is one record left by Heron in Alexandria around 150 BCE as sketched in Fig. 1.6 [3]. He generated steam by heating water using fire. The steam was blown out from nozzles that were attached to a sphere tangentially, and then the sphere rotated. It was really the boiler and reaction turbine system itself. However, there is no record of how the system was used. Since then until the 18th century, a human being never used fire to get power in reality although several ideas had been proposed.

In ancient Greek era, the four fundamental elements were fire, air, water, and soil. The fire was one of the elements. This understanding continued until the 18th century. The fire had not been related to energy. The word "energy" was first used by Young in 1807. Joule showed that heat is one form of the energy in 1843 of the mid of the industrial revolution. That was far after acquiring the steam engine that was the way to get power from fire.

The Newcomen steam engine was remarkably improved by Watt. He introduced an external condenser instead of condensing steam in the piston-cylinder in 1776. It contributed to improve thermal efficiency remarkably; or that is less use of coal and more power getting.

He tried to show that his steam engine was efficient and excellent. In his days, water was pumped up from a coal mine by using a horse. The horse was connected to an edge of a 12 ft. long arm. Another edge of the arm was fixed at the rotating axis that was linked to a pump at a coal mine. Watt let the horse walk around the axis and measured the force that the horse pulled the arm with as shown in Fig. 1.7.

Figure 1.6 Heron steam turbine [3].

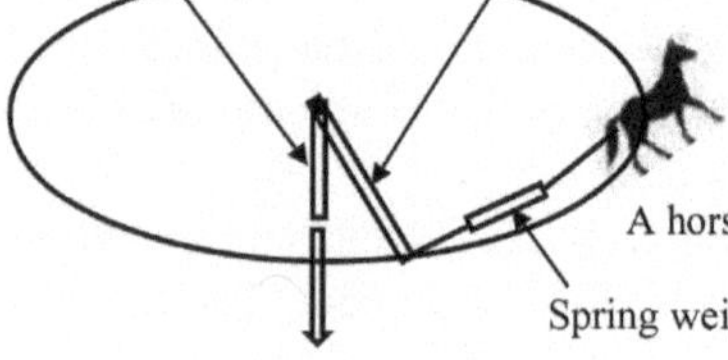

Figure 1.7 Definition of unit of power.

He found that the force measured by the spring weighing was 175 lb_f and the horse rotated 2.5 times per minute. Thus, the horse did the work L of

$$L = 2\pi \times 12 \times 2.5 \times 175 = 33,000 \text{ ftlb}_f. \tag{1.3}$$

Then, the work rate W of the horse per second became

$$W = \frac{L}{60} = \frac{33,000}{60} = 550 \text{ ftlb}_f/\text{s}. \tag{1.4}$$

He realized that the horse did the work by the work rate of 550 $ftlb_f$/s. He decided that the horse power is 550 $ftlb_f$/s. This work rate became horsepower. This unit was quite useful for him to show that the steam engine developed by him could do how many horses. The unit was quite useful and valuable for not only his machine but also other machines to express how the machine was efficient and large. The horsepower unit of 550 $ftlb_f$/s became used widely in general.

The horsepower unit was in the English ft-lb_f unit system. It was not convenient for people in other unit systems. A new horsepower unit was developed at the beginning of the 20th century. The new horsepower was 75 kg_fm/s in the metric unit system. The value itself is physically very close to the value of 550 $ftlb_f$/s in the English ft-lb_f unit system. The former and the latter are usually expressed as PS and HP, respectively.

Both the horsepowers of 550 $ftlb_f$/s and 75 kg_fm/s are in the gravitational unit system. The force in that system depends on gravity. Since gravity varies place by place on the earth; the unit that includes the gravity is not universal and convenient. Thus, the absolute unit system is used widely nowadays for the work rate; the power. The power is expressed as J/s = W in the absolute unit system. Thus, the horsepower PS of 75 kg_fm/s is equal to 0.735 kW.

Here, it should be remembered that power is produced from fire nowadays. The thermal energy of the fire is expressed using the unit of J in the absolute unit system. The thermal energy and the work are the same in the first law of thermodynamics. Thus, the work rate is also expressed using the unit of J. Work rate is J/s. This work rate; the power, is expressed as W. The unit came from Watt in honor of the modern steam engine founder and the definer of the horsepower.

Then, steam engines became widely used in coal mines. Steam engines in those days were reciprocating engines. Thus, it was hard to apply those to machines that required rotating motion such as spinning machines. The spinning industry was one of the leading industries at that time. Watt tried to develop many rotating steam engines; however, finally he introduced a crank-connecting rod device to his reciprocating engines in 1794 to turn the reciprocating motion of the steam engine to the rotating motion. It trigged the widely spreading-out of the steam engines originated from Watt to many industry fields as a general steam engine.

Replacing hydro, wind, and horsepower for spinning machines, milling machines, and so on with steam power was initiated. Steam power originated from fire; coal burning was used easily, efficiently, and greatly, which led to the explosive development of machine civilization: the industrial revolution from the 18th century to the 19th century. Since then, the energy consumption of a human being has begun to increase and is still increasing nowadays as shown in Fig. 1.4. A human being keeps seeking for ease, comfort, and convenience.

1.5 Full maturity of modern civilization

Getting power from thermal energy started from using coal as a heat source. As time proceeded, oil and then gas have been naturally also used as the heat source since those are easy to convey and handle, and also have higher calorific values than coal. It made mankind possible to get the higher growing rate of using power in the 20th century as shown in Fig. 1.4. Civilization has greatly improved the food situation, the medical circumstances, and the living environment of humankind. It has led to the increase in population in the world. It also has given an impetus to the power consumption increase.

Electricity that is one kind of energy has also been used for many usages; light, electric devices such as a radio or a television, generating heat, motors producing power, and so on. The electricity is very useful, clean environmentally, easy to handle, easy to transmit, and so on. The electricity was initially produced by using hydro energy. Soon, the electricity has been produced from the thermal energy of a steam engine. It started commercially in 1882 at the 120 kW Pearl Street Power Plant in the United States, and also in 1887 at 50 kW Kayabacho Power Plant in Japan.

The thermal electric power generation turned from the steam engine system to the boiler-turbine system in the end of the 19th century. Electricity has been efficiently and greatly generated from thermal energy by the boiler-turbine system. The great increase in power usage by mankind in the 20th century has been promoted by the electrical power generation growth.

Becquerel discovered radiation in 1896. Radiation has led to another source of power: nuclear power. Einstein pointed out that mass M and energy E are the same thing and have different forms and that the relation is expressed using light speed c as

$$E = Mc^2. \tag{1.5}$$

When atomic nucleus fission occurs, a small amount of mass turns to huge energy following Eq. (1.5), and the energy is released. This is nuclear energy. When 1 kg U^{235} is collapsed, the released energy is equivalent to the heat of combustion of coal of 3000 ton. Fermi succeeded in collapsing the nucleus in a chain reaction way in 1942. It was quite natural to use nuclear energy as a heat source for power generation. If the nucleus fission chain reaction could be occurred steadily, continuously, and safely under controlled conditions, humans acquired a new way to create heat for power generation. EBR-1 of an experimental fast breeder reactor in Idaho in the United States succeeded in the electric generation from heat of the nuclear fission in 1951. Four 200 W electric bulbs were lighted on December 20th.

Nuclear energy was naturally introduced into commercial electric power generation: the nuclear power generation. The commercial nuclear power generation was started in 1954 at the Obninsk Nuclear Power Plant of the 5000 kW channel-type boiling water reactor in Russia. Soon after, the pressurized water reactor (PWR) at the 100,000 kW Shipping Port Nuclear Power Plant was initiated in 1957 in the United States, and then the boiling water reactor (BWR) at the 210,000 kW Dresden Nuclear Power Plant in 1960 in the United States. In Japan, the commercial gas-cooled reactor power generation started at the 12,500 kW Tokai Nuclear Power Plant in 1966, the BWR power generation started at the 320,000 kW Tsuruga Nuclear Power Plant, and the PWR power generation started at the 340,000 kW Mihama Nuclear Power Plant in 1970.

It is interesting to note that thermodynamics that has been used to explain phenomena in a thermal-energy system and to develop the thermal-energy technology of the present day was evolved following the evolution of the thermal-energy system. That discussion is outside the scope of this book.

1.6 Rule for power generation systems

It is necessary for current civilization to supply power massively, steadily, economically, and safely. To achieve it, humankind has developed sophisticated intelligent operating system, rules, codes, standards, and laws learning from many lessons experienced. One example is the pressure of steam engines and boilers. It is obvious instinctively and theoretically from thermodynamics that elevating working pressure enhances the efficiency of the steam power generation systems. However, it caused the rupture of pressure vessels in the systems, which resulted in many injuries and deaths in the 19th century. Learning from those experiences, sophisticated and highly advanced rules and standards for designing them have been elaborated in the 20th century. Since the steam power generation plants are big and costly, the effect of those construction on society is quite large. Thus, it is important to assess how and what are influenced in advance and to prepare the way to avoid negative things. These are essential to keep civilized mankind daily life comfortable. These are briefly treated in this volume.

1.7 Future prospect of power

Humans have the technology to get power easily and greatly from heat. The daily life of the humankind has been greatly improved and becomes cultural. Humans are also reaching into cosmic space by using power. Humans enjoy a developed civilization by using power from heat. Looking into the future, it is hard to say that a current state will continue as it is. Several problems have been already pointed out. Among them, big issues are exhaustion of energy resources, environmental problems, and security. These will be briefly discussed later in this book.

1.7.1 Energy resource exhaustion

The main sources of power that humankind is leaning on are coal, oil, gas, and uranium. These are natural resources that had been stored in the earth for a quite long time before human beings appeared on the earth. Regardless of whether they are more or less, the existing amount of them is naturally limited. It is a matter of course that keeping on taking them out from the earth, those will be exhausted in the future.

Fig. 1.8 shows prospects for the future. Before humans discovered ways to get power from the thermal energy that came from natural resources, they walked by themselves and used horses and natural energy such as wind or hydro energy at the most. After getting power easily and sufficiently by using natural resources, humankind has been released from self-walking to use a car and even reached outer space. If humans lose natural resources for getting power, they may be forced to walk by themselves again in the future.

One possible way to avoid this situation is to get the technology of a fast breeder reactor as a reality. In that reactor, the fuel of the uranium is turned into more fuel of the plutonium than the used uranium, and the produced plutonium is used in the reactor. Humans will find ways to get enough power to enjoy civilization for quite a long time; several hundreds of years, as shown in Fig. 1.9.

Another possibility to avoid natural resource exhaustion is to use solar energy. The earth is getting large amounts of energy from the sun and releasing it back into outer space. The amount of energy that the earth is getting is several tens of thousands of times more than what humans are using on earth. Energy that comes from the sun can be used in many forms such as a solar battery and solar power generation, although wind and hydro power also have origins from the sun.

1.7.2 Environmental problems

As already mentioned, the earth is getting huge amounts of energy of 1.8×10^{14} kW from the sun. Humans are dissipating 1.6×10^{10} kW on the earth.

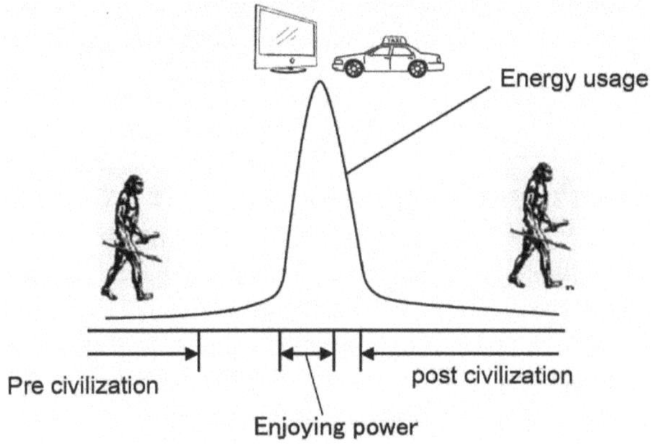

Figure 1.8 Dangerous possibility of future mankind situation.

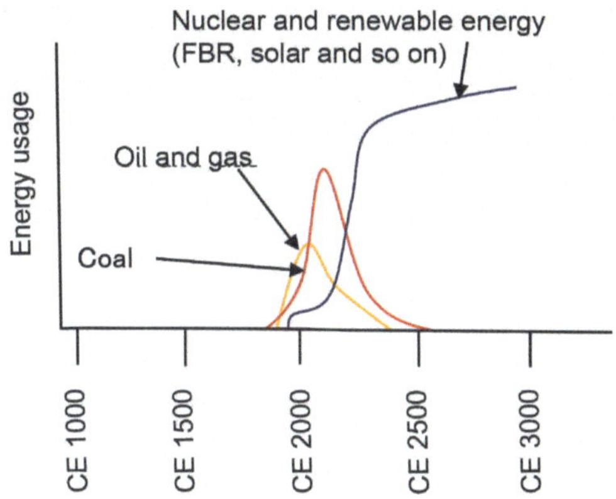

Figure 1.9 Provision for future energy exhaustion problems.

The dissipation that mainly originates from the natural resources stored in the earth is approximately one-ten-thousandth and quite smaller than solar energy coming to the earth. The solar energy coming to the earth and the human-dissipated energy are released radiantly to outer space. Thus, the temperature of the earth is kept constant. However, the air temperature of the earth started going gradually up in the 19th century. It is supposed that the gradual increase in air temperature has been caused by mismatching between the coming energy to the earth and the releasing energy to outer space.

The density of carbon dioxide in the air was 280 ppm until the 18th century. The density started to increase gradually in the 19th century. This timing well corresponds to the initiation of using the fossil resources of coal, oil, and gas to get power. It is supposed that the carbon dioxide density going up was initiated by starting to use fossil resources to get power. The density has been approximately 400 ppm now. The carbon dioxide has a radiation absorption band at the temperature of the air on the earth. Thus, the radiant release of energy from the globe to outer space has been suppressed by an increase in the absorption of the carbon dioxide in the air and stored in the air. The suppression has led to the gradual increase in the thermal energy in the air; the gradual temperature increase of the air. It is global warming.

It is obvious that global warming has been the result of burning fossil fuels such as coal, oil, and gas to get power. It seems that the temperature of the globe that is the average air temperature on earth has become approximately 0.8K higher than that in the 18th century. The increase seems small, however, even the small change is fatal to our natural world, leading to climate change, living environment changes for plants, animals, and fish, and so on. For example, just 1°C change of sea water causes sardine evacuation. Humankind is also of course affected by changes to the natural world situation. The environmental problems caused by the consumption of the fossil fuels has to be solved and calmed down. How to solve is closely related to what mankind relies on to get power.

1.7.3 Security

One risk that may be caused by the lack of natural resources is trouble between areas or countries. In some case, the trouble may become a war since securing the supply of energy is related to maintaining own lives. Needless to say, a political response is necessary, however, to solve the energy exhaustion problem is the most certain.

Another risk is accidents related to power generation. Since power is vital for modern civilized society nowadays, the risk is significant to society. If power is lost, a serious situation comes out physically and socially; factory operation will be stopped, transportation system will be suffered, and so on. It is not only important and useful to settle rules, code, standards, and organizations to secure safety but also to make social and technical system robust and resilient to the risk.

References

[1] K. Takemura, Japanese culture and energy, J. Jpn. Soc. Energy Resour. 40 (3) (2019) 221–224.

[2] Japan Atomic Energy Relation Organization, 1-1-1 Relation of human being and energy. <https://www.ene100.jp/zumen/1-1-1>, 2017.

[3] M. Conrad, Die Entwicklung der Dampfmaschine, Verlag von Julius Springer, Berlin, 1908, p. 282.

Development in power technology

2

Mamoru Ozawa[1] *and Yasushi Saito*[2]
[1]Kansai University, Takatsuki, Osaka, Japan, [2]Kyoto University, Kumatori, Osaka, Japan

2.1 Development of thermal power generation

2.1.1 Dawn of steam engine

Technology development is greatly dependent on the surrounding technology, social and economical environment, and natural environment. Although Heron's rotating machine being the first trial of the power from steam never contributed to human life and welfare, the machine consisted of principal elements, for example, a steam generator and a reaction turbine. Probably based on the same concept, G. Branca [1] constructed the image of substitution of boiler and primitive turbine as the power source for a water-driven flour mill in 1629, but the practical application might not be conducted. Heron's rotating machine was composed of submerged-type boiler and pure reaction turbine with reference to the current technology, and Branca's one was a submerged-type boiler, nozzle and impulse turbine. The unit output of Branca's turbine was unknown and supposed rather small.

In 1690, Denis Papin published a power generation concept [2] consisted of piston and cylinder where power was generated by the vacuum induced by the steam condensation and the atmospheric pressure. Such a concept was realized in a practical steam engine by Thomas Newcomen. A new technology does not appear suddenly without any precursors, while industrial success was achieved first by Newcomen.

Fig. 2.1 shows the illustration of Newcomen's engine in 1712. Coal-fired submerged boiler was installed on the brick work. Generated steam was introduced into the cylinder in accordance with the beam and piston movement. Then the fresh water was injected into the cylinder, and rapid condensation of steam took place leading to vacuum condition in the cylinder. The outside of piston, in the present case upper side, was opened to the atmosphere, then the piston was pushed downward by the atmospheric pressure. This downward movement turned to upward motion of the pump rod. The success of this Newcomen engine owed to coal refuse from coal mines, even though the first model produced only 4 kW with 0.5% thermal efficiency.

The main cause of rather low efficiency is heat loss owing to direct-contact condensation in the cylinder. The direct contact is not serious for the heat transfer, while the cylinder wall is cooled by the water injection. Then the initial stage of the

Fundamentals of Thermal and Nuclear Power Generation. DOI: https://doi.org/10.1016/B978-0-12-820733-8.00002-9

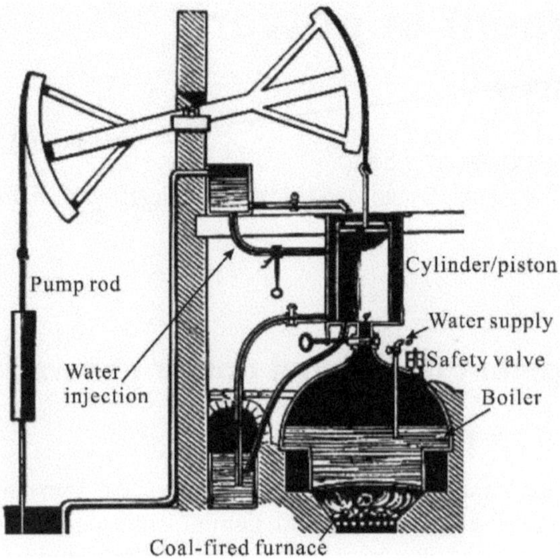

Figure 2.1 Newcomen's steam engine (1712) [3].

preceding cycle thermal energy of steam from the boiler is uselessly supplied for warming-up the cylinder.

James Watt found out this problem, and improved Newcomen's engine, that is, steam condenser was introduced (the first patent, 1769). In addition, he introduced planetary gear, by which the reciprocating motion was converted to rotating motion (1781), and control systems, that is, speed control with a governor, pressure regulation with pressure gauge and damper at the chimney inlet and water level control. Such an example is shown in Fig. 2.2.

The innovative aspect of Watt's engine is on the functional differentiation, that is steam condenser from cylinder-piston system, and control systems. Such an aspect is leading to the current steam power technologies. An appearance of rotating engine brought about various applications in the field of transportations on land and in the sea, factory prime movers for looms, and spinning machines.

2.1.2 Appearance of high-pressure engine

Before the discussion of steam power development, it is worthwhile to overview the long history of technology and the society. Fig. 2.3 shows such historical feature.

The origin of the industrial patent in England is the Statute of Monopolies in 1624, and the foundation of the Bank of England 1694. Such social surrounding encouraged public investments in industrial systems and equipment, for example, coal mines, maritime trade, and so on. An increase in coal production essentially brought about deepening coal mines, then the pumping of groundwater and lifting of coal prepared the market of steam power. In accordance with the development of cotton industry, the factory system became common, leading to mass production,

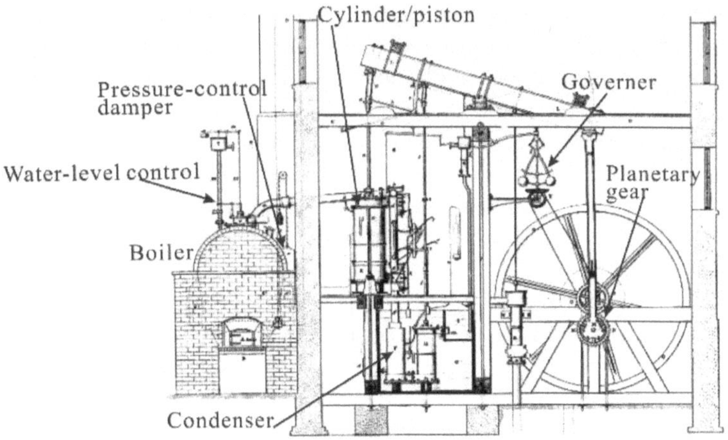

Figure 2.2 Boulton & Watt engine of 10 HP (1787) [4].

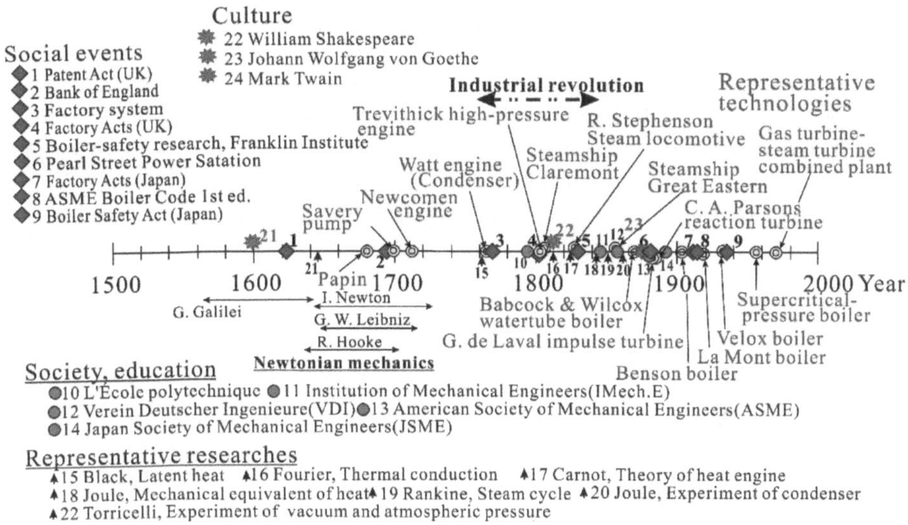

Figure 2.3 Historical overview of steam power and the society.

which need large power source for looms and spinning machines in factories. This is the base of steam engine development, production technology, iron-making technology. From the middle of 1700s to the middle of 1800s, technology innovation and economic development were drastically enhanced in Britain, which was the so-called industrial revolution. During such a situation, Watt's patent of atmospheric engine expired in 1800. Watt's patent strictly limited the steam engine operation at around atmospheric pressure. Thus, limitation prevented the boiler explosion, while might hinder the development of steam engine.

In relation to the steam power, Evangelista Torricelli conducted vacuum experiment and demonstrated the intensity of atmospheric pressure. This might largely affect the construction of Papin's concept, and later Thomas Savery's water commanding engine [5], that is, pump. At this stage, thermodynamics was not constructed, but probably based on the technological experience Newcomen developed his steam engine. Joseph Black discovered the latent heat being an important fundamental of steam engine. Around the same time, Watt conducted various experiments including condenser to improve the efficiency of steam engine. In 1824, N.L. Sadi Carnot proposed heat engine cycle [6], being the present thermodynamics principle. Around 1860, W.J.M. Rankine [7] discussed heat engine theory, so-called Rankine cycle. Then J.P. Joule measured condensation heat transfer and discussed various influencing parameters [8] (Fig. 2.4).

In 1800, Watt's patent expired and soon after, O. Evans developed high-pressure engine [10] as shown in Fig. 2.5.

Evans' boiler was installed on the brickwork and coal was fired on a grate. The flue-gas heated boiler-water outside the drum and at the end of furnace turned into relatively large-sized smoke tube through the drum. Evans' engine was installed on the steamship Oruktor Amphibolis in 1804. Around the same time, Richard Trevithick also developed high-pressure engine as shown in Fig. 2.6.

In the Trevithick engine, furnace and flue tube was U-shaped and installed within the boiler drum. This type of boiler was later referred to as Cornish boiler. By removing the brickwork, Trevithick's engine opened a wide application especially for transportation on the land, that is, steam locomotive.

As is easily understand, high-pressure means high temperature and high thermal efficiency of the engine. Practical application of steam engine to steamship, however, was first conducted with low-pressure engine in Britain. Typical example was the steamship Great Eastern in 1859 shown in Fig. 2.7 of displacement capacity

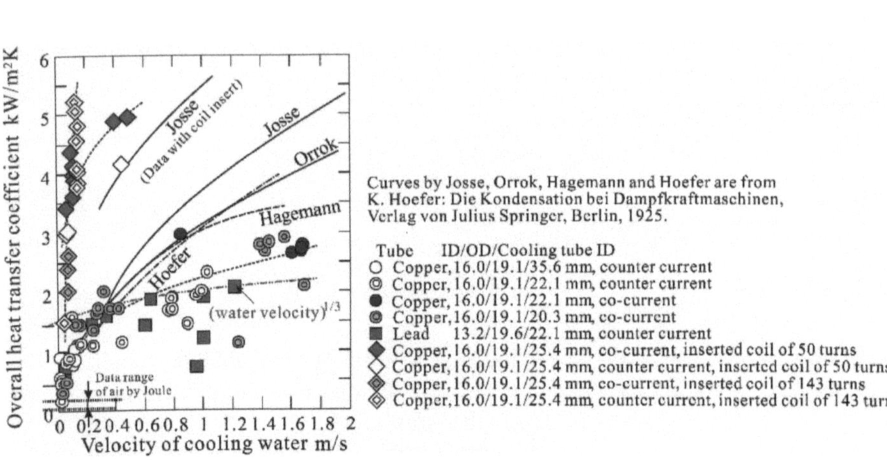

Figure 2.4 Condensation heat transfer by Joule and others.
Source: Drawn by the author based on the data of J.P. Joule, On the surface-condensation of steam, Philos. Trans. R. Soc. Lond. 151 (1861) 133–160 [8] and K. Hoefer, Die Kondensation bei Dampfkraftmaschinen, Verlag von Julius Springer, Berlin, 1925[9].

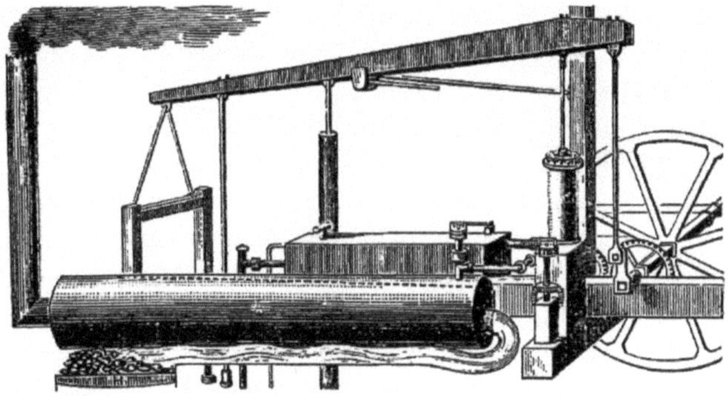

Figure 2.5 Evans' high-pressure engine (1800) [10].

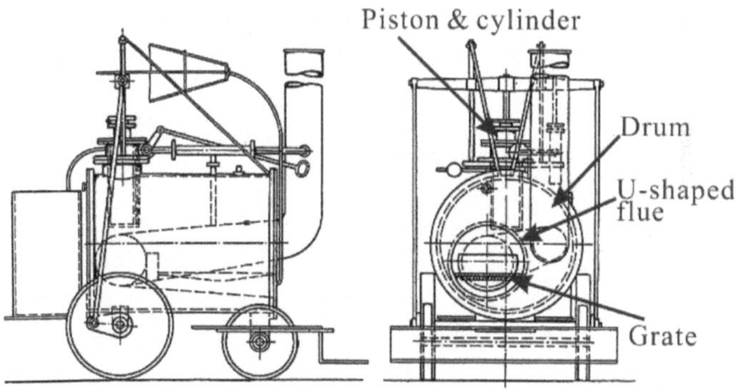

Figure 2.6 Trevithick's high-pressure steam engine [11].

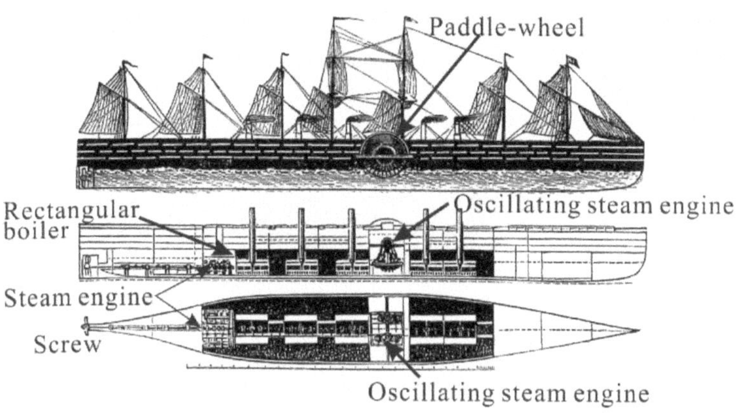

Figure 2.7 Steamship Great Eastern [10].

32,000 ton. The Great Eastern had 10 low-pressure rectangular boilers which supplied steam to four oscillating paddle-wheel engines and four screw engines of total 10,000 HP with navigation speed 13 knot.

The production of steamship was highly enhanced in the Clyde Side of Glasgow since 1850. Fig. 2.8 indicates the annual tonnage of shipbuilding in Clyde Side being the most famous shipbuilding area in the word at that time. Sailing ships were still constructed but the number of steamships increased steadily.

The power produced by one engine system is referred to as the unit power, being a suitable measure for technological development. As described below, the unit power of the steam engine of mercantile and naval ships far greater than that of land-use engines. Thus, the technological development in the late 1880s was driven by the steamship. In the course of development in marine steam engines, the boiler type went through transitions from low-pressure rectangular boiler, Flue and smoke-tube boilers, typically Scotch boiler. Substantial increase in steam pressure and temperature needed to introduce watertube boilers. Thus, watertube boilers like Thornycroft boiler, Belleville boiler, Yarrow boiler and so on were developed. Of course, the transition was not straight forward but was going back and forth. In accordance with such development in steamships, the steam pressure as well as the steam temperature increased as shown in Fig. 2.9. By around 1890, the steam engines were operated with saturated steam, and since then superheated steam has been used for steam power.

Fig. 2.10 shows one of typical cylindrical boilers for marine use, that is, Scotch boiler, which was extensively installed on steamships including well-known Titanic. Titanic was navigated with 29 boilers of 1.48 MPa steam, two sets of reciprocating engines of 15,000 HP and 16,000 HP steam turbine.

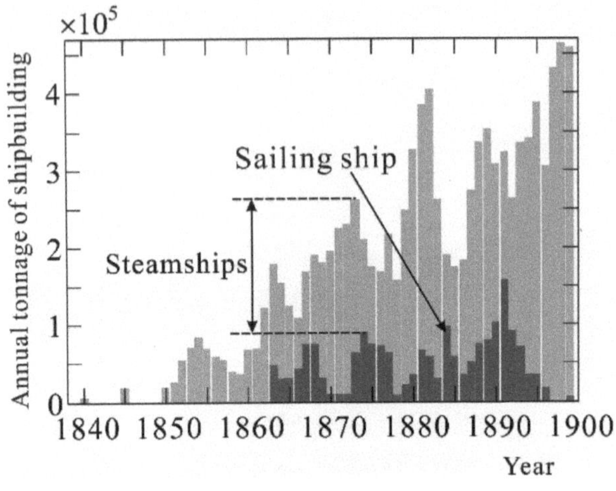

Figure 2.8 Annual tonnage of shipbuilding at Clyde Side.
Source: Drawn by the author based on the data of A.M. Robb A.M. Robb, Chapter 16: Shipbuilding, in: C. Singer, E.J. Holmyard, A.R. Hall, T.I. Williams (Eds.), A History of Technology, Vol. V, Oxford University Press, Oxford, 1958, pp. 350–390 [12].

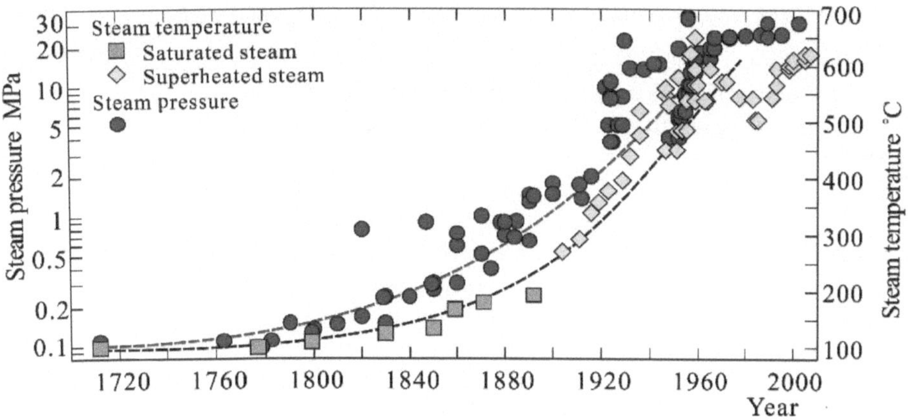

Figure 2.9 Transition of steam pressure and temperature.
Source: Drawn by the author based on the data of S. Ishigai, Design Principles of Steam Boiler, Sankai-do, Tokyo, 1961, pp. 58–59; Report from the Select Committee on Steam Boiler Explosions; together with the Proceedings of the Committee, Minutes of Evidence, and Appendix, The House of Commons, 1870.; The Manchester Steam Users' Association (MSUA), A Sketch of the Foundation and of the Past Fifty Years' Activity of the Manchester Steam Users' Association for the Prevention of Steam Boiler Explosions and for the Attainment of Economy in the Application of Steam, Taylor, Granett, Evans, & Co., Manchester, 1905.; K. Akagawa, Thermal and hydraulic design of steam-generation systems, in: S. Ishigai (Ed.), Steam Power Engineering, Cambridge University Press, New York, 1999, pp. 255–283 [13–16], Central Research Institute of Electric Power Industry, and Thermal and Nuclear Power Society.

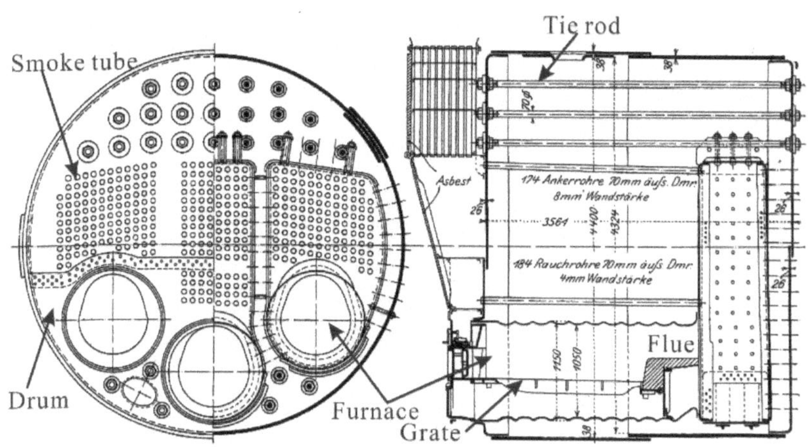

Figure 2.10 One-side firing Scotch boiler (1900) [11].

2.1.3 Watertube boiler development to the present

The first sectional watertube boiler was put into the market by Babcock and Wilcox Co. The water separated upper horizontal section flows downward through the left-sided vertical channel, that is, downcomer, and then goes up through gently inclined steam generating tubes. The right-sided vertical channel is a riser. These channels composed one unit, and by arranging multiple units, the steam generation rate is increased so as to meet the requirement, which is referred to as a sectional boiler.

When the Babcock and Wilcox's boiler shown in Fig. 2.11 was developed, the boiler-water treatment was not well established, and thus frequent clean-up of watertube was an essential task for removing scale. This might be the reason why the straight tubes and both ends plugs were used. In accordance with the progress in boiler-water chemistry, curved, and steeply inclined watertubes were installed as shown in Fig. 2.12. By applying such design, natural circulation rate was highly enhanced and boiler capacity increased.

An enhancement in boiler performance was further conducted with the installation waterwalls in the furnace, that is, so-called radiant boiler as shown in Fig. 2.13. With such radiant boiler, the natural circulation boiler reached a certain established level.

The overall feature in the boiler development was described by S. Ishigai as shown in Fig. 2.14.

Further increase in the boiler capacity necessitates an increase in the circulation rate together with an increase in the heat release rate in the furnace. Then one of the solutions was to assist water circulation by installing pump in the water circulation line, which is a forced-circulation boiler like La Mont boiler. The increase in the boiler efficiency and capacity to meet social and economic demand essentially raised up the steam pressure and temperature. The pressure increase brings about a decrease in the density difference between water and steam, being inappropriate to the natural circulation. This inappropriateness was broken through with the once-through boiler, typically Benson boiler shown in Fig. 2.15, Sulzer monotube boiler, and Ramzin boiler around 1920.

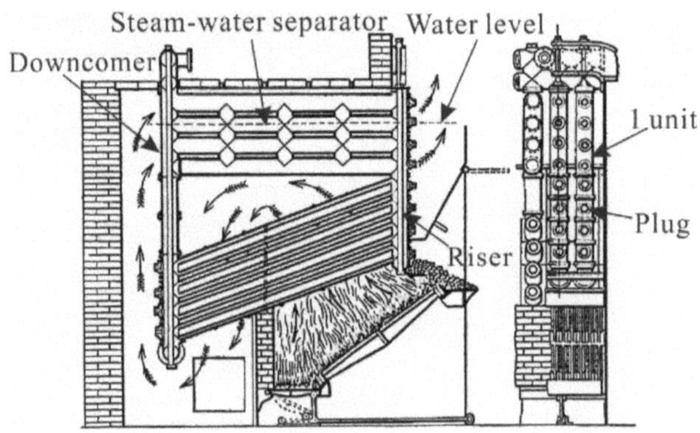

Figure 2.11 Babcock & Wilcox's sectional watertube boiler (1867) [17].

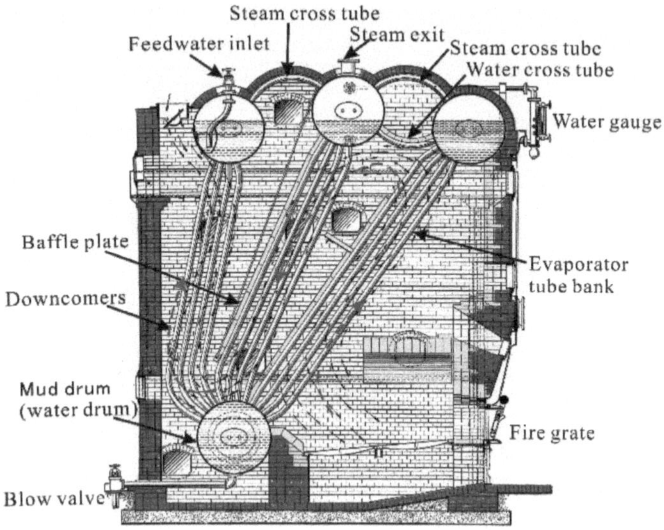

Figure 2.12 Stirling boiler (1905) [18].

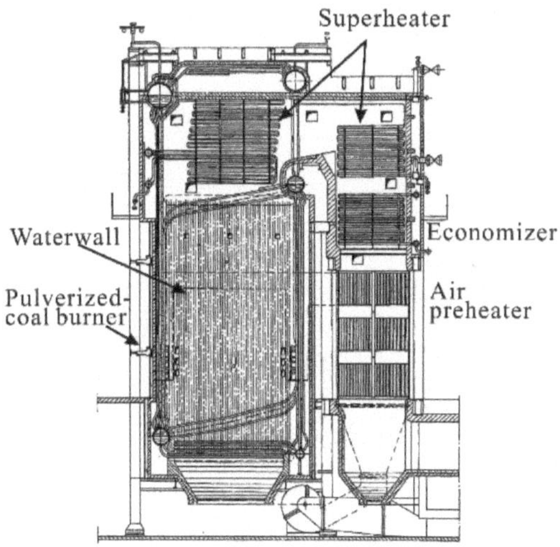

Figure 2.13 Example of 4-drum radiant boiler (1936) [19].

The once-through concept was extended to the supercritical pressure boilers. The first supercritical pressure plant in Germany was Hüls Block I unit of 260 t/h steam generation rate, 29.4 MPaG, 600/560/560°C steam temperature with 2-stage reheat in 1956 [20]. The first unit in the US was Philo #6 Unit of 306 t/h, 31.0 MPaG, 621/566/538°C in 1957 [21].

A present once-through supercritical pressure boiler is exemplified in Fig. 2.16.

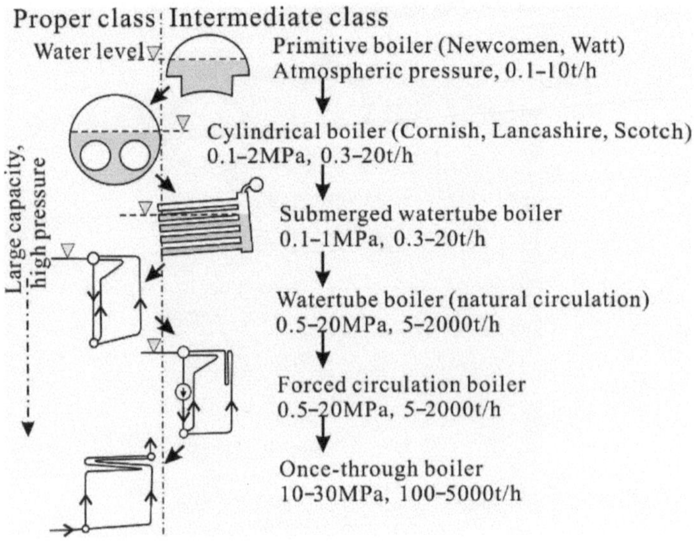

Figure 2.14 Principal types of boilers and development sequence.
Source: Drawn by the author based on the Ishigai's classification, Courtesy Prof. Emeritus Seikan Ishigai.

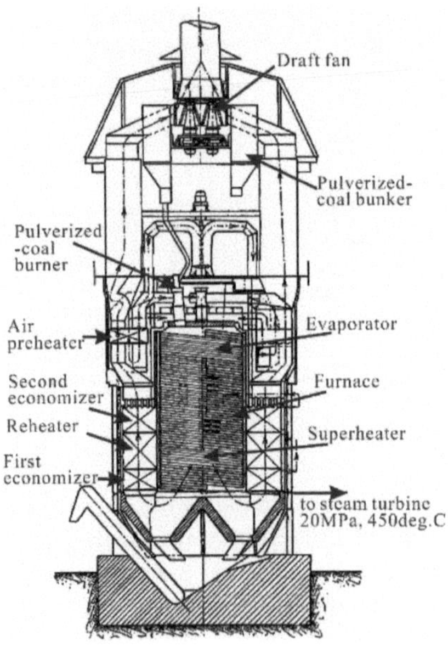

Figure 2.15 Benson boiler of Siemens-Schuckert cable factory in Berlin (1927) [19].

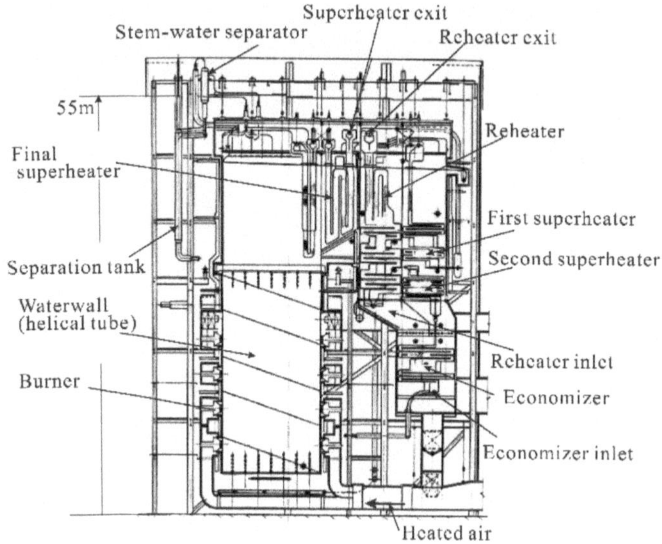

Figure 2.16 Once-through supercritical pressure boiler.
Source: Courtesy IHI Co.

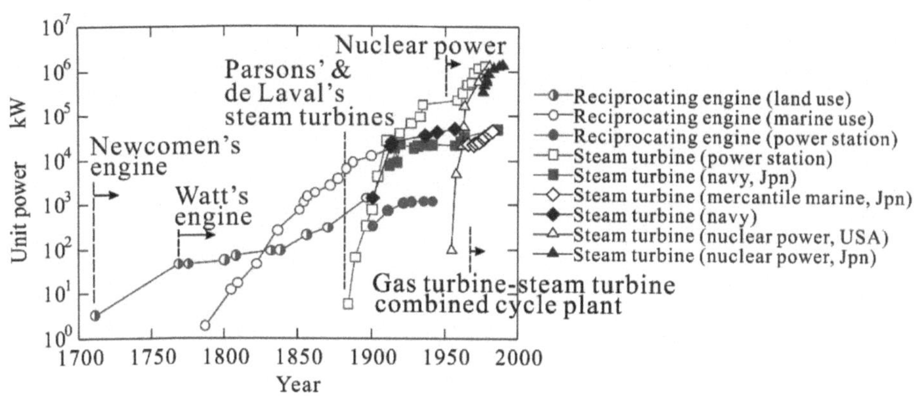

Figure 2.17 Development of unit power.
Source: Courtesy Prof. Emeritus Koji Akagawa.

2.1.4 History of steam engine and turbine

Fig. 2.17 shows unit power of steam engine and power generation systems. The smallest one located left end is the Newcomen engine mentioned above. The unit power of Watt's engine was about one order higher, compared with the Newcomen's engine. The steam engine started in its application from land use, while the marine use started about 70 years behind. The unit power of marine use grew rapidly up and became comparable around 1830 and further grew up beyond the land use. This is mainly because the factory system development was relatively

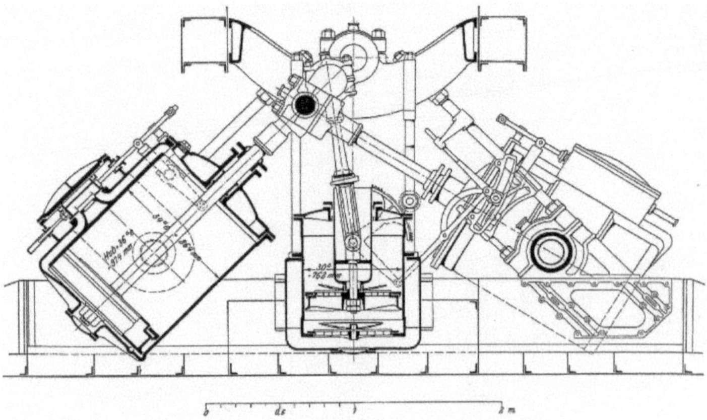

Figure 2.18 Oscillating engine for paddle-wheel steamship 1855 [11].

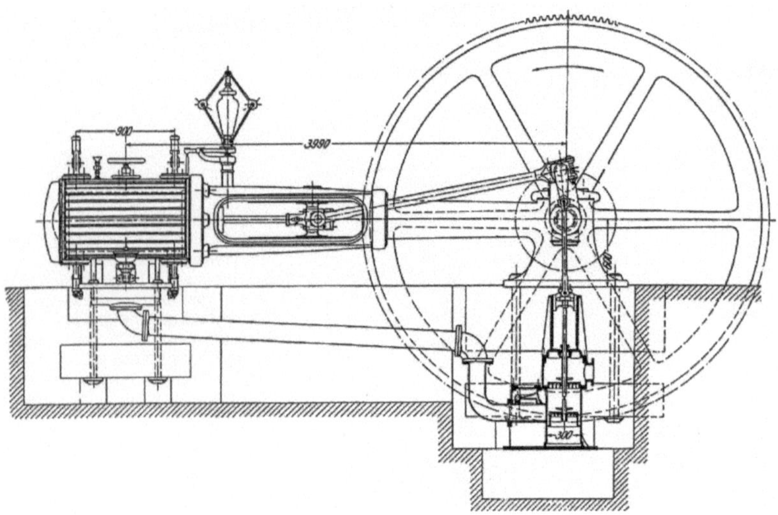

Figure 2.19 Sulzer steam engine 1866 [11].

slow compared with the marine use. Not only mercantile transportations but also the development in battle ships of navy enhanced the growth in the unit power of steam engine. They needed large capacity and high-speed navigation. These demand in the market was covered by the reciprocating engines at that time.

Typical reciprocating engines are shown in Figs. 2.18 and 2.19.

In Fig. 2.20, three cylinders are arranged in line, that is, tandem compound. The cylinder at the right end is high-pressure cylinder, then intermediate-pressure and low-pressure cylinder are aligned for driving same shaft. The exhausted steam from high-pressure cylinder flows into the intermediate cylinder for the next stage expansion, and finally followed by low-pressure cylinder.

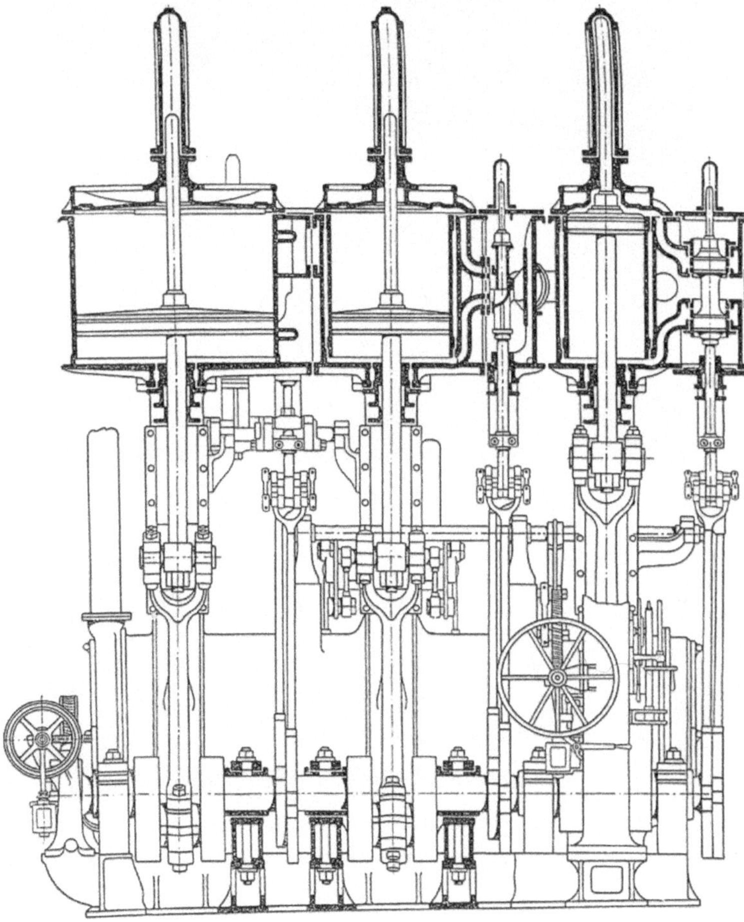

Figure 2.20 Triple expansion steam engine (1886) [11].

Fig. 2.21 shows one of main engines installed German battle-ship Kaiser Wilhelm II in 1902. Each engine is 4-stage expansion with 1.47 MPaG steam and 80 rpm. Each cylinder has the diameter 750, 1250, 1900, and 2850 mm, and 1800 mm stroke. The generated power was around 7400 kW.

The development of the reciprocating engine is drastic as shown in Fig. 2.17, and around 1900, over 10,000 kW engine appeared in marine use, while the reciprocating engine was taken over by the later-appeared steam turbine.

It is worthwhile to note the marine-use steam power was dominated by steam turbine since 1900, while the prevalent of steam turbine in marine field ceased around 1970s by substituting with high-efficiency large-capacity diesel engines. At present, steam-turbine navigation is limited only in LNG tanker.

The first reaction turbine was probably Heron's one, while no output of power. Then the above-mentioned Branca's concept appeared while probably no practical

Figure 2.21 Main steam engine of Kaiser Wilhelm II (1902) [11].

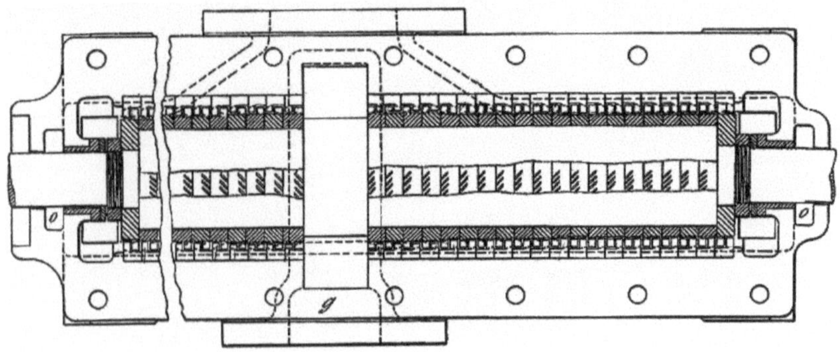

Figure 2.22 Parsons' turbine (1884).
Source: From US patent No. 328710 C.A. Parsons, Rotary Motors, US Patent No. 328710, Official Gazette of the United State Patent Office, Vol. 33, 1885, pp. 303–304 [22] corresponding to UK patent 6735.

application. The reason why the Branca's concept was realized is mainly because of the technological and economical situations underdeveloped to accept his idea. Waterwheels or windmills generated sufficient power for flour mills and coal mines, and the factory system was not developed at that time.

The first type of steam power was reciprocating engine, being suitable for small-capacity fields. This is also true at present as can be easily seen in motorcycles and automobiles. The practical application of steam turbine started from Charles A. Parsons of his UK patent No. 6735 in 1884, "Improvements in rotary motors actuated by elastic fluid pressure and applicable also as pump" as shown in Fig. 2.22. H.W. Dickinson [23] described that Parsons' patent was comparable to the Watt's patent of 1769 for the separate condenser.

Table 2.1 Development in Parsons turbine up to 1910.

Year	Output	Steam consumption/kWh	Steam pressure	Steam temperature
	kW	kg	MPaG	°C
1885	4	90.7	0.41	153
1888	75	24.9	0.69	170
1892	100	12.2	0.69	180
1900	1250	8.26	0.90	232
1902	3000	6.70	0.95	294
1907–10	5000	5.99	1.38	247
1907*	7500	6.94	1.22	242

Source: Made by the author based on the data of R.L. Hill [24]; and the data with * is the Westinghouse-Parsons turbine of New York Edison Co. [25].

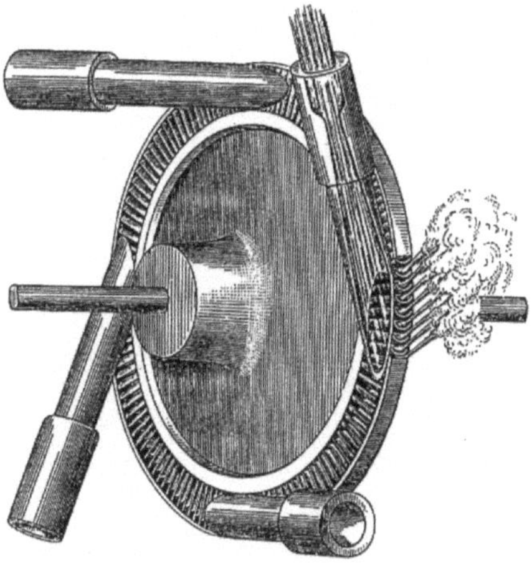

Figure 2.23 de Laval's steam turbine (1889) [25].

This turbine has 14 fixed and 14 moving blades and generated 7.5 kW at 18,000 rpm with 0.55 MPa saturated steam [23]. The Parsons' turbine is classified as a reaction turbine, moved half by impulse forced and half by reaction force, that is, degree of reaction is 0.5 (Table 2.1).

Around the same time, C.G.P. de Laval developed simple impulse turbine shown in Fig. 2.23 while the unit power was rather small compared with reciprocating engines at that time. This de Laval turbine is a typical impulse turbine.

The de Laval turbine covered in the range 1.1–225 kW, and the mean diameter of wheel was 75–750 mm, being corresponding 40,000–10,600 rpm.

Curtis turbine and Rateau turbines are also the impulse turbine, while the former is the velocity compound, and the latter is the pressure compound, shown in Figs. 2.24 and 2.25, respectively.

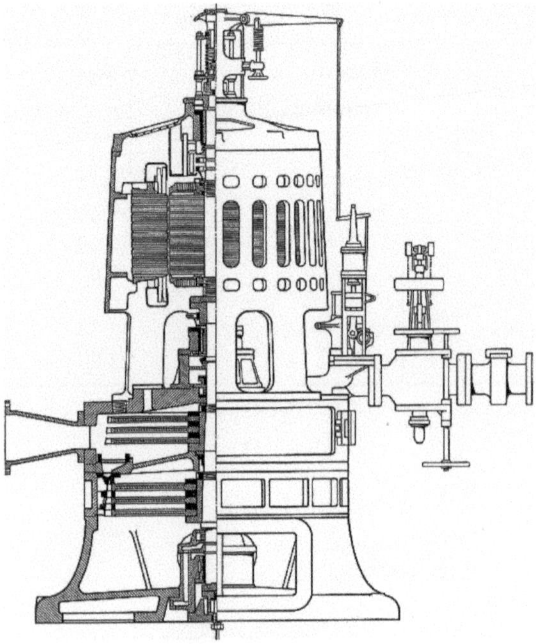

Figure 2.24 Curtis turbine of GE Co. at around 1900 of 750 kW, 1800 rpm [25].

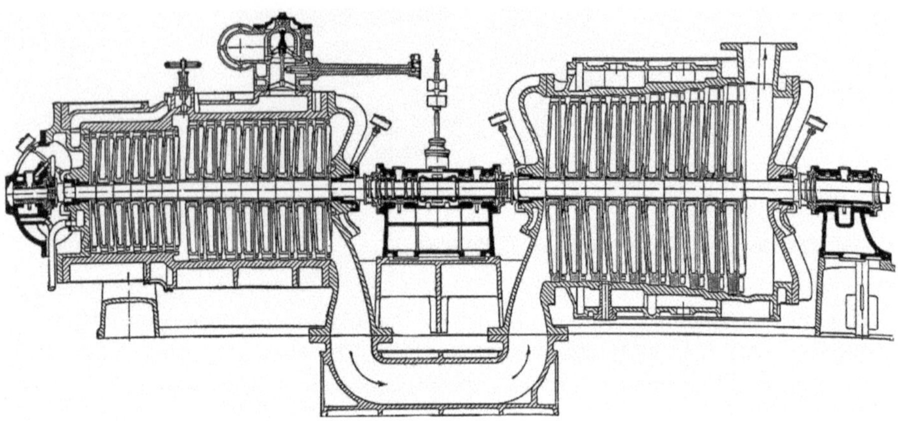

Figure 2.25 Rateau turbine at around 1900 [25].

The original US patent No. 591822 was issued in 1897 as "Elastic fluid turbine" by Charles G. Curtis [26], and in 1903 as "Multicellular turbine" (US Patent Office, No. 748216) by A.C.E. Rateau and G. Sautter [27].

Based on the Sautter's experiment, Rateau turbine with 58–463 kW output range, 0.22–1.05 MPa, 142°C–194°C of steam temperature showed 17.16 to 9.96 kg/kWh of steam consumption, and 74%–93.3% efficiency [25].

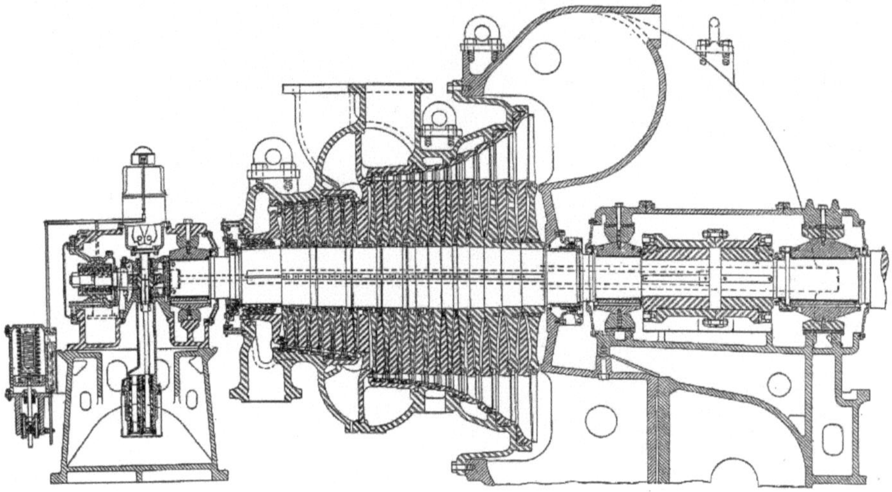

Figure 2.26 GE 40,000 kW resuperheating turbine [28].

The growth in steam turbine output was rather drastic and around 1907 over 7500 kW system was used for power station, and since then the unit power increased further and further. On the other hand, in the marine use, a 26,000 kW-turbine was installed in 1907. In 1935, 33,400 kW-turbine appeared. In a naval ship, 48,000 kW turbine was installed in 1954, which was the highest output [13].

One of the examples of steam turbine for power station is shown in Figs. 2.26, 40,000 kW GE resuperheating turbine.

Current large capacity steam turbine is illustrated in Fig. 2.27. This 4-cylinder tandem-compound turbine is constructed for 600 MW unit. This turbine is composed of one high-pressure cylinder to the left, one intermediate-pressure cylinder, and two low-pressure cylinders.

2.1.5 Dawn of electric power generation

The first power station, the Perl Street station of Edison Electric Illuminating Co. of New York, started direct-current electricity-distribution service in 1882, and at that time wiring was in place for 400 lamps. By 1884, the wiring increased to 10,164 lamps. As such, the electric supply started as lighting business. Six reciprocating steam engines of 120 kW were supplied saturated steam from four Babcock and Wilcox boilers. At the early stage of power station development, reciprocating engines were used as the prime movers, while in accordance with the development in steam turbine technology the reciprocating engines were replaced successively by steam turbines This is mainly because the turbine is much more appropriate to larger unit power.

The development in boiler technologies does not keep step with steam turbines. Thus, in the Pearl Street stations, the number of boilers was 0.7 per one prime

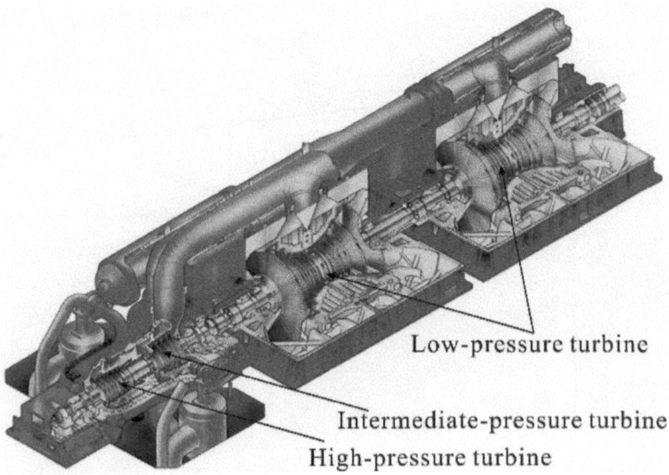

Figure 2.27 Tandem-compound steam turbine for 600 MW thermal power plant.
Source: Courtesy Mitsubishi Hitachi Power Systems Co.

mover. This number increased first up to 8, that is, 80 boilers supplied steam to 10 prime movers of 1200 kW (total 120 MW) in 1903, while successively decreased to 3 in 1929, and then finally 1.0, that is, one turbine correspond to one boiler, at around 1940–1950 in land-use field [13].

The development process in the United State is shown in Fig. 2.28 together with that in Japan. As mention above, the Pearl Street station was practically the first commercial power station for electric lightning in the world, while such power generation system rapidly spread all over the world. Japan was not the exception. The first power station in Japan was Kayaba-cho plant of the Tokyo Electric Light Company constructed in 1987. The 25 kW steam engines were installed to generate direct current as in the Pearl Street. The development process in Japan is almost similar except slightly lower capacity than the US.

Fig. 2.29 shows an example of power generation efficiency. Around 1900, the efficiency was rather low, several %, since then the efficiency increased steadily to over 40% at around 1960.

Fig. 2.30 shows overview of power station with reciprocating engines. The right side is a boiler room and the engines are located at the lower floor on the left. The electric generators are located at the 2nd floor connected with horizontal engines via belts. By introducing steam turbine with larger capacity, the plant size extended, and design and arrangement became different as shown in Fig. 2.31 from that of reciprocating engines

It is worthwhile to note that exhaust-gas treatment equipment is not installed before entering a chimney in both power stations shown in Figs. 2.30 and 2.31. On the other hand, in the power station at around 1960 shown in Fig. 2.32, mechanical dust collector, that is multicyclone type, for fly ash and electrostatic precipitator are installed to prevent the diffusion of dust and fly ash. In the present coal-fired plants,

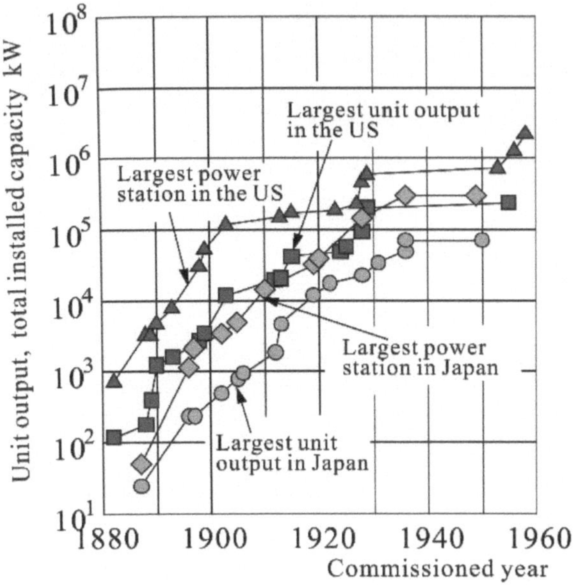

Figure 2.28 Trend of installation capacity in the US and Japan.
Source: Drawn by the author based on the data of S. Ishigai, Design Principles of Steam Boiler, Sankai-do, Tokyo, 1961, pp. 58–59 [13].

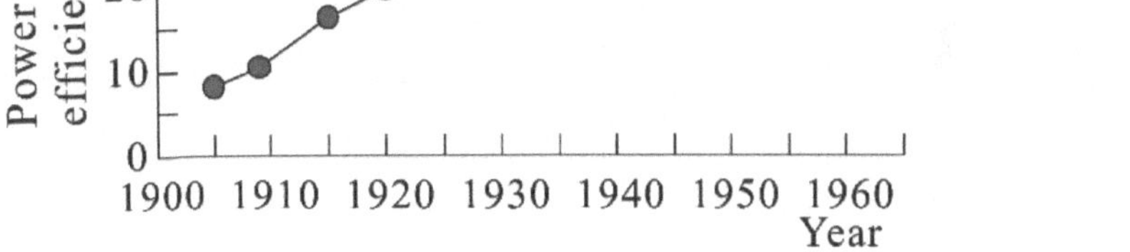

Figure 2.29 Trend of power generation efficiency of Philadelphia Electric Co.
Source: Drawn by the author based on the data of Gubler, Eddystone steam power station, Sulzer Tech. Rev. 41 (1) (1959) 3–20 [29] and S. Miyaoka, Supercritical unit and its economic potential, Therm. Power 1 (11) (1965) 902–916 [30].

electrostatic precipitator or bag-filter, desulfurization equipment, and denitration equipment were installed for preventing diffusion of sulfur oxide (NOx) and nitrogen oxide (SOx). In the oil-fired power stations, denitration equipment, electrostatic precipitator and desulfurization equipment are arranged in series in the exhaust line

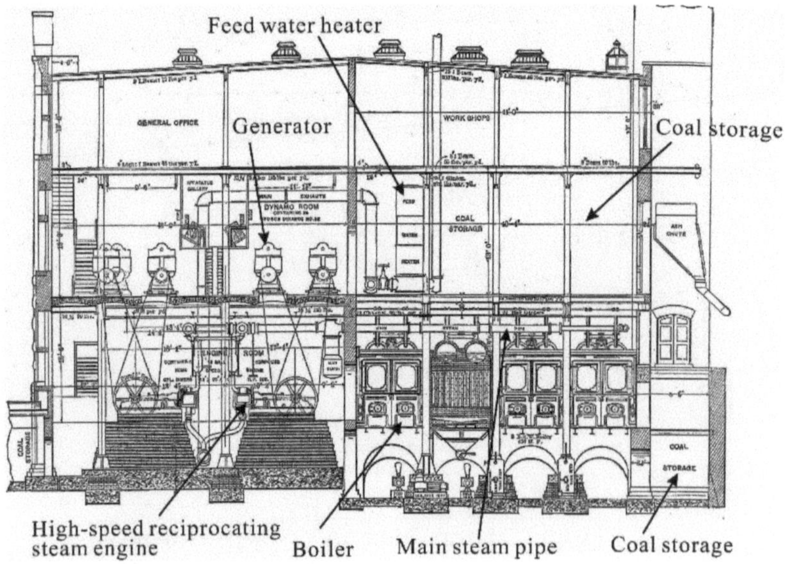

Figure 2.30 Coal-fired power station with reciprocating steam engines of Edison Electric Illuminating Co. in New York (1889) [17].

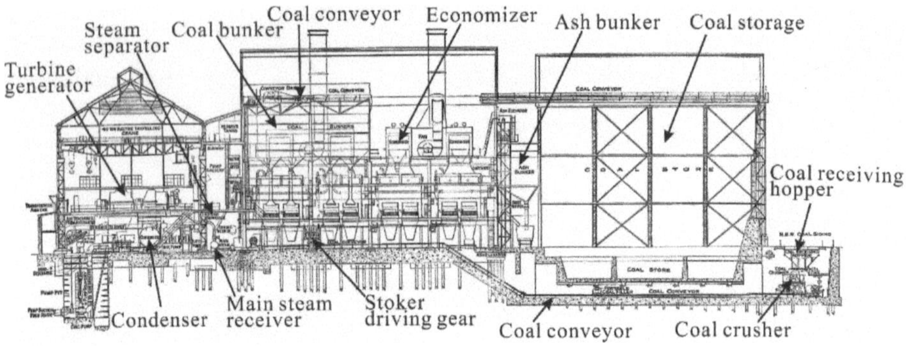

Figure 2.31 Coal-fired power station with steam turbines of Newcastle-Upon-Tyne Electric Supply Co. at around 1910 [31].

to chimney. Gas-fired power stations needs only denitration equipment to suppress NOx emission to the environment. Further detailed discussions on the power plant will be found in Vol.2 Advances in Power Boilers.

2.1.6 The road to modern steam power generation

Fig. 2.33 shows transition of fuels for power generation in this six decades. The fuel for fossil-fired power stations was grain coal since the Newcomen's engine.

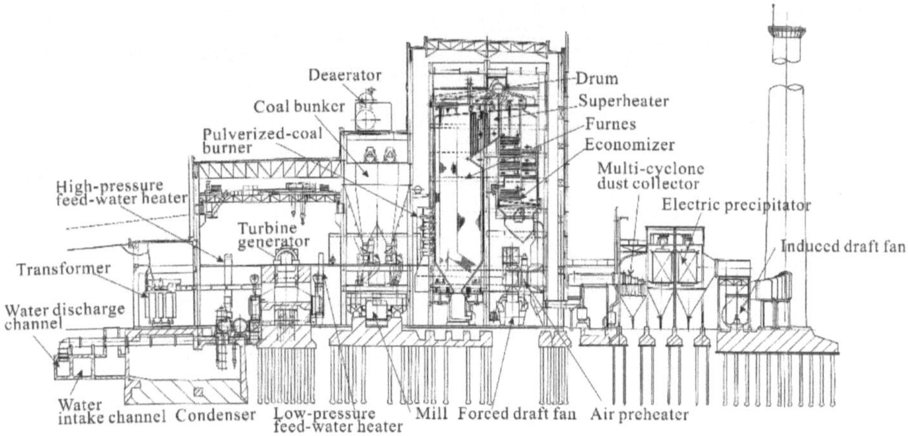

Figure 2.32 Coal-fired power station with radiant boiler commissioned in 1959 (12.5 MPa, 538/538°C, 430 t/h, 125 MW) [32].
Source: Courtesy Kansai Electric Power Co.

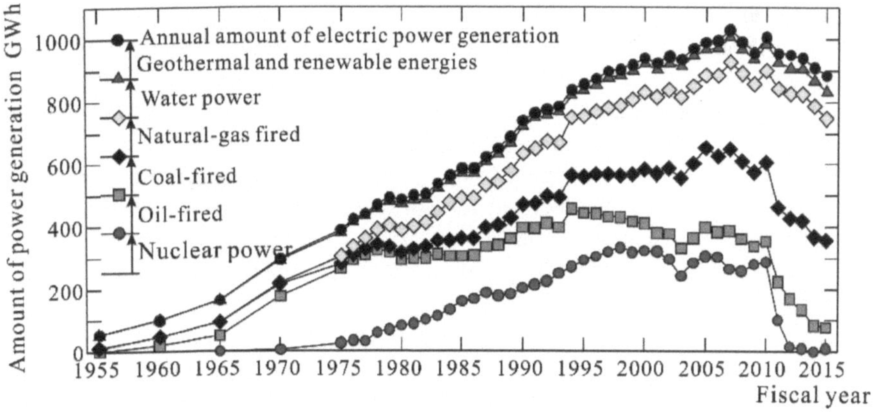

Figure 2.33 Transition of fuels for power generation in Japan.
Source: Drawn by the author based on the statistical data of the Federation of Electric Power Companies of Japan.

Then the pulverized coal technologies were developed around the end of 1880s. Since then pulverized coal was extensively used for power stations. Figure shows the transition of fuels for power generation in Japan. Up to around 1950, coal-fired plant dominated in Japan, while around 1955 oil utilization started to increase, and around 1960 became dominant fuel for power generation in Japan. Nuclear power station with gas-cooled reactor was first commissioned at Tokai-mura in 1965. Substantial increase in the nuclear power started at 1970 with light-water nuclear reactor of Tsuruga and Mihama. The number increased steadily as shown in Fig. 2.33, up to the middle of 1990s.

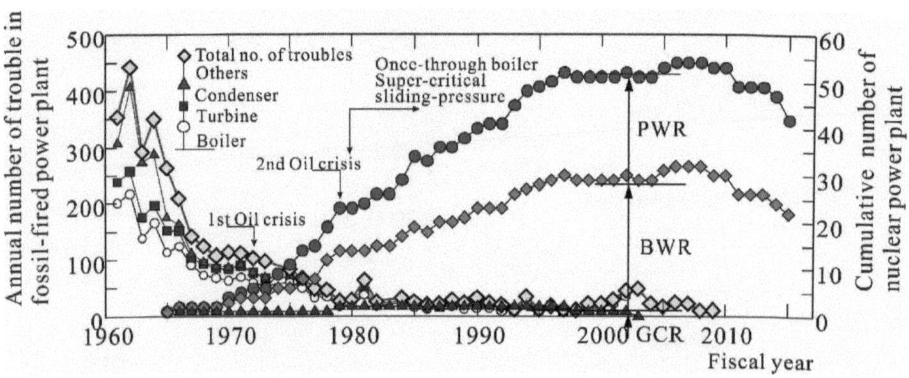

Figure 2.34 Trend of troubles in fossil-fuel-fired power stations and cumulative installation number of nuclear power plants in Japan.
Source: Drawn by the author based on the data from the Ministry of Economy, Trade and Industry, Japan, and WNA World Nuclear Association, Reactor Database.

The troubles in fossil-fuel-fired power stations are shown in Fig. 2.34. The number at around 1960 was very high but decreased drastically toward the end of 1960s.The number further decreased in 1970s. The first drastic decrease was due to the boiler technology development together with the computer development. Nuclear power plants have been applied for base-load power generation, then the large variation of electricity demand has been covered by fossil-fuel power plants. The plant operation below the rated power generally causes a decrease in plant efficiency. Thus to suppress the decrease the variable-pressure-operation technology as well as the daily-start and stop (DSS) operation were introduced.

One of the typical examples of the flow system of supercritical variable-pressure once-through boiler and turbine plant is shown in Fig. 2.35.

A few centuries backward, when coal was only the fuel for heating and steam generation, so-called London fog frequently took place. The highest number of occurrences was beyond 80 times in 1880s [33], which was the first air pollution problem. Acid rainfalls were also the severe problems at that time. The main reason might be the emissions of dust and SOx from personal and industrial chimneys. To achieve complete burning of coal, relatively large residence time or large furnace are needed. In accordance with the development in combustion technology, such pollution problems were successively solved, while the total emission steadily increase together with an increase in the number of power generation units and automobiles by 1960s. Owing to severe air pollution, “Clean Air Act” of 1963 in the US and similarly in Japan “Air Pollution Control Act” of 1968 were enforced. Almost the similar time, the Club of Rome reported “The Limits of Growth” [34] in 1972. This report suggested the growth of the world economy was limited first by fuel resources, while turned around to be limited by the environment. Such a social movement as well as the oil cries in 1973 and 1979 in Japan results in a decrease in number of coal- and oil-fired power stations, instead nuclear and

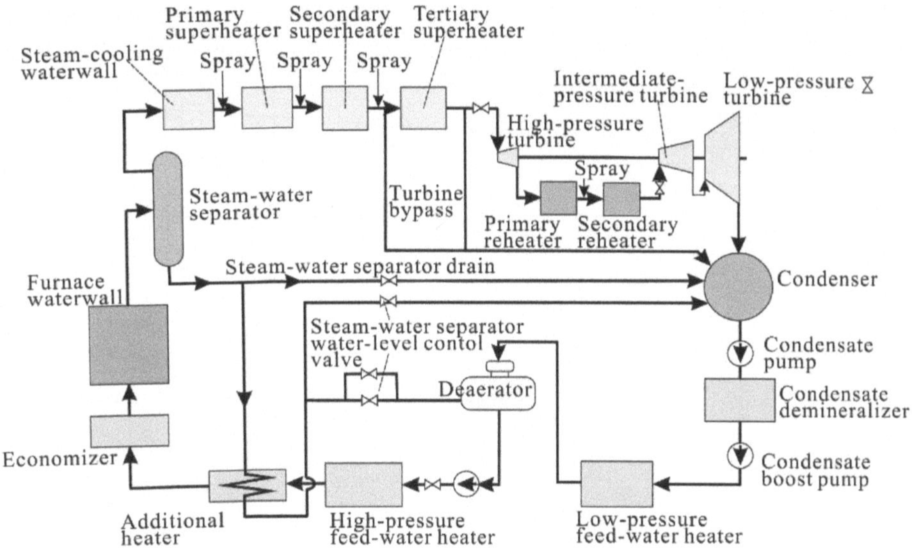

Figure 2.35 Flow diagram of supercritical variable-pressure once-through boiler.
Source: Drawn by the author referring to the figure of K. Akagawa, Thermal and hydraulic design of steam-generation systems, in: S. Ishigai (Ed.), Steam Power Engineering, Cambridge University Press, New York, 1999, pp. 255–283. [16]

natural-gas-fired power plants increased in number, being effective strategies for energy security especially in Japan. Natural-gas-fired plants have a merit of no emission of SOx and less-emission of carbon dioxide relative to coal and oil.

The United Nations Framework Convention on Climate Change since 1994 claimed to prevent global warming via a reduction in carbon-dioxide (CO_2) emission. One of the solutions is the development in energy-saving technology, and alternatively the extension of nuclear power is a prevailing option, while the safety and radioactive waste issues were fully unsolved especially in Japan. The general public has a tend to believe that an extension of renewable-energy utilization seems a powerful and promising option, while the renewable energy, for example, solar and wind power, is unstable, and thus some backup systems, for example, thermal power plant, are essential for stable electricity supply.

Development of high-efficiency power plant is also very important solution for the reduction of CO_2, while simple power plant efficiency is at most 40 several percent. To utilize high-temperature energy of combustion gas in the furnace, it is effective to substitute a topping cycle, that is gas-turbine cycle, on a conventional steam cycle, which is referred to as gas turbine/steam turbine combined plant. Natural gas is suitable for such combined plant. The combined plant started 1980s with relatively small capacity, while an increase in gas turbine inlet temperature (TIT) has been successively conducted and in the middle of 1980s over 1000 MW unit with 1154°C TIT was commissioned. Currently, the TIT has increased further to around 1600°C, and then the plant efficiency becomes over 60% (LHV base).

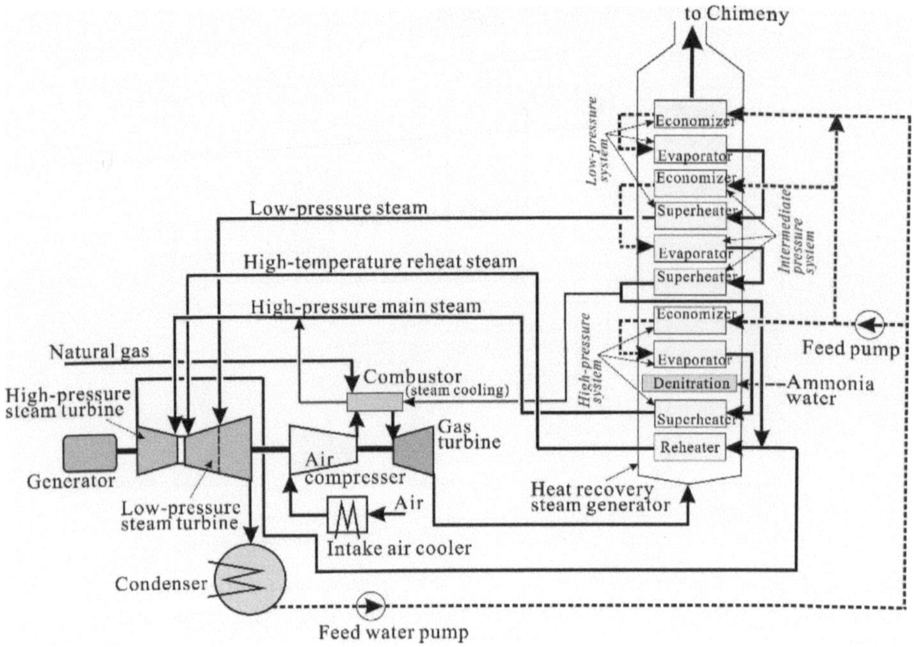

Figure 2.36 Example of natural-gas-fired gas-turbine steam-turbine combined power plant. *Source:* Drawn by the author referring to several articles of The Thermal and Nuclear Power and Catalogue of the Kansai Electric Power Co.

Typical example of natural-gas-fired gas-turbine/steam-turbine combined plant is shown in Fig. 2.36. In this example, steam turbine with high-, intermediate- and low-pressure cylinder and compressor/gas turbine are arranged to form tandem-compound. The heat recovery steam generator is composed of reheater and three-stage steam generators, that is high-, intermediate- and low-pressure units consisted of superheater, evaporator and economizer. To reduce NOx, the denitration equipment with ammonia water is installed.

2.2 Development of nuclear power generation

2.2.1 Dawn of nuclear energy

In 1938 nuclear fission was discovered by Otto Hahn and Fritz Strassmann [35]. The following year in 1939, Lise Meitner and Otto Robert Frisch gave a theoretical explanation to the nuclear reaction and named it "nuclear fission." which can release large amounts of energy both as electromagnetic radiation and as kinetic energy of the fragments. Scientists considered the possibility that the fission chain reaction could be employed as a source of power. However, World War II began in 1939, and scientists focused their thoughts on development of atomic bomb using the chain reaction.

One of the "fissile materials" is Uranium-235 (^{235}U), which exists at 0.7% in natural uranium. Without isotopic enrichment available, only a handful of configurations could sustain a chain reaction. The Manhattan Project included work on uranium enrichment to procure ^{235}U in high concentrations and research on reactor development. Two main goals of this project were to investigate the chain reaction for atomic bomb development and to develop a method of producing a new element, plutonium, which was expected to be fissile and could be chemically isolated from uranium.

2.2.1.1 First artificial chain reaction at Chicago Pile No.1

Under the Manhattan Project, Enrico Fermi supervised reactor development, and began his project at Columbia University. Fermi and his colleagues constructed a graphite-moderated reactor, later called Chicago Pile No.1 (CP-1), using pieces of natural uranium metal dispersed carefully in a lattice of high-purity graphite blocks in a Chicago squash court as shown in Fig. 2.37. In 1942 they succeeded to attain the chain reaction in the reactor as shown in Fig. 2.38.

2.2.1.2 Hanford B reactor

After the success of CP-1 experiments, the first high-power reactors were designed and built to produce plutonium as fuel for nuclear explosives. The B reactor in Hanford site was a graphite-moderated reactor with metallic natural uranium shown in Fig. 2.39. It was designed to operate at 250 MW (thermal) cooled with ordinary water, based on experimental designs by Enrico Fermi and test from the X-10 Graphite Reactor at Oak Ridge National Laboratory. It was designed to operate 250 MW, cooled by ordinary water. Since water is a neutron absorber, the reactor

Figure 2.37 Chicago Pile No.1.

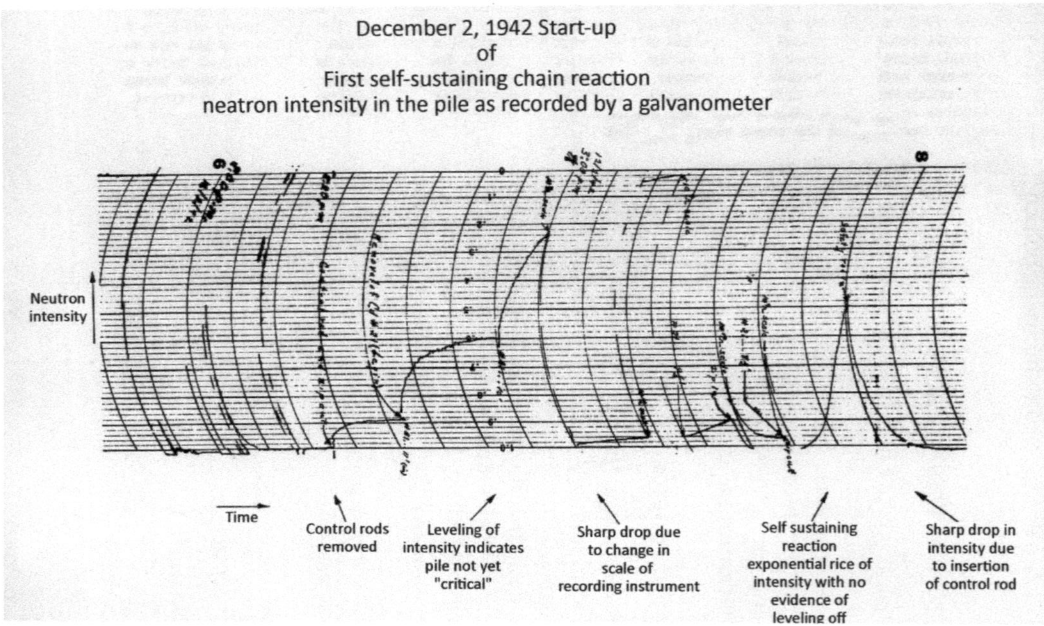

Figure 2.38 First self-sustaining chain reaction at CP-1.

Figure 2.39 Front face of the B reactor at the Hanford site.

had to be large, and it required extra-pure graphite with minimal neutron-absorbing impurities. It also needed a lot of metallic natural uranium. The reactor produced plutonium-239 (^{239}Pu) by irradiating uranium-238 (^{238}U)with neutron generated by the nuclear reaction. Generated plutonium was delivered to Los Alamos and used in world's first explosion (Trinity Test Site, New Mexico) and also for Fat man nuclear bomb dropped on Nagasaki, Japan.

On the morning of April 26, 1944, Enrico Fermi, Leo Szilard, Eugene Wigner, Alvin Weinberg, and others gathered to discuss the possibilities for using nuclear fission t heat and light cities. However, it was unclear at that time whether there was sufficient uranium for producing highly enriched uranium and plutonium for a significant number of nuclear weapons. They recognized that some reactor configuration might permit the conversion of ^{238}U to fissile plutonium at a rate faster than the fissile ^{235}U was consumed. This was promoting the development of "breeder reactor."

2.2.1.3 Clementine reactor

The world's first fast-neutron reactor was Clementine, a 25 kW (thermal), mercury-cooled experimental reactor built at Omega Site at Los Alamos. It was proposed and approved in 1945. High intensities of fission-spectrum neutrons were needed by the bomb designers. Also, operation of the reactor would supply information about fast reactors that would be relevant to their possible use for production of power and fissile materials. Construction began in August 1946, criticality was achieved in late-1946, and full power in March 1949. The fuel was plutonium metal with natural uranium slugs at each end of the steel-clad rods. The rods were installed in a steel cage through which the liquid-mercury coolant flowed, driven by an electromagnetic pump.

The experience and data provided by operating the Clementine reactor was particularly useful for both military and civilian applications. One of the notable achievements of the Clementine project included measurements for the total neutron cross sections of 41 elements to a 10% accuracy. Additionally, Clementine provided invaluable experience in the control and design of fast neutron reactors. It was also determined that mercury was not an ideal cooling medium for this type of reactor due to its poor heat transfer characteristics.

After Clementine, Los Alamos developed and briefly operated one additional fast reactor, LAMPRE-I. This sodium-cooled reactor was fueled with molten plutonium. It achieved initial criticality in early-1961 and operated successfully for several thousand hours until mid-1963.

2.2.1.4 Experimental breeder reactor I

As part of the National Reactor Testing Station, a construction of Experimental Breeder Reactor I (EBR-I) started in late 1949, which was a world's first sodium-cooled breeder reactor. The reactor design and construction were led by a team of Walter Zinn at the Argonne National Laboratory, Idaho site, known as Argonne-West [36]. The design purpose of EBR-I was not to produce electricity but instead to validate nuclear physics theory that suggested that a breeder reactor should be possible. It began power operation on August 24, 1951. On December 20 of that year, atomic energy was successfully harvested at EBR-1 for the first time (Fig.2.40).

The following day, the reactor produced enough power to light the whole building. The power plant produced 200 kW of electricity out of 1.4 MW of heat generated by the reactor [37].

In 1953, experiments revealed the reactor was producing additional fuel during fission, thus confirming the hypothesis. On November 29, 1955, the reactor at EBR-I suffered a partial meltdown during a coolant flow test as shown in Fig.2.41. The flow test was trying to determine the cause of unexpected reactor responses to changes in coolant flow. It was subsequently repaired for further experiments, which determined that thermal expansion of the fuel rods and the thick plates supporting the fuel rods was the cause of the unexpected reactor response [38].

2.2.1.5 Power generation at AM-1 in the Soviet Union

In 1953 President Eisenhower proposed his "Atoms for Peace" program, which reoriented significant research effort toward electricity generation and set the course

Figure 2.40 The first production of usable nuclear electricity occurred on December 20, 1951, when four light bulbs were lit with electricity generated from the EBR-1 reactor.

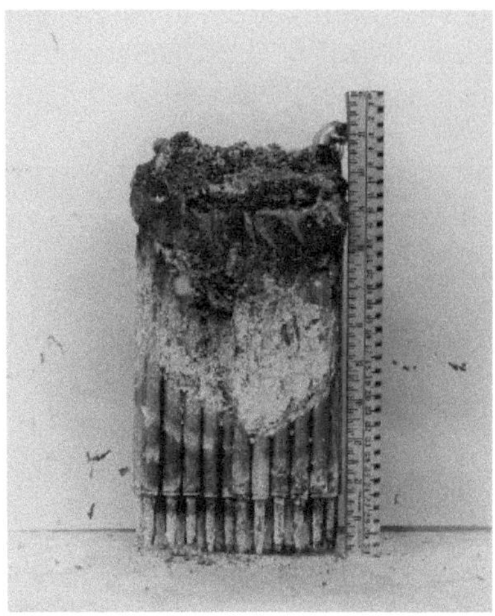

Figure 2.41 Part of the core after the 1955 partial meltdown.

for civil nuclear energy development in the USA [39]. In the Soviet Union, work was under way at various centers to refine existing reactor designs and develop new ones. The Institute of Physics and Power Engineering (FEI) was set up in May 1946 at the then-closed city of Obninsk to develop nuclear power technology. The existing graphite-moderated channel-type plutonium production reactor was modified for heat and electricity generation and in June 1954 the world's first nuclear powered electricity generator began operation at the FEI in Obninsk. The AM-1 (Atom Mirny: peaceful atom) reactor was water-cooled and graphite-moderated, with a design capacity of 30 MWt or 5 MWe. It was similar in principle to the plutonium production reactors in the closed military cities and served as a prototype for other graphite channel reactor designs including the Chernobyl-type RBMK (reaktor bolshoi moshchnosty kanalny – high power channel reactor) reactors. AM-1 produced electricity until 1959 and was used until 2000 as a research facility and to produce isotopes. In the 1950s FEI at Obninsk was developing fast breeder reactors (FBRs) and lead-bismuth reactors for the naval use.

2.2.1.6 Naval reactor

The main US effort was under Admiral Hyman Rickover, which developed the nuclear reactor for naval use [40]. The prototype naval reactor started up in March 1953 in Idaho (Fig. 2.42), and the first nuclear-powered submarine, USS Nautilus (Fig. 2.43), was launched in 1954. A schematic of naval reactor is shown in Fig. 2.44, which can generate electricity and propulsion energy. Nautilus was powered by the submarine thermal reactor (STR), a pressurized water reactor (PWR) produced for the US Navy by Westinghouse Electric Corporation. In addition, the nuclear propulsion allowed the Nautilus to remain submerged much longer time than diesel-electric submarines and to realized 1800-mile travel under the Arctic ice in the summer of 1958 from the Bering Strait to the eastern coast of Greenland.

During the same period, a submarine intermediate reactor (SIR) had been developed. The propulsion system was more technologically advanced. SIR was designed

Figure 2.42 STR construction for USS Nautilus.

Figure 2.43 USS Nautilus, first nuclear submarine.

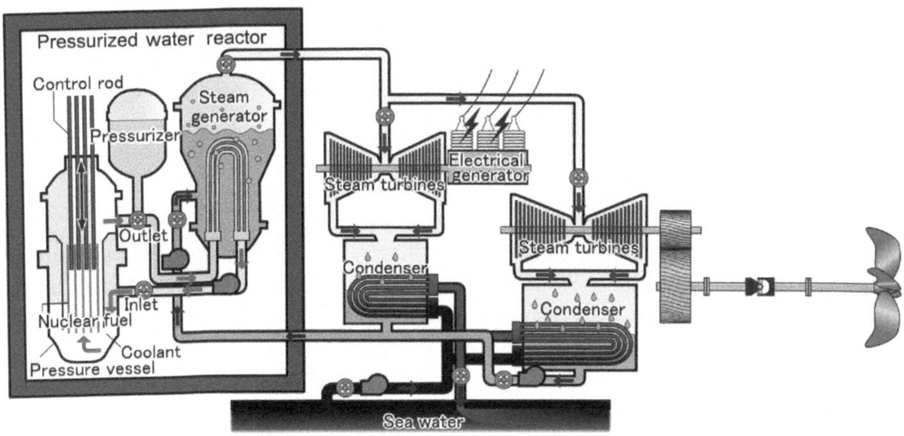

Figure 2.44 Schematic of submarine thermal reactor.

by General Electric's Knolls Atomic Power Laboratory. SIR was a liquid-sodium cooled epithermal reactor, and more powerful rather than Nautilus's alternative light-water reactor, reduced size of machinery in the engineering spaces nearly 40%. The liquid-sodium cooled reactor was more thermally efficient than a light water-cooled reactor, quieter, and presumably better system, but posed several additional safety hazards for the ship and crew over already hazardous submarine and naval warcraft conditions. Primary system pressure was 100 kPa and the only moving part in the primary system was the liquid sodium which was magnetically pumped by electromagnetic pump. There was only one coolant leak ever noted, and that was during fitting out in the yards. However, the superheaters suffered from tube sheet welding cracks with allowed high-pressure steam to leak into the low-pressure primary system and react with the sodium. USS Seawolf was the only US submarine with SIR or liquid-metal cooled reactor. In 1959 both USA and USSR launched their first nuclear-powered surface vessels.

During the same period, a nuclear-powered aircraft had been studied. The intention was to produce a jet engine that would heat compressed air with heat from fission reactions, instead of heat from burning fuel. The United States and Soviet Union researched such nuclear-powered aircraft, but neither country succeeded in development of such nuclear-powered aircraft. In the United States, Aircraft Reactor Experiment (ARE) was performed, which was a 2.5 MWt thermal-spectrum nuclear reactor experiment designed to attain a high power density and high output temperature for use as an engine in a nuclear-powered bomber aircraft. The ARE was the first molten salt reactor (MSR) to be built and operated. It used the molten fluoride salt NaF-ZrF4-UF4 as fuel, was moderated by a hexagonal-configuration beryllium oxide (BeO), and had a peak temperature of 860°C. While the ARE successfully demonstrated operation of a MSR concept, the program was canceled in 1961 citing the outrageous cost.

Various nuclear reactors with different concepts have been tested in 1950's, including the graphite-moderate water-cooled, gas-cooled, light/heavy water-cooled, and liquid-metal cooled. However, due to the huge success of the Nautilus and PWR, the nuclear power plant design has been developed mainly with light (or heavy) water-cooling system.

Along with the development of naval reactor, developments of large-scale nuclear power plants became the next directions in 1950's and 1960's.

2.2.1.7 BORAX experiments

In boiling water reactor (BWR), steam passes directly from the reactor to the turbine as shown in Fig. 2.45 [41], but in PWR the heat generated in the core is transferred to the primary pump and the secondary coolant is boiled in the intermediate heat exchanger (boiler) shown in Fig. 2.46. The pressure in the reactor core is much higher than that in BWR. In BWR, the same operating steam pressure can be achieved with a reactor vessel pressure. In the direct cycle only the reactor feed pumps are necessary, and since the size of these units is considerably smaller than the coolant circulating pumps in PWR or in boiling cycles using forced circulation, the pumping requirements of natural circulation boiling reactors are comparatively small.

The BORAX Reactor Program owed its beginning to a nuclear criticality accident that had occurred earlier on June 2, 1952 at the Argonne National Laboratory near Chicago. Some tests were being conducted on a Critical Assembly at Argonne to measure the characteristics of various control rods used in the reactor power plants on nuclear submarines. The Critical Assembly was a mockup of the Nautilus

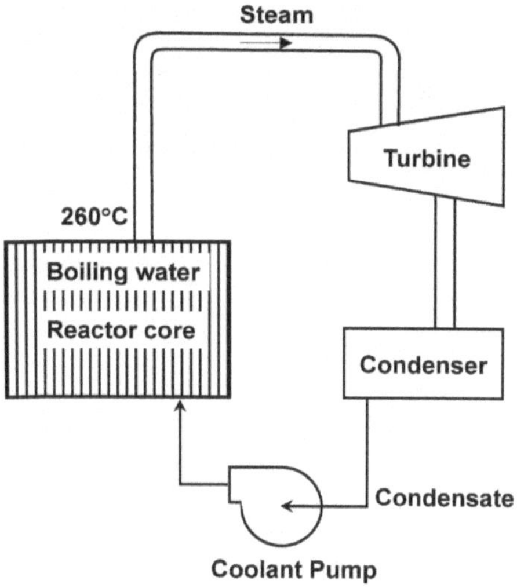

Figure 2.45 Direct cycle (BWR) system.

nuclear submarine reactor. It had real fuel elements and could be operated something like the real reactor but only at extremely low power. During these tests there occurred an operator error which caused the Critical Assembly to go prompt critical and resulted in a steam explosion within the assembly. The Critical Assembly was damaged, and four operators received substantial radiation exposure and a shower bath as the water in the Critical Assembly was expelled from the assembly tank.

The Argonne staff marveled that the incident was so comparatively mild. It had previously been assumed that a water moderated reactor would be very unstable if allowed to go into a boiling state. The criticality incident gave indication that, perhaps, a water moderated reactor could be made to operate in a stable boiling mode. The objective of the BORAX experiments was to confirm the stability of the BWR under the operating and the accidental conditions.

The first BORAX-I was built in 1952 to determine the stability of a reactor using boiling water both as a heat transfer agent and moderator. It used aluminum-uranium alloy MTR type fuel elements. The reactor characteristics of other experiments are shown in Table 2.2.

The BORAX experiment can be divided roughly into four groups, in term of the differences in the reactor arrangements. The first of these experiments, BORAX-I (Fig. 2.47), was conducted to investigate whether or not the formation of steam in the reactor would reduce the reactivity and so cause the system to become self-regulating. BORAX-I experiment proved that a reactor using direct boiling of water would be practical, rather than unstable, because of the bubble formation in the core. Subsequently the reactor was used for power excursion tests which showed that rapid conversion of water to steam would safely control the reaction as shown in Fig. 2.48. The final, deliberately destructive test in 1954 produced an unexpectedly large power excursion that "instead of the melting of a few fuel plates, the test

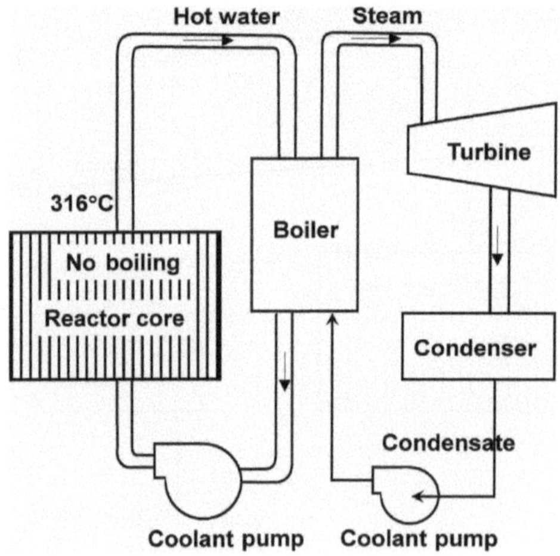

Figure 2.46 Indirect cycle (PWR) system.

Table 2.2 Reactor characteristics [41].

Reactor	Maximum pressure, psig	Fuel material	Fuel conductivity	Operator
BORAX-I	130	Al + U235	High	Argonne
BORAX-II	300	Al + U235	High	Argonne
BORAX-III	300	Al + U235	High	Argonne
BORAX-IV	300	Al + U235	Low	Argonne
SPERT-I	0	Al + U235	High	Phillips Petroleum
EBWR	600	U-Zr-Nb	Intermediate	Argonne
Vallecitos	1000	SS + U235	High	General Electric

melted a major fraction of the entire core." However, this core meltdown and release of nuclear fuel and nuclear fission products provided additional useful data to improve mathematical models. The tests proved key safety principles of the design of modern nuclear power reactors. Design power of BORAX-I was 1.4 megawatts thermal. Although BORAX-I was a non-pressurized reactor and tests were performed with water at various temperatures. The BORAX-I design was a precursor to the SL-1 plant, which was sited nearby and began operations in 1958. The principles discovered in the BORAX-I experiments helped scientists understand the issues that contributed to the fatal incident at SL-1 in 1961 .

After additional tests in the summer of 1954, BORAX-I was destroyed in a violent subcooled power excursion by the sudden formation of steam in the reactor vessel. Shortly after the destructive experiment of BORAX-I, BORAX-II was constructed, utilizing the same control equipment and many of the actual parts of BORAX-I which had been salvaged after the explosion. This new reactor was capable of operation at 300 psi and with twice the BORAX-I core volume, delivered about 6 MW.

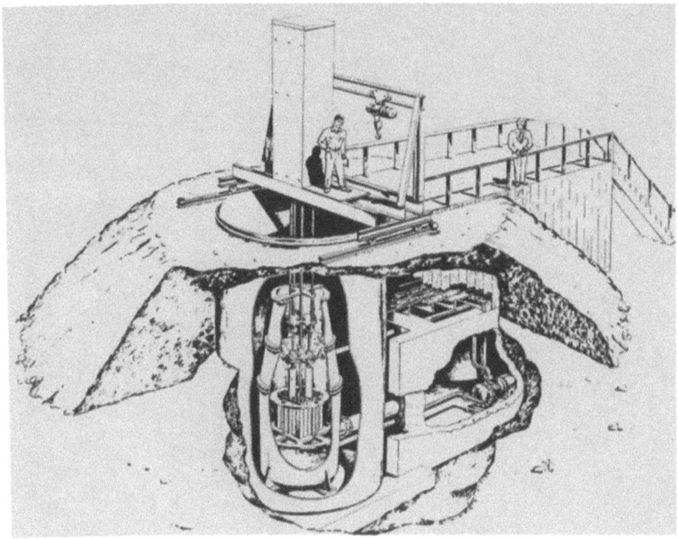

Figure 2.47 Schematic of BORAX-I.

Figure 2.48 Excursion experiment at BORAX-I.

In 1955, BORAX-II was installed. This was essentially a new core in the same pressure vessel used for BORAX-II and at the same time a turbine and generator were added to provide a complete power generating system. In March 1955 BORAX-II was intentionally destroyed by taking the reactor "prompt critical."

Figure 2.49 BORAX-III reactor provided all the electricity for Arco, Idaho, in 1955.

The design of BORAX-II was modified into BORAX-III with the addition of a turbine, proving that turbine contamination would not be a problem. It was linked to the local power grid for about an hour on July 17, 1955.

BORAX-III provided 2000 kW to power nearby Arco, Idaho (500 kW), the BORAX test facility (500 kW), and partially powered the National Reactor Testing Station (after 2004, the Idaho National Laboratory) (1000 kW). On July 17, 1955, BORAX III reactor provided all the electricity for Arco, the first time any community's electricity was provided entirely by nuclear energy (Fig. 2.49). The reactor continued to be used for tests until 1956.

BORAX-IV, built in 1956, explored the thorium fuel cycle and uranium-233 fuel with a power of 20 MW thermal. This experiment utilized fuel plates that were purposely full of defects in order to explore long-term plant operation with damaged fuel plates. Radioactive gases were released into the atmosphere.

BORAX-V continued the work on BWR designs, including the use of a superheater. It operated from 1962 to 1964.

2.2.2 Development of nuclear power plant

2.2.2.1 Power plants in Soviet Union

Many efforts had been made to use nuclear energy for peaceful applications [42]. One of those was to build a large-scale commercial nuclear power plant. In Soviet

Union, AM-1 nuclear power plant was built in the Science City of Obninsk, about 110 km southwest of Moscow and connected to power grids in June 1954. Obnisk was the first grid-connected nuclear power plant in the world. The single reactor unit at the plant, AM-1, had a total electrical capacity of 6 MW and a net capacity of around 5 MWe. Thermal output was 30 MW. It was a prototype design using a graphite moderator and water coolant, which was a forerunner of the RBMK reactors.

2.2.2.2 Power plants in United Kingdom

The first full-scale nuclear reactor in the United Kingdom was the Windscale Pile in Sellarfield. The reactor consists of a huge cube of highly purified graphite made up of many smaller blocks and drilled through horizontally to make a large number of fuel channels. The system was designed to work at low temperatures and power levels and was air-cooled with large fans.

However, graphite is flammable and presents a serious safety risk. This was demonstrated on October 10, 1957 when Unit 1 caught fire.

In order to be economically useful, the plant would have to run at much higher power levels, and in order to efficiently convert that power to electricity, it would have to run at higher temperature. Therefore, carbon dioxide (CO_2) was chosen as the coolant for large-scale nuclear power "Magnox" plant in UK.

The first Magnox plant in Calder Hall was first connected to the grid on August 27, 1956 and was the world's first power generate electricity on an industrial scale (four 60 MWe plants) from nuclear energy.

2.2.2.3 Power plants in United States

In the United States, the Shippingport Atomic Power Station began supplying electricity to the Duquesne Light Company, serving the Greater Pittsburgh, Pennsylvania on December 18, 1957. It was a PWR, generating electrical capacity of 60 MWe. A that time, there had been a growing realization that planning for future nuclear power plants of two or three hundred megawatts electrical output could be made with greater confidence if operating experience from a reactor plant of intermediate power were available. At same time, another project was going on to develop a large military PWR for a naval ship. Therefore, it was better way to construct PWR type power plant extrapolated from the preliminary work carried out under naval project. In addition, it would speed nuclear power development both in the United States and abroad.

From inception, the main purpose of the Shippingport plant has been to advance the technology of pressurized water reactors rather than to generate electricity at costs competitive with ordinary fuels. Fig. 2.50 shows the pressure vessel of the Shippingport power plant. A schematic diagram of the cooling system of the Shippingport reactor is shown in Fig. 2.51. The reactor coolant system consists of four closed piping loops connected in parallel to the reactor vessel. In the primary loop, high purity water under 136 atm serves as both coolant and moderator. At full

Figure 2.50 Pressure vessel installation.

power, with three loops operating the coolant temperature is 264°C at inlet and 281°C at outlet of the reactor vessel. Each coolant loop contains two inlet stop valves, a steam generator, a coolant circulating pump, a check valve, a venturi flowmeter, temperature and pressure instrumentation, and two outlet stop valves. Each loop also has a bypass line connecting the inlet piping to the outlet piping, bypassing the reactor vessel.

Each of the four coolant loops contains a steam generator for use as a heat exchanger. Two of the loops use a straight-tube heat exchanger, and the other two use a U-bend heat exchanger as shown in Fig. 2.52. Because of this difference in steam generator design, two different loop arrangement were used. Experience gained in this design, manufacture, and operation of both types was helpful in future reactor plant design. This reactor, the largest of the power-reactor prototypes, went online in 1957.

It was considered probable that future PWR plants would tend toward higher steam pressure to increase steam plant efficiency and reduce operating costs.

Saturated steam was selected since to superheat 600 psi steam with the core maximum surface temperature available was not considered feasible. This maximum temperature was set at 636°F on the basis of the current Zircaloy corrosion data and a further design requirement that coolant boiling in the core during normal steady state conditions must be prevented.

It was recognized that for a fixed steam temperature, steam generator heat transfer surface and cost decrease, but that core heat transfer surface cost increase as the

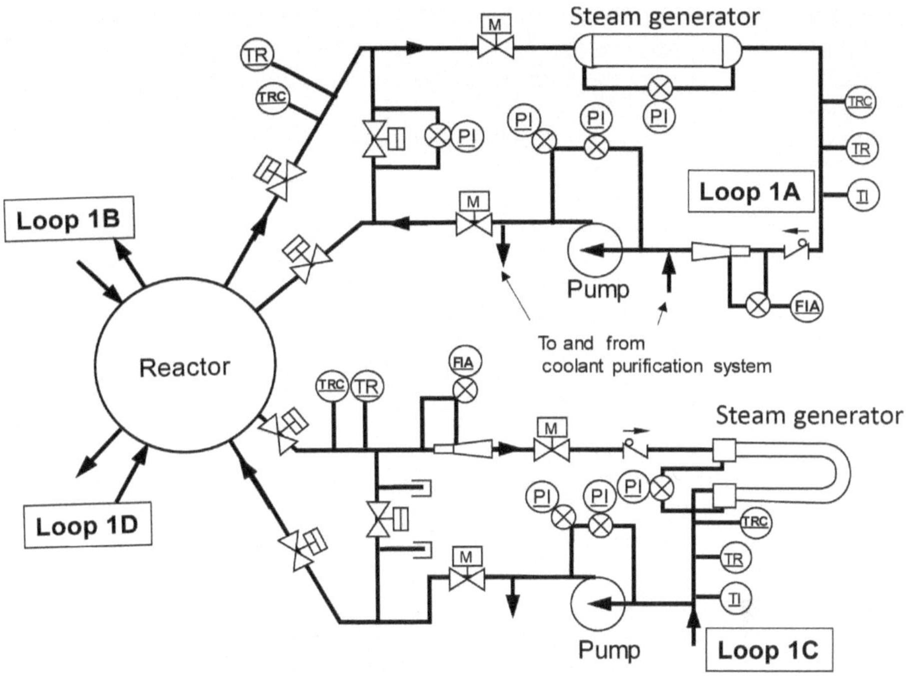

Figure 2.51 Cooling system of shippingport reactor.

Figure 2.52 U-bend heat exchanger.

coolant temperature increases. However, because of the much higher cost of core heat transfer surface, the average coolant temperature at full power was et while still maintaining a practical steam generators size. The average temperature so chosen was 523°F.

The main criteria forming the basis of the thermal design of the PWR core were as follows:

1. Local boiling was not to occur in the core under normal full load conditions, because of a lack of burnout date for flow geometries other than flow in single round tubes.
2. To avoid any possibility of burnout, bulk boiling was not to occur in the water leaving the hottest channel during a loss-of-coolant-flow accident
3. The maximum temperatures in the core components were not to exceed permissible values based on the mechanical and metallurgical properties of the material involved. This criterion did not impose a severe limitation on the design except insofar as temperature differences induce thermal stresses in the materials.

In addition to the aforementioned factors, the thermal and hydraulic operating characteristics of the PWR core were dependent on the power plant parameters and on the detailed power distribution pattern and associated cooling within the core.

Table 2.3 shows nuclear power reactors planned to operate before 1963 in the United States. As can be seen here, many water-cooled power plants were constructed including various types power plants in the 1960s. It can be said that most of the basics of water-cooled nuclear power plants had been established in the 1960s.

Table 2.3 Nuclear power reactors planned for operation before 1963 [42].

Project	Reactor type	Power, electric kW	Startup
Experimental Boiling Water Reactor ANL	Boiling water	5000	1956
Sodium Reactor Experiment NAA SCE	Sodium graphite	6000	1957
General Electric & Pacific Gas & Elec.	Boiling water	5000	1957
Shippingport Pressurized Water Reactor WAPD Duquesne	Pressurized water	60,000	1957
Homogeneous Reactor Experiment ORNL	Aqueous homogenous	1000	1957
Experimental Breeder Reactor No. 2 ANL	Sodium fast reactor	20,000	1960
Rural Coop. Power Association	Boiling water	22,000	1960
Commonwealth Edison G.E.	Boiling water	180,000	1960
Consolidated Edison Co. B. & W.	Pressurized water	275,000	1960
Power Reactor Development Co. APDA	Sodium fast breeder	100,000	1960
Yankee Atomic Electric. Co. West.	Pressurized water	134,000	1960
City of Piqua	Organic moderated	12,500	1961
Consumers Public Power District	Sodium graphite	75,000	1962
Northern States Power Co.	Boiling water	66,000	1962
East Central Florida	Heavy water, gas-cooled	50,000	1963
Pacific Gas and Electric.	Boiling water	50,000	1962
Chugach Electrical Association, Inc.	Sodium, heavy water	10,000	1962
Total		1,071,500	

Of the prototype commercial nuclear power plants had been built during the late 1950s and early 1960s, the most successful types used the light-water reactor system, either PWR or BWR. From the mid-1960s, larger units were ordered by the early 1970s, nuclear plant orders were coming in at such a rapid pace.

2.2.3 Growth of nuclear power plants and nuclear accidents

The first fully commercial PWR in the United States was Yankee Rowe of 250 MWe, which started up in 1960 and operated to 1992. The first commercial BWR, Dresden-1 of 250 MWe, was started up earlier in 1960. By 2020, more than 100 nuclear power plants have been constructed in the United States. In Canada, different type of nuclear reactor had been developed using natural uranium fuel and heavy water as a moderator and coolant, which is called CANada Deuterium Uranium reactor (CANDU). In 1964 the first two Soviet nuclear power plants were commissioned. A 100 MW boiling water graphite channel reactor (RBMK) began operating in Beloyarsk. In Novovoronezh, a small Russian PWR known as a VVER was built. The first large RBMK of 1000 MW started up at Sosnovy Bor near Leningrad in 1973, and in the Arctic northwest a VVER with a rated capacity of 440 MW began operating. In Kazakhstan the world's first commercial prototype fast neutron reactor (the BN-350) started up in 1972 with a design capacity of 135 MWe, producing electricity and heat to desalinate Caspian seawater.

Nuclear energy has been deployed for over five decades, accounting for 17% of global electricity production during the 1990s and early 2000s from 440 operating units in over thirty countries. Such nuclear power plants can be found in all continents except Australia and Antarctica. Fig. 2.53 shows variation of net electricity generated by nuclear [43].

Nuclear energy differs from other energy options such as oil, coal, natural gas, hydro, and renewables because of concerns with safety, long-term storage of nuclear waste, high capital costs, and nuclear weapons proliferation. Several severe accidents have occurred since 1970s, such as Three Mile Island Unit 2 (TMI-2) accident in 1979, Chernobyl disaster in 1986, and Fukushima Daiichi Disaster in 2011. Two of them,

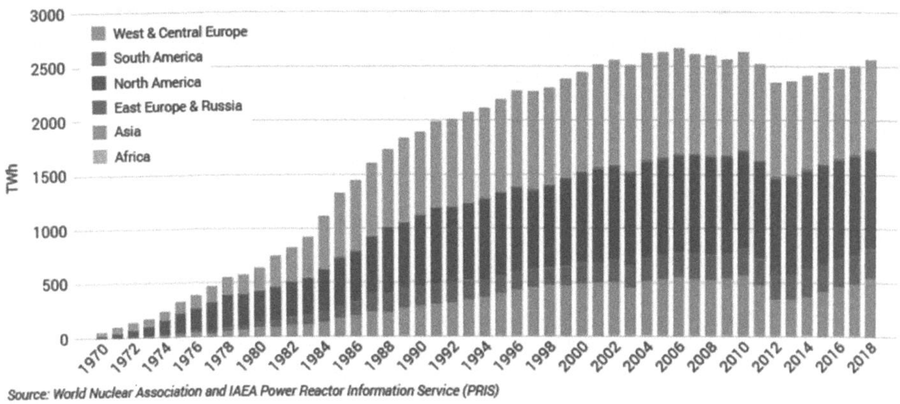

Figure 2.53 History of net electrical generation in the world [43].

TMI-2 accident and Fukushima Daiichi Disaster took place due to Loss-Of-Coolant Accident (LOCA) caused by human error or natural disaster. Nuclear power plants are equipped with the emergency core cooling systems (ECCS), which provide core cooling under LOCA conditions to limit fuel cladding damage including core melt. ECCS consists of the high-pressure coolant injection (HPCI) system, the automatic depressurization system, the low-pressure coolant injection (LPCI) mode of the residual heat removal (RHR) system and some additional cooling systems.

Nuclear power plants are cooled under operation conditions to drive a steam turbine. If something goes wrong or in case of some accidents, the nuclear power plants can be immediately shut down similar to other kinds of power plants. However, nuclear fuels generate decay heat which needs to be cooled by RHR system to avoid fuel damage, even after the reactor shutdown. Despite of safety system, above-mentioned severe accidents took place. Each accident will be briefly mentioned as follows.

2.2.3.1 Three Mile Island-2 pressurized water reactor plant and its accident

Three Mile Island Nuclear Generating Station (TMI) is a closed nuclear power plant located on Three Mile Island in Londonderry Township, Pennsylvania, on the Susquehanna River just south of Harrisburg. It had two separate units, TMI-1 and TMI-2. Fig. 2.54 shows a schematic of TMI-2 unit, which generated 906 MWe. The plant was one of the Gen-II PWR and is widely known for having been the site of the most significant accident in United States commercial nuclear energy, when TMI-2 suffered a partial meltdown. On March 28, 1979, there was a cooling system malfunction that caused a partial meltdown of the reactor core. Fig. 2.55 shows NRC graphic of TMI-2 core end-state configuration.

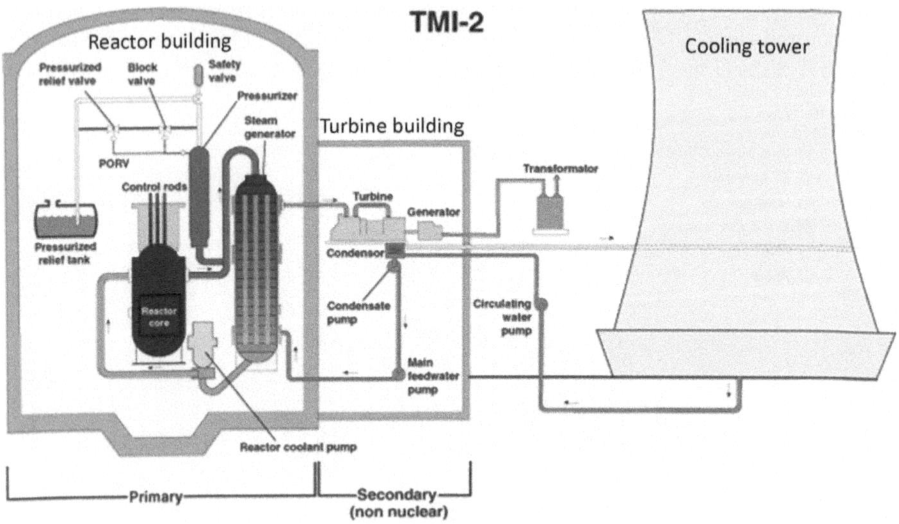

Figure 2.54 Schematic of Three Mile Island No.2 power plant.

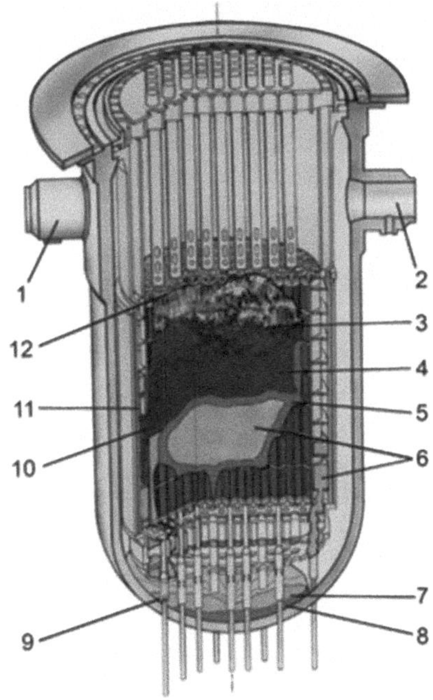

1. 2B inlet
2. 1A inlet
3. Cavity
4. Loose core debris
5. Crust,
6. Previously molten material,
7. Lower plenum debris,
8. Possible region depleted in uranium,
9. Ablated incore-instrument guide
10. Hole in baffle plate,
11. Coating of previously-molten material on bypass region interior surfaces,
12. Upper grid damage

Figure 2.55 TMI-2 core end-state configuration.

The accident began with failures in the non-nuclear secondary system, followed by a stuck-open pilot-operated relief valve in the primary system, which allowed large amounts of nuclear reactor coolant to escape. The mechanical failures were compounded by the initial failure of plant operators to recognize the situation as a LOCA due to inadequate training and human factors, such as human-computer interaction design oversights relating to ambiguous control room indicators in the power plant's user interface. In particular, a hidden indicator light led to an operator manually overriding the automatic emergency cooling system of the reactor because the operator mistakenly believed that there was too much coolant water present in the reactor and causing the steam pressure release.

This LOCA resulted in the release of an amount of radioactivity, estimated at 1.59 PBq of radioactive krypton-85 gas but less than 740 GBq of the especially hazardous iodine-131, into the surrounding environment. According to the Nuclear Regulatory Commission (NRC) report, the accident resulted in no deaths or injuries to plant workers or members of nearby communities. Follow-up epidemiology studies have linked no incidents of cancer to the accident.

2.2.3.2 Reaktor bolshoi moshchnosty kanalny reactor and chernobyl disaster

The RBMK was the culmination of the Soviet nuclear power program to produce a water-cooled power reactor with dual-use potential based on their graphite-moderated plutonium production military reactors as shown in Fig. 2.56. The first of these,

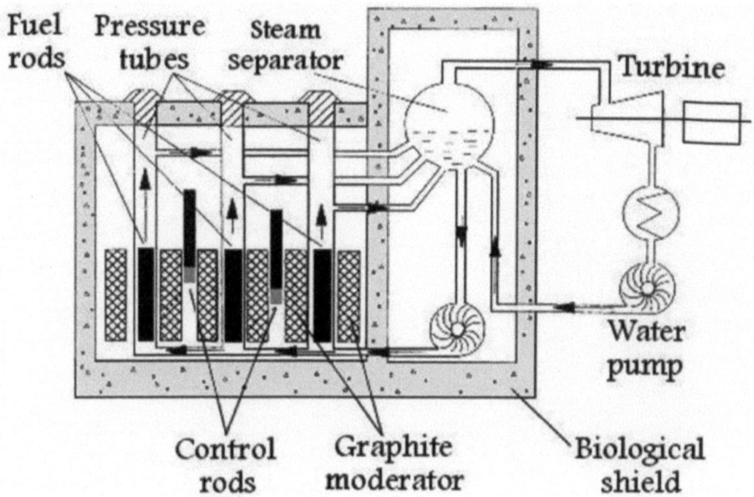

Figure 2.56 Schematic of RBMK type nuclear reactor.

Obninsk AM-1 generated 5 MW of electricity from 30 MW thermal power. By using a minimalist design that used regular (light) water for cooling and graphite for moderation, it was possible to use natural uranium for fuel (instead of the considerably more expensive enriched uranium). This allowed for an extraordinarily large and powerful reactor that was also cheap enough to be built in large numbers and simple enough to be maintained and operated by local personnel. For example, the RBMK reactors at the Ignalina Nuclear Power Plant in Lithuania were rated at 1500 MWe each, a very large size for the time and even for the early 21st century.

The accident started during a safety test on an RBMK-type nuclear reactor, which was commonly used throughout the Soviet Union. The test was a simulation of an electrical power outage to aid the development of a safety procedure for maintaining reactor cooling water circulation until the backup electrical generators could provide power. This gap was about one minute and had been identified as a potential safety problem that could cause the nuclear reactor core to overheat. It was hoped to prove that the residual rotational energy in a turbine generator could provide enough power to cover the gap. During the planned decrease of reactor power in preparation for the electrical test, the power unexpectedly dropped to a near-zero level. Upon test completion, the operators triggered a reactor shutdown, but a combination of unstable conditions and reactor design flaws caused an uncontrolled nuclear chain reaction instead. A large amount of energy was suddenly released, vaporizing superheated cooling water and rupturing the reactor core in a highly destructive steam explosion (Fig. 2.57).

2.2.3.3 Fukushima Daiichi nuclear power plant accident

The Fukushima Daiichi nuclear disaster was a nuclear accident at the Tokyo Electric Power Company (TEPCO) Fukushima Daiichi Nuclear Power Plant in

Figure 2.57 Steam plumes continued to be generated days after the initial explosion.

Table 2.4 Reactor data of units in Fukushima Daiichi Nuclear Power Plant [44].

Unit	Type (containment)	Net power	Commercial operation	NSSS
No.1	BWR-3 (Mark-I)	439 MWe	March 26, 1971	General Electric
No.2	BWR-4 (Mark-I)	760 MWe	July 18, 1974	General Electric
No.3	BWR-4 (Mark-I)	760 MWe	March 27, 1976	Toshiba
No.4	BWR-4 (Mark-I)	760 MWe	October 12, 1978	Hitachi
No.5	BWR-4 (Mark-I)	760 MWe	April 18, 1978	Toshiba
No.6	BWR-5 (Mark-II)	1067 MWe	October 24, 1979	General Electric

Ōkuma, Fukushima Prefecture. It was the most severe nuclear accident since the 1986 Chernobyl disaster, and the only other disaster to receive the Level 7 event classification of the International Nuclear Event Scale.

In the plant, there were six units and their reactor data are summarized in Table 2.4. All the units are Gen-II BWRs, with Mark-I type containment. A schematic of Mark-I containment vessel and its cut view are shown in Figs. 2.58

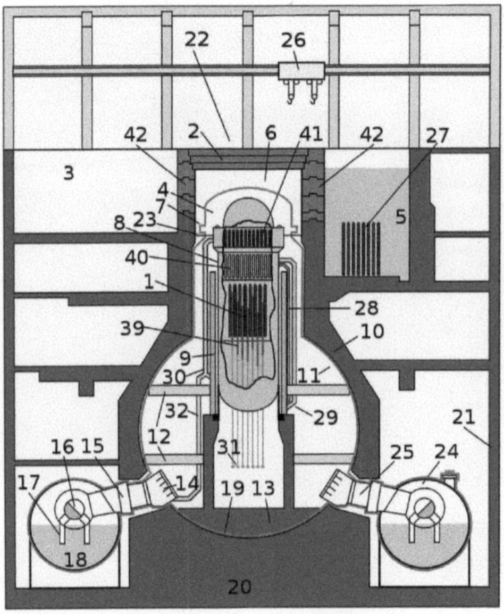

Figure 2.58 Mark I BWR.

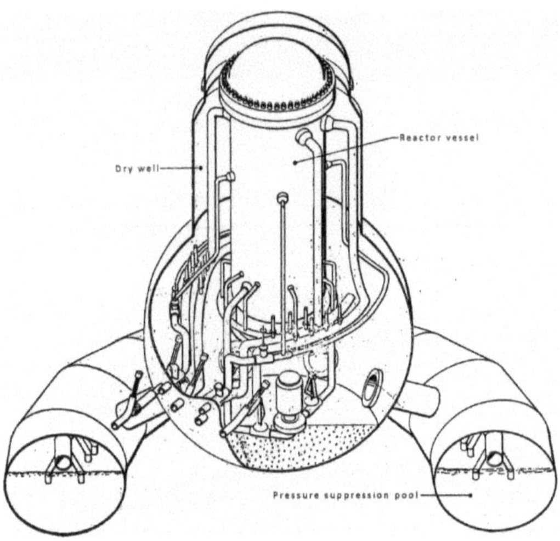

Figure 2.59 Cut view of Mark-I.

and 2.59. The Mark-I containment design consists of several major components as follows:

- The dry well ("10" in Fig. 2.58), which surrounds the reactor vessel and recirculation loops.
- A suppression chamber ("24"), which stores a large body of water (suppression pool).

- An interconnecting vent network ("15" or "25") between the drywell and the suppression chamber.
- The secondary containment, which surrounds the primary containment and houses the spent fuel pool and ECCS.

Nuclear power plants should be designed to be safe even in severe accident conditions. A reactor protection system is designed to immediately terminate the nuclear reaction. By breaking the nuclear chain reaction, the source of heat is eliminated. Other systems can be used to remove decay heat from the core. All nuclear plants have some form of such reactor protection system. The essential service circulates the water that cools the plant's heat exchangers and other components before dissipating the heat into the environment. Because this includes cooling the systems that remove decay heat from both the primary system and the spent fuel rod cooling pods, the ESWS is a safety-critical system. Since the water is frequently drawn from an adjacent river, the sea, or other large body of water, the system can be fouled by seaweed, marine organisms, oil pollution, ice, and debris. Total loss of ESWS occurred during the Fukushima Daiichi nuclear accidents.

The accident was started by the Tōhoku earthquake and tsunami on Friday, March 11, 2011. On detecting the earthquake, the active reactors automatically shut down their fission reactions. Because of the reactor trips and other grid problems, the electricity supply failed, and the reactors' emergency diesel generators automatically started. Critically, they were powering the pumps that circulated coolant through the reactors' cores to remove decay heat, which continues to be produced after fission has ceased. The earthquake generated a 14-m-high tsunami that swept over the plant's seawall and flooded the plant's lower grounds around the Units 1–4 reactor buildings with sea water, filling the basements and knocking out the emergency generators.

The resultant LOCAs led to three nuclear meltdowns, three hydrogen explosions, and the release of radioactive contamination.

Based on Fukushima Daiichi Disaster, Lessons learned to enhance the nuclear safety would be as follows [45]:

- To enhance the capability to maintain plant safety during a prolonged loss of electrical power.
- To provide a reliable hardened containment vent system for BWRs with Mark-I or Mark-II containment designs.
- To provide a reliable wide-range indication of water level in spent fuel storage pools.
- To reanalyze potential seismic effects using present-day information to determine if safety upgrades are needed.
- To reanalyze potential flooding effects using present-day information to determine if safety upgrades are needed.
- To inspect existing plant protection features against seismic and flooding events, and correct any degraded conditions.
- To assess staffing needs and communications capabilities to effectively respond to an event affecting multiple reactors at a site.
- To enhance the capability to maintain plant safety during a prolonged loss of electrical power.
- To strengthen and integrate different types of emergency procedures and capabilities at plants.
- To evaluate potential strategies that may further confine or filter radioactive material if core damage occurs.

Nuclear technology has been developed for five decades, but it may still not mature technology, which could be modified to meet the above listed tasks to enhance the safety and reliability.

2.2.4 Advanced nuclear power generation

Nuclear power plant designs are usually categorized by generation; that is, Generation I, II, III, III + , and IV as shown in Fig. 2.60 [45]. The key attributes characterizing the development and deployment of nuclear power plants illuminate the essential differences between the various generations of reactors.

- Generation I refers to the prototype and power reactors that launched civil nuclear power plants. This generation consists of early prototype reactors from 1950s and 1960s, such as Shippingport, Dresden-1 (BWR) in the United States, and Calder Hall-1 in the United Kingdom. All the Generation I power plants have been already shut down.
- Generation II denotes a class of commercial reactors built until the end of the 1990s designed to be economical and reliable. Designed for typical operational period of 40 years, including PWR, CANDU, BWR, advanced gas-cooled reactors, and VVER/RBMK. Most of the existing nuclear power plants in the world belong to Generation II.
- Generation III is a development of Generation II nuclear reactor designs incorporating evolutionary improvements in design developed during the lifetime of the Generation II reactor designs. These include improved fuel technology, superior thermal efficiency, significantly enhanced safety systems including passive nuclear safety, and standardized designs for reduced maintenance and capital costs, such as AP1000, Advanced BWR (ABWR), Evolutionary Power Reactor (EPR), VVER-640 and Pebble Bed Modular Reactor (PBMR).
- Generation IV systems are innovative nuclear power reactors ready for construction during the 2020s and in operation by 2030.

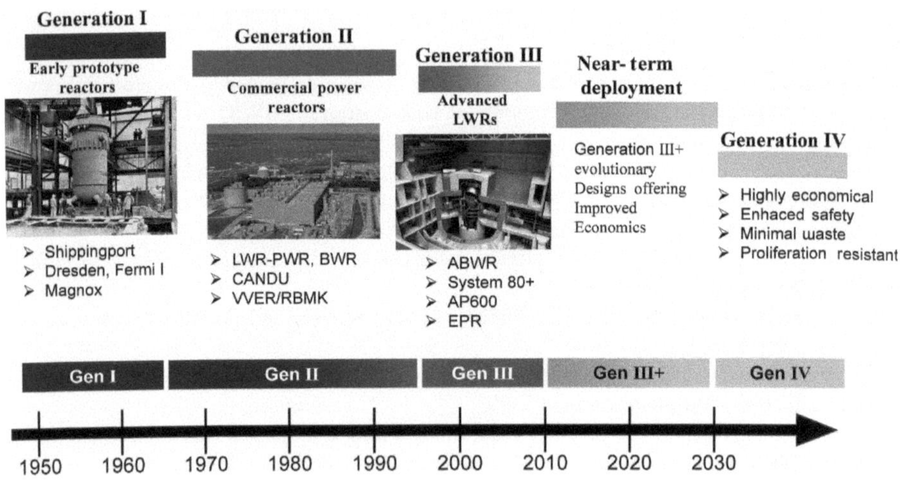

Figure 2.60 The evolution of nuclear power.
Source: Redrawn based on U.S. Department of Energy, Office of Nuclear Energy.

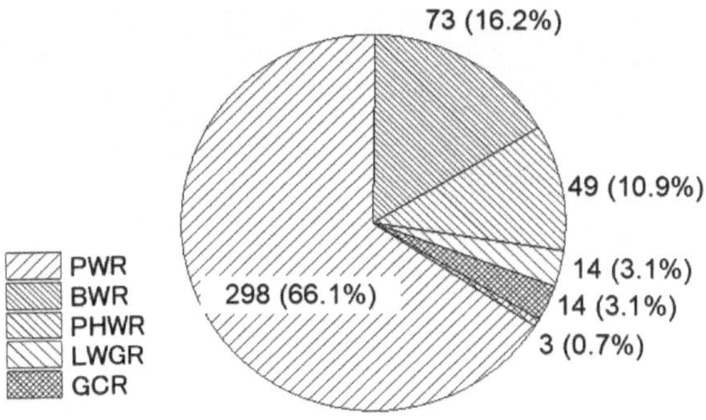

Figure 2.61 Number of operating plants in OECD.

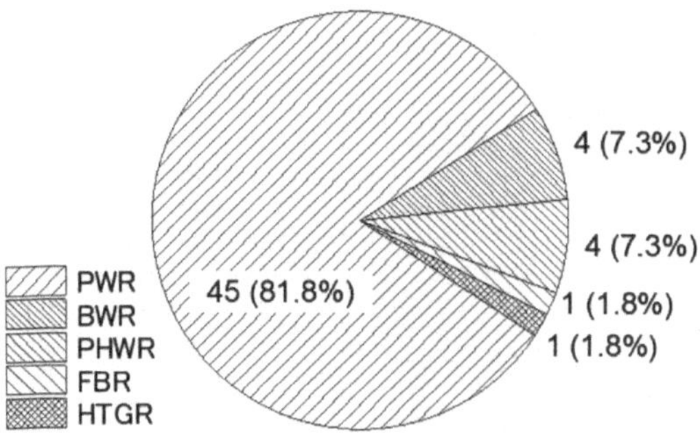

Figure 2.62 Number of planning plants in OECD.

Figs. 2.61 and 2.62 show the number of operating plants and planning plants in OECD countries [46]. More than 80% of operating/planning nuclear power plants are Light-Water Reactor (BWR, PWR). PHWR denotes Pressurized Heavy Water Reactor, or CANDU, the Light Water-Cooled Graphite-Moderated Reactor (LWGR), or RBMK.

As mentioned, nuclear power plants are equipped with various safety systems including ECCS, which needs normally electricity to work for Generation II systems. In Fukushima Daiichi Disaster, most of safety systems correctly functioned at the earthquake and the units were safely shut down. However, the tsunami destroyed some of the cooling systems and diesel generator, resulting in a Station Black Out (SBO) without final heat sink (see water). Electrical driven cooling system or active cooling system cannot be utilized without electricity. Generation III/III + incorporate lessons learned during 1980s and 1990s from the construction, commissioning and mainly of LWRs throughout the world. Gen III systems include passive cooling systems, which do not require any active intervention on the part of the operator or electrical/electronic feedback in order to bring the reactor to a safe shutdown state in the event of a particular type of emergency. Such design features tend to rely on the engineering of components, which typically take advantage of natural forces or phenomena such as gravity, buoyancy, pressure differences, conduction or natural heat convection to accomplish safety functions without requiring an active power source.

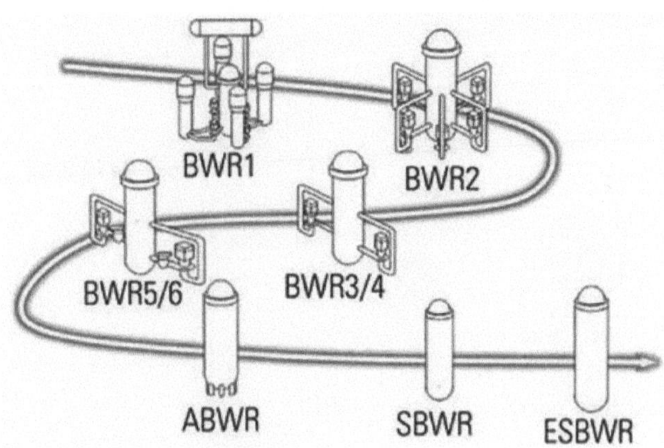

Figure 2.63 Development of BWR design [47].

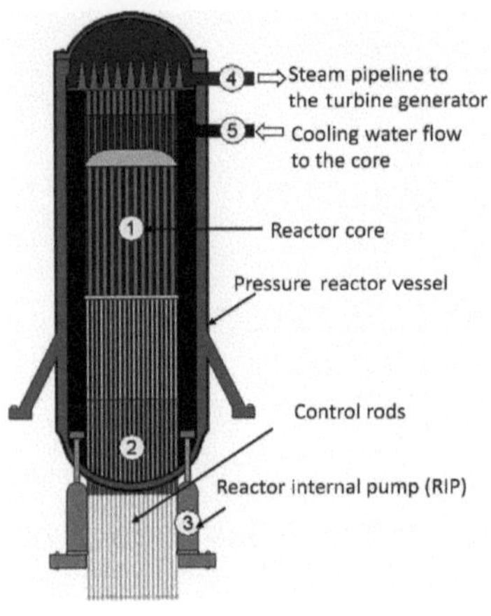

Figure 2.64 Schematic of ABWR RPV.

Due to the prolonged period of stagnation in the construction of new reactors and the continued (but declining) popularity of Generation II/II + designs in new construction, relatively few Gen III reactors have been built. Generation IV designs are still in development as of 2017, and are not expected to start entering commercial operation until 2020–30.

Some examples of Gen III designs are shown as follows.

2.2.4.1 Advanced boiling water reactor

The ABWR represents an evolutionary route for the BWR family (Fig. 2.63) [47], with numerous changes and improvements to previous BWR designs. Major improvements include the addition of Reactor Internal Pumps (RIP) mounted on the bottom of the reactor pressure vessel (RPV) as shown in Fig. 2.64. which achieve

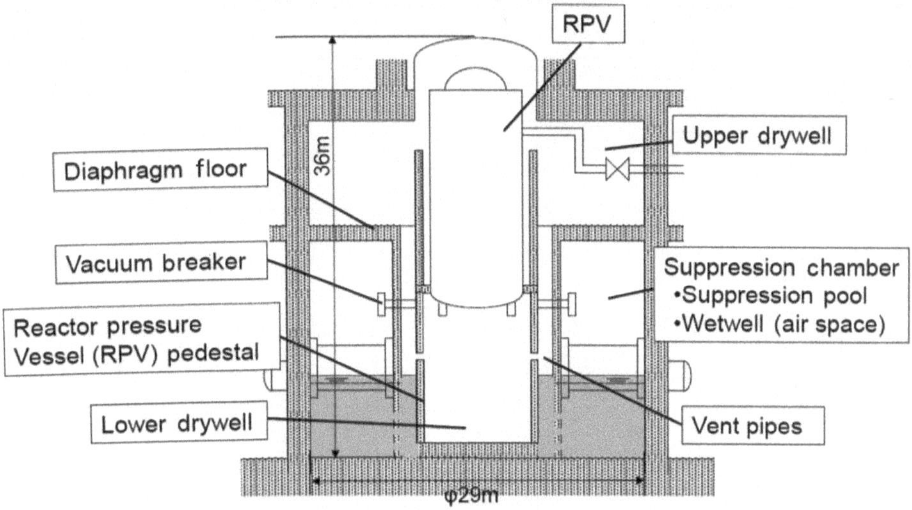

Figure 2.65 Cross section of ABWR design reinforced concrete containment vessel.

improved performance while eliminating large recirculation pumps in containment and associated large-diameter and complex piping interfaces with the RPV. In the recirculation loop found in earlier BWR models. Only the RIP motor is located outside of the RPV in the ABWR. The first ABWR plant began at Kashiwazaki-Kariwa unit 6 in Japan, which is considered the first Generation III reactor in the world.

Fig. 2.65 shows the cross section of ABWR design Reinforced Concrete Containment Vessel (RCCV), which has been significantly improved over the conventional Mark-I type. Like the conventional Mark-I type, it is of the pressure suppression type, designed to handle evolved steam in the event of a transient, incident, or accident by routing the steam using pipes that go into a pool of water enclosed in the wet well, the low temperature of which will condense the steam back into liquid water. This will keep containment pressure low. Notably, the typical ABWR containment has numerous hardened layers between the interior of the primary containment and the outer shield wall and is cubical in shape.

2.2.4.2 Economic simplified boiling water reactor

The Economic Simplified BWR (ESBWR) is a passively safe generation III+ reactor design derived from its predecessor, the Simplified BWR (SBWR) and from the Advanced BWR (ABWR). All are designs by GE Hitachi Nuclear Energy (GEH) and are based on previous BWR designs.

Fig. 2.66 shows the schematic of the ESBWR. Its passive safety systems in an ESBWR operate without using any pumps, which creates increased design safety, integrity, and reliability, while simultaneously reducing overall reactor cost. It also uses natural circulation to drive coolant flow within the RPV; this results in fewer

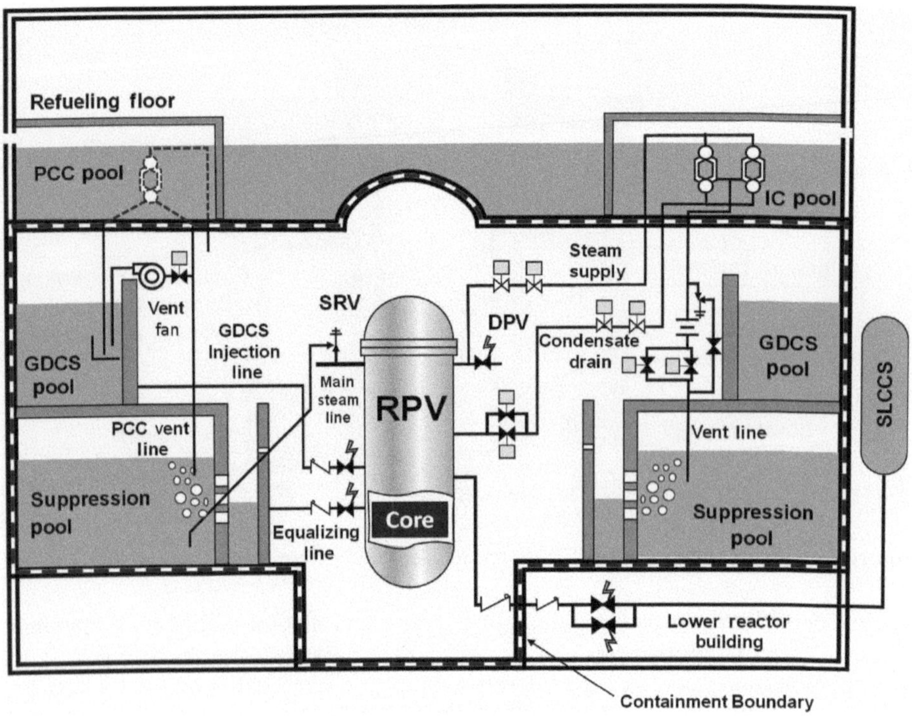

Figure 2.66 Schematic of cooling system for ESBWR.

systems to maintain, and precludes significant BWR casualties such as recirculation line breaks. There are no circulation pumps or associated piping, power supplies, heat exchangers, instrumentation, or controls needed for these systems.

2.2.4.3 AP1000

The AP1000 is a nuclear power plant designed and sold by Westinghouse Electric Company [48]. The plant is a pressurized water reactor with improved use of passive nuclear safety as shown in Fig. 2.67 and many design features intended to lower its capital cost and improve its economics.

Six AP1000s are currently in operation or under construction. Four are located at two sites in China, two at Sanmen Nuclear Power Station and two at Haiyang Nuclear Power Plant. Two are under construction at the Vogtle Electric Generating Plant in the US, while a further two at the Virgil C. Summer Nuclear Generating Station were canceled in 2017. As of 2019, all four Chinese plants have completed construction and are at various stages of connecting to the grid. Construction at Vogtle has suffered numerous delays and Unit 3 is now expected to be completed in 2021. Cost overruns at Vogtle and the cancellation of summer led to Westinghouse's bankruptcy in 2017.

The first AP1000 began operations in China at Sanmen, where Unit 1 became the first AP1000 to achieve criticality in June 2018 and was connected to the

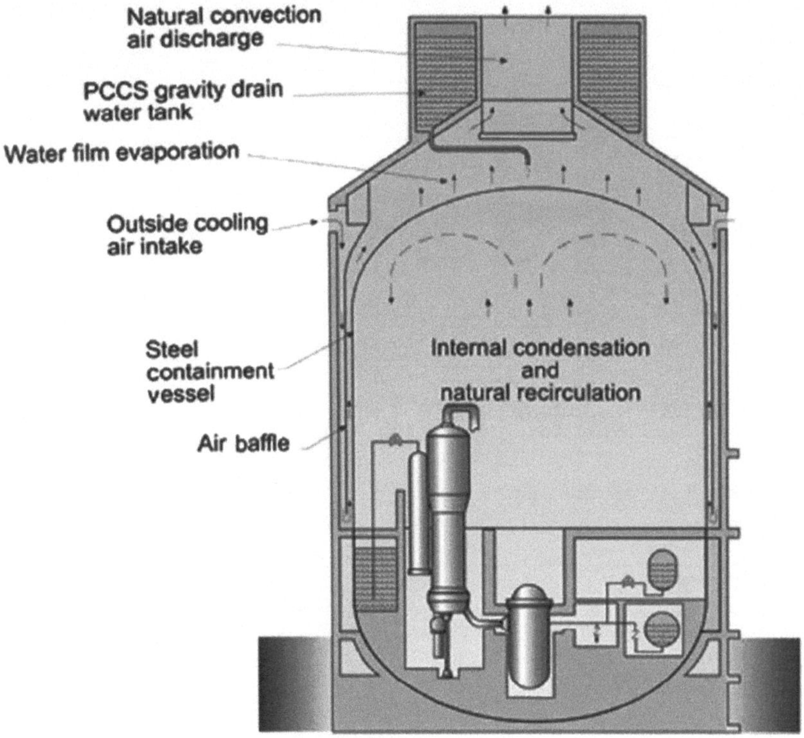

Figure 2.67 Passive cooling system for AP1000.

grid the next month. Further builds in China will be based on the modified CAP1400 design.

2.2.4.4 Evolutionary power reactor

The EPR is a Gen-III PWR design. It has been designed and developed mainly by Framatome (part of Areva between 2001 and 2017) and Électricité de France in France, and Siemens in Germany. In Europe this reactor design was called European Pressurized Reactor, and the internationalized name was Evolutionary Power Reactor, but it is now simply named EPR. Fig. 2.68 shows the schematic of EPR safety system. The EPR design has several active and passive protection measures against accidents: (1) Four independent emergency cooling systems, each providing the required cooling of the decay heat that continues for one to three years after the reactor's initial shutdown (i.e., 300% redundancy), (2) Leak-tight containment around the reactor, (3) An extra container and cooling area if a molten core manages to escape the reactor (see containment building and core catcher), (4) Two-layer concrete wall with total thickness 2.6 m, designed to withstand impact by airplanes and internal overpressure.

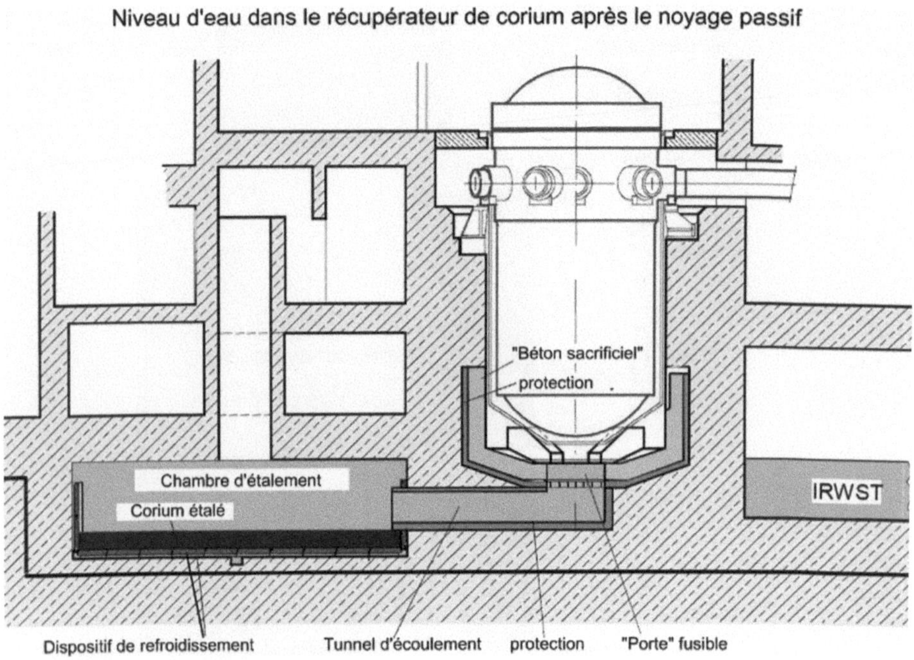

Figure 2.68 Safety system for EPR.

The first operational EPR unit was China's Taishan 1, which started in December 2018. Taishan 2 started operations in September 2019. The first two EPR units to start construction, at Olkiluoto in Finland and Flamanville in France, are both facing costly delays (to 2021 and 2023, respectively). Two units at Hinkley Point in the United Kingdom received final approval in September 2016 and are expected to be completed around 2026.

Safety and reliability of Generation III designs are much higher than those of Generation II, because they have been designed based on the long operating experiences of Generation II. Generation III designs make use of two important principles such as simplification and modularization. Simplification leads to lower construction costs and the modularization leads to lower construction time alongside other improvements. However, recently the requirements for nuclear power plants are increasing such as against aircraft collision or terrorism. Therefore, there is still room for further improvement, including safety enhancement.

2.2.5 *Road to future nuclear power generation*

Existing or existed nuclear power plants belong to Generation I to Generation III [49]. However, there exist more challenging nuclear technologies, which will be realized in the future. The next stage of the nuclear technology would be Generation IV. The Generation IV International Forum (GIF) was initiated by the

US Department of Energy in 2000, which is an international collective representing governments of 13 countries. The original charter members of GIF are Argentina, Brazil, Canada, France, Japan, South Korea, South Africa, the UK and the USA and later they have been joined by Switzerland, China, Russia, and Australia.

The GIF have discussed the hopeful candidates for Generation IV and selected originally six technologies, such as Sodium-cooled Fast Reactor (SFR), Gas-cooled Fast Reactor (GFR), Very High-Temperature Reactor (VHTR), Lead-cooled Fast Reactor (LFR), Super Critical Water-cooled Reactor (SCWR), and MSR. Among above-mentioned candidates, only one candidate, SCWR, employs water as the coolant, and others employ liquid metal, gas, or molten salt. The working temperature of Gen IV reactors is much higher than that of existing water-cooled reactor to attain higher thermal efficiency.

Three of six technologies for Gen IV are fast reactor, not FBR. At the beginning of nuclear development, it was believed that the fissile material was scarce in the world. The possibility of a plutonium-fueled nuclear reactor that could produce more fuel than it consumed was first raised during World War II in the United States by scientists in the atomic bomb program. In the following two decades, several countries followed the United States in establishing national plutonium breeder reactor programs. In these programs, the main objective had been to solve the long-term energy supply problem using the large-scale deployment of nuclear energy for electric power. However, uranium proved to be much more abundant than originally imagined. The urgency of development of FBR has abated. In contrast, the needs are increasing to reduce plutonium stockpile generated from existing/existed nuclear power plants. Therefore, "breeding" is not necessary at present stage, but "burner" fast reactor is needed to reduce the plutonium stockpile and minor actinides which are the major elements in high level nuclear wastes.

2.2.5.1 Sodium-cooled fast reactor

The SFR uses liquid sodium as the primary coolant like EBR-I/II developed in the United States, allowing high power density at low pressure. High temperature of liquid sodium enables high thermal efficiency to generate electricity. Since the start of the nuclear development, its initial object had been plutonium production from uranium-238, resulting in the FBR. However, as mentioned above the breeder is not necessary at the present stage.

Although the SFR has been one of the most hopeful candidates for next generation nuclear technologies, there still remains at the forefront despite needing a sealed coolant system because of the possibility of sodium-water reaction, This was one of reason why the sodium-cooled reactor has not fully employed by submarine technology for USS Seawolf, which was initially intended a sodium-cooled nuclear reactor.

The coolant temperature of the SFR is 500°C–550°C. The cooling system consists of two circuits: the primary sodium circuit and the secondary sodium circuit, to avoid the sodium-water reaction as possible. Two typical arrangements have been proposed such as pool type and loop type, as shown in Fig. 2.69. The loop

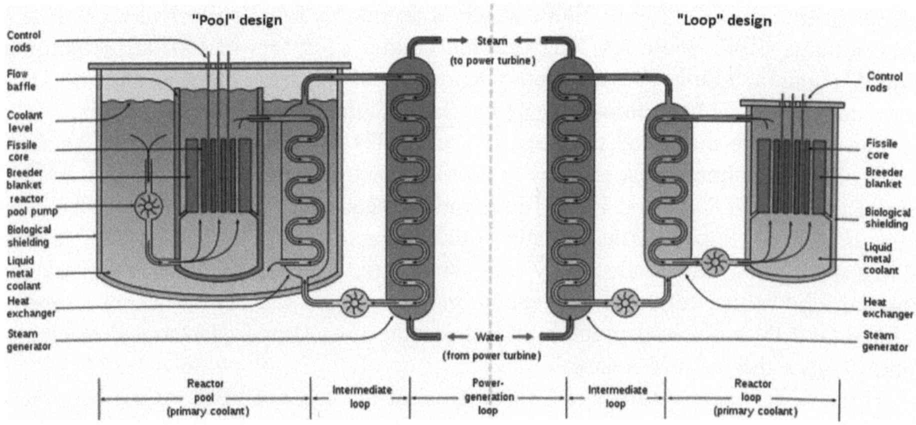

Figure 2.69 Schematics of pool design (*Left*) and loop design (*Right*) of SFR.

design has been employed by Monju FBR prototype in Japan, and the pool design has been employed by Phenix in France and others. The output of the SFR would be 300−1500 MWe for pool type and 600−1500 MWe for loop type. In addition to these, a 50−150 MWe modular type SMR has been also proposed.

2.2.5.2 Lead-cooled fast reactor

Liquid metal of (Pb or Pb-Bi) can be utilized for fast reactor. The LFR is a flexible fast neutron reactor which can use depleted uranium or thorium fuel matrices and burn actinides from LWR fuel. Its decay heat can be cooled at atmospheric pressure by the natural convection of the liquid metal. Fuel is metal or nitride, with full actinide recycle from regional or central reprocessing plants. A wide range of unit sizes is envisaged, from factory-built "battery" for small grids or developing countries, to modular 300−400 MWe units and large single plants of 1400 MWe. Operating temperature of 550°C is readily achievable but the liquid metal (Pb or Pb-Bi) shows corrosive/erosive nature to structural materials at higher temperature. A two-stage development program leading to industrial deployment is envisaged: by 2025 for reactors operating with relatively low temperature and power density, and by 2040 for more advanced higher-temperature designs.

This corresponds with Russia's BREST LFR technology which is based on 80 reactor-years' experience of Russian submarine reactors. However, these propulsion reactors were small, operated at low capacity factors, featured an epithermal neutron spectrum and operated at significantly lower temperatures than those anticipated in Generation IV reactors.

Initial development work on the LFR was focused on two pool-type reactors: SSTAR—Small Secure Transportable Autonomous Reactor of 20 MWe in USA; and the European Lead-cooled SYstem or European Lead Fast Reactor of 600 MWe in Europe. In 2014 the leading developments were Russia's SVBR-100 and BREST-300, Europe's 300 MWt ALFRED, and, with R&D focus on fuels and materials corrosion.

In addition, substantial advances in the design and use of powerful accelerators make it possible to consider accelerator driven systems (ADS), which consist of subcritical fuel assembly, proton accelerator, and spallation target. The interest in ADS is strongly connected with the transmutation of high level nuclear waste, but it is possible that such subcritical reactors may have advantages over critical fast reactors, especially with regard to the fuel composition. ADS project are going on for Belgium's MYRRHA (Multi-purpose hYbrid Research Reactor for High-tech Application).

2.2.5.3 Very high-temperature gas reactor

These are graphite-moderated, helium-cooled reactors, based on substantial experience. The core can be built of prismatic blocks such as the Japanese HTTR and General Atomics' (GA) earlier GTMHR design and others in Russia, or it may be pebble bed such as the Chinese HTR-10 or HTR-PM and the PBMR. Outlet temperature of over 900°C and aiming for 1000°C enables thermochemical hydrogen production via an intermediate heat exchanger, with electricity cogeneration, or direct high-efficiency driving of a gas turbine (Brayton cycle). At lower outlet temperatures, the Rankine steam cycle may be used for electricity generation, and this is the focus for demonstration projects. Modules of 600 MW thermal are envisaged.

There is some flexibility in fuels, but no recycle initially. Fuel is in the form of TRISO (tristructural-isotropic) particles less than a millimeter in diameter. Each has a kernel (c. 0.5 mm) of uranium oxycarbide with the uranium enriched up to 20% U-235, though normally less. This is surrounded by layers of carbon and silicon carbide, giving a containment for fission products which is stable to over 1600°C. The particles may be incorporated in billiard ball sized pebbles, or in prismatic graphite blocks.

2.2.5.4 Gas-cooled fast reactor

Like other helium-cooled reactors which have operated or are under development, GFRs will be high-temperature units up to 850°C. They employ similar reactor technology to the VHTR, suitable for power generation, thermochemical hydrogen production or other process heat. The reference GFR unit is 2400 MWt/1200 MWe, large enough for breakeven breeding, with thick steel RPV and three 800 MWt loops. For electricity, an indirect cycle with helium will be on the primary circuit, in the secondary circuit the helium gas will directly drive a gas turbine (Brayton cycle), and a steam cycle will comprise the tertiary circuit.

An alternative GFR design has lower temperature (600°C–650°C) helium cooling in a primary circuit and supercritical CO_2 at 550°C and 20 MPa in a secondary system for power generation. This reduces the metallurgical and fuel challenges associated with very high temperatures. The GA Energy Multiplier Module (EM2) design is a 500 MWt, 240 MWe helium-cooled fast-neutron HTR operating at 850°C and fueled with used PWR fuel or depleted uranium, plus some low-enriched uranium as starter.

2.2.5.5 Supercritical water-cooled reactor

This is a very high-pressure water-cooled reactor which operates above the thermodynamic critical point of water (374°C, 22 MPa) to give a thermal efficiency about one-third higher than today's light-water reactors, from which the design evolves. The supercritical water (25 MPa and 510°C–550°C) directly drives the turbine, without any secondary steam system, simplifying the plant. Two design options are considered: pressure vessel and pressure tube. Passive safety features are similar to those of simplified BWR. Fuel is uranium oxide, enriched in the case of the open fuel cycle option. The core may use thermal neutron spectrum with light or heavy water moderation or be a fast reactor with full actinide recycle based on conventional reprocessing. Since the SCWR builds both on much BWR experience and that from hundreds of fossil-fired power plants operated with supercritical water, it can readily be developed, and the operation of a 30–150 MWe technology demonstration reactor is targeted for 2022.

2.2.5.6 Molten salt reactor

The MSR is a nuclear reactor with fissile material dissolved in the circulation fuel salt. The uranium fuel is dissolved in the fluoride salt coolant which circulates through graphite core channels to achieve some moderation and an epithermal neutron spectrum. The reference plant is up to 1000 MWe. Fission products are removed continuously, and the actinides are fully recycled, while plutonium and other actinides can be added along with ^{238}U, without the need for fuel fabrication. Coolant temperature is 700°C–800°C at low pressure. A secondary coolant system is used for electricity generation, and thermochemical hydrogen production is also feasible.

Compared with solid-fueled reactors, MSR systems have lower fissile inventories, no radiation damage constraint on fuel burn-up, no requirement to fabricate and handle solid fuel or solid used fuel, and a homogeneous isotopic composition of fuel in the reactor. These and other characteristics may enable MSRs to have unique capabilities and competitive economics for actinide burning and extending fuel resources. The Molten Salt Fast Neutron Reactor (MSFR), which will take in thorium fuel cycle, recycling of actinides, closed Th/U fuel cycle with no U enrichment, with enhanced safety and minimal wastes.

Up to now, the unit size of nuclear power plants has grown from 60 MWe to more than 1600 MWe since the Shippingport construction. From the viewpoint of scale merit, larger units must be economically competitive. However, due to the high capital cost of large-scale power reactors, smaller and more flexible nuclear design may be one of the future options for nuclear technology. Recently there has been a move to develop Small Modular Reactors (SMRs). SMRs are defined as nuclear reactors generally less than 300 MWe, designed with modular technology using module factory fabrication, pursuing economies of series production and short construction times. These may be built independently or as modules in a larger complex, with capacity added incrementally as required. Economies of scale depends on the numbers produced. Some examples of SMR are described below.

Fig. 2.70 shows a schematic of BWRX-300, which is a 300 MWe water-cooled, natural circulation SMR with passive safety systems [50]. BWRX-300 leverages the design and licensing basis of the US NRC certified ESBWR, which is a 1520 MWe Generation III + unit (Fig. 2.66). By employing proven technologies from conventional BWR design, it can become cost-competitive with power generation from combined cycle gas and renewable sources.

Fig. 2.71 shows a schematic of NuScale's SMR, an integral pressurized water reactor, which has been designed based on Multi-Application Small Light-Water Reactor developed at Oregon State University in the early 2000s [51]. NuScale is a natural circulation light-water reactor with the reactor core and helical coil steam generators located in a common reactor vessel in a cylindrical steel containment. The reactor vessel containment is intended to be kept in an underground pool. The reactor building is designed to uphold 12 SMR units. Each unit has a thermal output of 160 MWt and electrical output of 50 MWe, resulting a total capacity of 600 MWe. Under severe accident conditions, the nuclear fuels in the core can be cooled without any electricity and any human input. Due to the advantages such as their size, construction efficiency, and passive safety system, they can lead to easier financing compared to that for larger plants.

Both above-mentioned SMRs are designed based on existing experiences of operating PWR and BWR. In addition, more innovative technology can be applied to SMR such as fast reactor technology. Nuclear power plants have been often

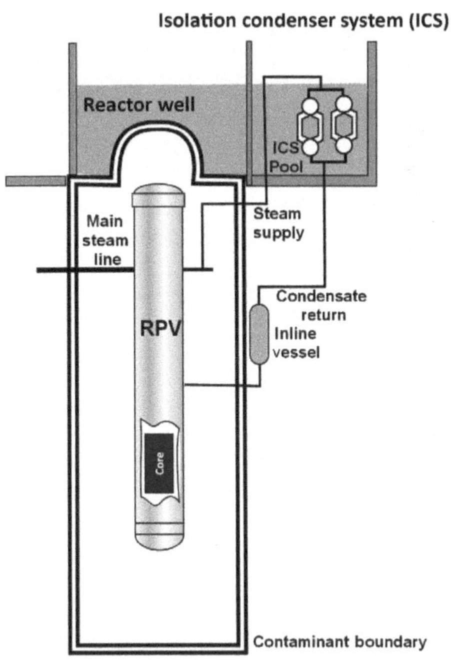

Figure 2.70 Schematic of BWRX-300.

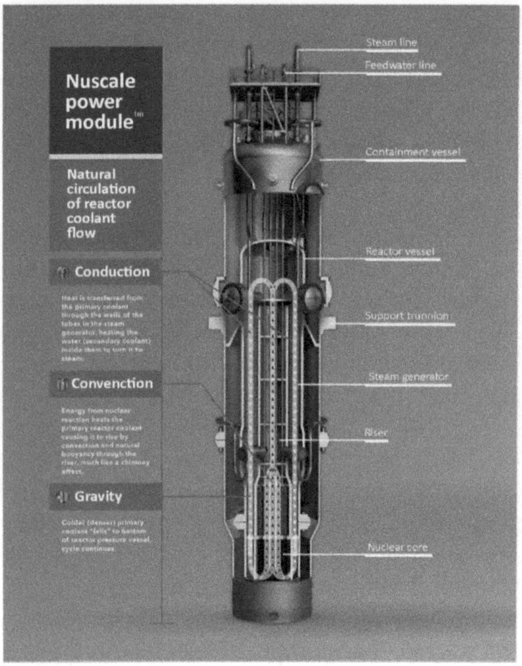

Figure 2.71 Schematic of NuScale [51].

considered base-load electricity sources and operating them at their rated power level is usually efficient. However, due to increasing renewable electric generation, it is necessary for nuclear power plants to have load-following abilities to compensate variations in the electricity generation. Most of the modern nuclear power plants are capable to operate in a load-following mode, to change their power level once or twice per day in the range of 100%–50% of their rated power. Future SMRs may open up possibilities for advanced load-following abilities to meet various requirements, which cannot be attained by large-scale nuclear power plants [52].

References

[1] G. Branca, Le Machine, Giacomo Manuci for Giacomo Mascardi, Rome, 1629, p. 55.

[2] D. Papin, Nova methodus ad vires motrices validissimas levi pretio comparandas, Acta Eruditorum, Leipzig Grosse & Gleditsch (1690), ss. 410–414.

[3] R. Stuart, Descriptive History of the Steam Engine, John Knight and Henry Lacey, London, 1824, p. 65.

[4] J. Farey, A Treaties on the Steam Engine, Printed for Longman, Rees, Orme, Brown, and Green, London, 1827, p. 444.

[5] T. Savery, The Miner's Friend, or an Engine to Rais Water by Fire, Described, and of the Manner of Fixing it in Mines, with an Account of the Several Other Uses it is Applicable unto and an Answer to the Objections Made against it, S. Crouch, London, 1702.

[6] N.L.S. Sadi Carnot, Reflections on the motive power of heat, original French version, in: R.H. Thurston (Ed.), English version (1897) includes "An Account of Carnot's Theory" by William Thomson, John Wiley & Sons, New York, 1824.
[7] W.J.M. Rankine, A Manual of the Steam Engine and Other Prime Movers, Richard Griffin and Co., London, 1859.
[8] J.P. Joule, On the surface-condensation of steam, Philos. Trans. R. Soc. Lond. 151 (1861) 133–160.
[9] K. Hoefer, Die Kondensation bei Dampfkraftmaschinen, Verlag von Julius Springer, Berlin, 1925.
[10] R.H. Thurston, A History of the Growth of the Steam-Engine, Cornell University Press, New York, 1907, 415, p. 156.
[11] C. Matschoss, Die Entwicklung der Dampfmaschine, Band I, Verlag von Julius Springer, Berlin, 1908, Band I: p. 677 & 766; Band II: p. 36, 511, 526 & 551.
[12] A.M. Robb, Chapter 16: Ship-building, in: C. Singer, E.J. Holmyard, A.R. Hall, T.I. Williams (Eds.), A History of Technology, Vol. V, Oxford University Press, Oxford, 1958, pp. 350–390.
[13] S. Ishigai, Design Principles of Steam Boiler, Sankai-do, Tokyo, 1961, pp. 58–59.
[14] Report from the Select Committee on Steam Boiler Explosions; together with the Proceedings of the Committee, Minutes of Evidence, and Appendix, The House of Commons, 1870.
[15] The Manchester Steam Users' Association (MSUA), A Sketch of the Foundation and of the Past Fifty Years' Activity of the Manchester Steam Users' Association for the Prevention of Steam Boiler Explosions and for the Attainment of Economy in the Application of Steam, Taylor, Granett, Evans, & Co., Manchester, 1905.
[16] K. Akagawa, Thermal and hydraulic design of steam-generation systems, in: S. Ishigai (Ed.), Steam Power Engineering, Cambridge University Press, New York, 1999, pp. 255–283.
[17] The Babcock & Wilcox Co, Steam – Its Generation and Use, 22nd ed., 1890, pp. 27–34, p. 108.
[18] The Engineering Staff of the Stirling Co., Stirling – A Book on Steam for Engineers, The Stirling Co., 1905, pp. 8–9.
[19] A. Loschge, Die Dampfkessel, Verlag von Julius Springer, Berlin, 1937, pp. 206, & 231, Fig. 179b.
[20] G. Noetzlin, Das neue Kraftwerk Hüls—eine Anlage mit 300 at/600 Frischdampfzustand, Mitteilungen der Vereinigung der Grosskesselbesitzer (VGB), Heft 55, 1958, ss. 230–255.
[21] S.N. Fiala, First commercial supercritical-pressure steam-electric generating unit for philo plant, Trans. ASME 79 (1957) 389–407.
[22] C.A. Parsons, Rotary Motors, US Patent No. 328710, Official Gazette of the United State Patent Office, Vol. 33, 1885, pp. 303–304.
[23] H.W. Dickinson, A Short History of the Steam Engine, The University Press, Cambridge, 1938, pp. 185–235.
[24] R.L. Hill, Power from Steam, Cambridge University Press, Cambridge, 1989, pp. 281–292.
[25] A. Stodola, Die Dampfturbinen, Verlag von Julius Springer, Berlin, 1910, p. 2, 379, 423 & 461, pp. 423–428, p. 587.
[26] C.G. Curtis, Elastic fluid turbine, US Patent Office, No. 591822, 1897.
[27] A.C.E. Rateau, G., Sautter, Multicellular turbine, US Patent Office, No. 748216, 1903.
[28] A.Stodola, Steam and Gas Turbine, McGraw-Hill, New York, 1927, p. 587.
[29] J. Gubler, Eddystone steam power station, Sulzer Tech. Rev. 41 (1) (1959) 3–20.
[30] S. Miyaoka, Supercritical unit and its economic potential, Therm. Power 1 (11) (1965) 902–916.

[31] Babcock & Wilcox Ltd., Steam – Its Generation and Use, 8th British ed., 1913, p. 200.
[32] Thermal Power Engineering Society, Data Book of Thermal Power Stations, 1961, p. 206.
[33] F.J. Brodie, On the Prevalence of Fog in London: During the 20 Years in 1871 to 1890, Q. J. R. Meteorol. Soc. 18 (81) (1982) 40–45.
[34] D.H. Meadows, D.L. Meadows, J. Randers, W.W. Behrens III, The Limits to Grows, Universe Books, New York, 1972.
[35] G.T. Mazuzan, J.S. Walker, Controlling the Atom: The Beginnings of Nuclear Regulation, 1946–1962, University of California Press, Berkeley and Los Angeles, CA, 1984.
[36] Argonne's Nuclear Science and Technology Legacy: Chicago Pile Reactors Create Enduring Research Legacy Part of the Argonne National Laboratory Highlights in the Period 1942–1949.
[37] History of Reactor Development, Britanica.com. <https://www.britannica.com/technology/nuclear-reactor/Growth-of-nuclear-programs>.
[38] R. Haroldsen, The Story of the BORAX Nuclear Reactor and the EBR-I Meltdown – How it All Came to be Destructive Test of the First Borax Lighting of Arco with Atomic Power the Russian Connection, Idaho Falls, Idaho, 2008. <https://www.ne.anl.gov/pdfs/reactors/Story-of-BORAX-Reactor-by-Ray-Haroldsen-v2.pdf>.
[39] I. Hore-Lacy, Nuclear Energy in the 21st Century: World Nuclear, University Press, Elsevier, 2010.
[40] C. Collins, Submarine Development: A Survey of Submarine Visionaries and Pioneering Boats, DefenseMediaNetwork, April 5, 2020. <https://www.defensemedianetwork.com/stories/submarine-development-a-survey-of-submarine-visionaries-and-pioneering-boats/>.
[41] A.W. Kramer, Boiling Water Reactor, Addison-Wesley, 1958.
[42] Division of Reactor Development United Personnel of the Naval Reactors Branch, Westinghouse Electric Corporation, Duquesne Light Company, The Shippingport Pressurized Water Reactor, Addison-Wesley, 1958.
[43] Nuclear Power in the World Today, March 2020, World Nuclear Association. <https://www.world-nuclear.org/information-library/current-and-future-generation/nuclear-power-in-the-world-today.aspx>.
[44] Japan: Nuclear Power Reactors. Power Reactor Information System – PRIS. IAEA. Archived from the original on May 28, 2011. Retrieved March 14, 2011.
[45] NRC Documents, Recommendations to Enhance U.S. Reactor Safety.
[46] Nuclear Power Reactors in the World, 2019 ed., International Atomic Energy Agency, Vienna, 2019.
[47] The Evolution of the ESBWR, <https://www.powermag.com/the-evolution-of-the-esbwr/>.
[48] AP1000, The Nuclear Renaissance Starts Here, Westinghouse AP1000 Brochure.
[49] T.B. Cochran, A.F. Harold, W. Patterson, G. Pshakin, M.V. Ramana, M. Schneider, et al., Fast Breeder Reactor Programs: History and Status, A Research Report of the International Panel on Fissile Materials, February 2010.
[50] <https://nuclear.gepower.com/build-a-plant/products/nuclear-power-plants-overview/bwrx-300>.
[51] <https://www.nuscalepower.com/>.
[52] M. Grubb, W.B. Janusz, Future Electricity Technologies and Systems, Cambridge University Press, 2006.

Fundamentals for power engineering

3

Tomio Okawa[1], Hitoshi Asano[2], Kei Ito[3], Shoji Mori[4], Hisashi Umekawa[5], Ryosuke Matsumoto[5], Cheol Ho Pyeon[3] and Daisuke Ito[3]
[1]The University of Electro-Communications, Chofu, Tokyo, Japan, [2]Kobe University, Kobe, Hyogo, Japan, [3]Kyoto University, Kumatori, Osaka, Japan, [4]Kyushu University, Fukuoka, Fukuoka, Japan, [5]Kansai University, Suita, Osaka, Japan

3.1 Fundamentals of thermodynamics

3.1.1 Basic concepts

3.1.1.1 Thermodynamic system

Closed systems and open systems

A thermodynamic system is defined as a quantity of matter or a region in space that is of interest. The mass or region outside the system is called the surroundings, and the surface that separates the system and the surroundings is called the boundary. For example, if the thermodynamic properties of the piston-cylinder device illustrated in Fig. 3.1 that is the simplified model of engine are of interest, the gas inside the cylinder is usually regarded as the thermodynamic system. Since the thickness of the boundary is zero, the inner surfaces of the cylinder and piston constitute the boundary. If the piston is not fixed in space, the boundary is movable in this case. Finally, the surroundings consist of the bodies of the cylinder and piston, and everything outside the piston-cylinder device.

Generally, mass and energy are exchanged between the system and the surroundings. The thermodynamic systems are classified to the closed and open systems. In the closed system, the exchange of mass is not allowed and consequently the mass in the system is always conserved. If the exchange of energy is also not allowed, the system is called the isolated system. In the piston-cylinder device shown in Fig. 3.1, the boundary is movable but no mass crosses it if gas leakage is assumed negligible. In this case, the device shown in Fig. 3.1 is classified as the closed system.

An example of the open system is depicted in Fig. 3.2. This device is similar to that shown in Fig. 3.1 but it is classified as the open system since the mass flow crossing the boundary is present. If the boundary is stationary and the inlet and outlet mass flow rates are equal and do not change with time, the system is call the

Fundamentals of Thermal and Nuclear Power Generation. DOI: https://doi.org/10.1016/B978-0-12-820733-8.00003-0

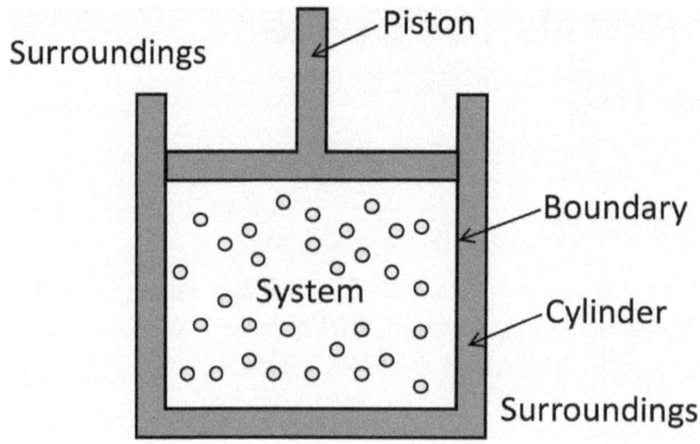

Figure 3.1 An example of thermodynamic system, boundary, and surroundings (piston-cylinder device).

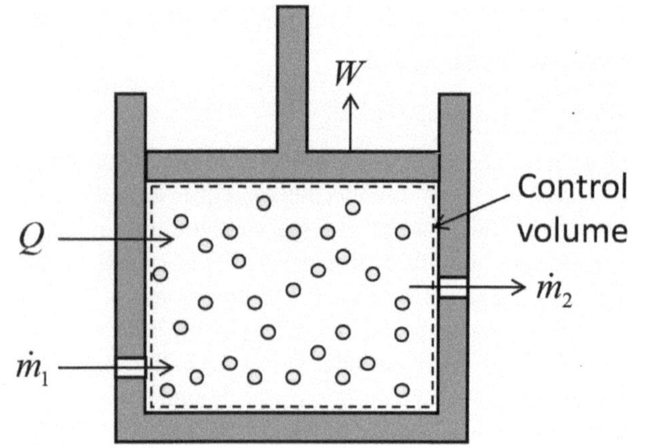

Figure 3.2 An example of open system.

steady flow system. For example, main components used in the thermal power plant such as boiler, turbine, condenser, and pump are often treated as the steady flow systems.

It is noted that the classification of the thermodynamic systems is dependent on the selection of the boundary. For example, if the boundary shown in Fig. 3.1 is fixed in space and the piston is moved upward, some amount of gas crosses the boundary. In this case, the system shown in Fig. 3.1 is not classified as the closed system but the open system. The proper choice of boundary is an important step to obtain valuable results through thermodynamic analysis.

Quantities of state

The quantities of state are the values representing the status of a system. Some important examples of the quantities of state are pressure p, temperature T, volume V, mass m, density ρ, internal energy U, enthalpy H, and entropy S. The quantities of state are classified to the intensive and extensive quantities. The intensive quantities of state are not dependent on the size of the system, such as pressure, temperature, and density. While the extensive quantities of state depend on the system size, such as mass, internal energy, enthalpy, and entropy. In consequence, if the system is divided into two equal parts, the intensive state quantities of each part are unchanged, but the extensive state quantities become half of the original values. As for the extensive quantities of state, the quantities per unit mass can be defined. Such quantities are called specific quantities of state. Several examples are specific volume v ($=V/m$), specific internal energy u ($=U/m$), specific enthalpy h ($=H/m$), and specific entropy s ($=S/m$).

Equilibrium state

Thermodynamics deal with the properties in the equilibrium state where the system does not undergo a change. When a system is in the equilibrium state, it experiences no change if it is isolated from its surroundings. If properties of system change under the influences of its surroundings, the state of the system will move to a different equilibrium state. There are several types of equilibrium. For example, if the temperature is uniform and there is no heat flow within a system, it is in thermal equilibrium. In mechanical equilibrium, no macroscopic motion is present in the system. If the chemical composition does not change with time, a system is in chemical equilibrium. If a system consists of multiple phases (two or three out of gas, liquid, and solid phases) and the mass of each phase does not change with time, it is in phase equilibrium. A system is in thermodynamic equilibrium when all the relevant equilibriums are established. Conversely, the system is not in thermodynamic equilibrium if even one equilibrium is not satisfied. For example, even when the temperature is uniform within a system and the thermal equilibrium is satisfied, macroscopic fluid flow is induced if mass concentration is not uniform within the system. Obviously, such system is not in the thermodynamic equilibrium since it undergoes a change.

3.1.1.2 Energy

Various forms of energy

The system can possess energy in various forms such as kinetic, potential, electric, magnetic, chemical, and nuclear energies. The sum of all the energies constitutes the total energy E of a system. The total energy can be regarded as the sum of macroscopic and microscopic energies. The macroscopic energies can be defined as the energies that are influenced by some outside reference frame, such as the kinetic energy E_k and the potential energy E_p. When the mass of a system is m, all parts of the system move with the same velocity w, and the elevation of the system in a gravitational field is z, E_k, and E_p are calculated, respectively, by

$$E_k = \frac{1}{2} m w^2 \tag{3.1}$$

$$E_p = mgz \tag{3.2}$$

where g is the gravitational acceleration, while the microscopic energies are related to the structure and activity of molecules constituting the system and they are not influenced by outside reference frames. The internal energy U refers the sum of all the microscopic energies of the system and particularly important form of energy in thermodynamics. In many cases. the total energy E consists of the kinetic, potential, and internal energies and is expressed by

$$E = E_k + E_p + U \tag{3.3}$$

If other components such as the magnetic and electric energies are not negligible, such components should also be included in the aforementioned equation.

Internal energy

From the microscopic point of view, the internal energy can be regarded as the sum of the kinetic and potential energies of the molecules constituting a system. The kinetic energy of molecules consists of those associated with the molecular translation, molecular rotation, molecular vibration, and so on. The motions of nuclei and electrons within molecules are also the components of the kinetic energy. Since temperature of the system increases with an increase in the kinetic energy, the kinetic energy of molecules is called the sensible energy. The potential energy of molecules is associated with the binding forces between the molecules, those between the atoms within a molecule, and those between the particles within an atom. The binding forces between molecules is stronger in solids and liquids than in gases. If sufficient energy is added to a solid or liquid, phase change into gas takes place. Due to this added energy, gas possesses higher internal energy than solid or liquid. The internal energy associated with the phase of a system is called the latent energy. Similarly, the internal energies associated with the binding forces between the atoms within a molecule and those between the particles within an atom are called the chemical and the nuclear energies, respectively.

As the simplest case, we consider the internal energy of an ideal monatomic gas in which the kinetic energy associated with the molecular translation is the only component of the total kinetic energy. Assuming that the mass of each molecule is m, the number of molecules in the system is N, and the position and velocity vectors of i-th molecule are denoted by $\boldsymbol{x}_i$ and $\boldsymbol{w}_i$, respectively, the mass of the system M and the position vector of the center-of-mass $\boldsymbol{X}$ are calculated, respectively, by

$$M = Nm \tag{3.4}$$

$$\mathbf{X} = \sum_{i=1}^{N} \mathbf{x}_i \tag{3.5}$$

In the center-of-mass coordinate system, the position vector $\boldsymbol{x}_i'$ and velocity vector $\boldsymbol{w}_i'$ of i-th molecule are expressed by

$$\boldsymbol{x}_i' = x_i - X \tag{3.6}$$

$$\boldsymbol{w}_i' = w_i - W \tag{3.7}$$

where $\boldsymbol{W}$ is the velocity vector of $\boldsymbol{X}$ (=d$\boldsymbol{X}$/dt). Using the aforementioned definitions, the total kinetic energy $E_{k,total}$ of the system is written by

$$E_{k,total} = \sum_{i=1}^{N} \frac{1}{2} m w_i^2 = \frac{1}{2} M W^2 + \frac{1}{2} m \sum_{i=1}^{N} w_i'^2 = E_k + U_k \tag{3.8}$$

The first and second terms in the aforementioned equation correspond to the macroscopic kinetic energy E_k and the internal kinetic energy U_k of the system, respectively.

Relation between microscopic and macroscopic properties

As shown in Fig. 3.3, let us consider the ideal monatomic gas enclosed in a cube of the volume $V = L^3$. The mass of each molecule is m and the number of molecules is N. For simplicity, elastic collision is assumed when the molecules collide with the inner wall of the cube, and interaction between the molecules is neglected. The velocity of i-th molecule is denoted by $\boldsymbol{w}_\mathrm{i} = (w_{i,x}, w_{i,y}, w_{i,z})$. When this molecule collides with the wall perpendicular to the x-axis, the velocity component in the x-direction becomes $-w_{i,x}$. Thus, the change in momentum per collision is $2mw_{i,x}$. The number of collision of this molecule with this wall within the time period Δt is calculated by $w_{i,x}\Delta t/2\,L$. The impulse received by the wall within Δt is hence estimated as

$$F\Delta t = \sum_{i=1}^{N} \left[2mw_{i,x} \times \frac{w_{i,x}\Delta t}{2L} \right] = \sum_{i=1}^{N} \frac{mw_{i,x}^2 \Delta t}{L} = \frac{Nm\overline{w_{i,x}^2}\Delta t}{L} \tag{3.9}$$

where $\overline{w_{i,x}^2}$ is the mean square velocity defined by

$$\overline{w_x^2} = \frac{1}{N} \sum_{i=1}^{N} w_{i,x}^2 \tag{3.10}$$

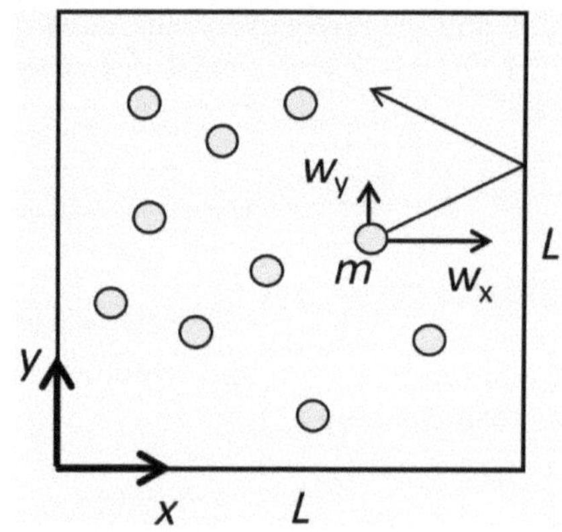

Figure 3.3 Ideal monatomic gas molecules enclosed in a cube.

Assuming that the molecular motion is random and there is no bias in any direction, the relation $\overline{w_{i,x}^2} = \overline{w_{i,y}^2} = \overline{w_{i,z}^2}$ can be expected. The mean square velocity of the molecules is

$$\overline{w^2} = \frac{1}{N}\sum_{i=1}^{N}(w_{i,x}^2 + w_{i,y}^2 + w_{i,z}^2) = 3\overline{w_x^2} \tag{3.11}$$

The pressure of the gas is hence calculated by

$$p = \frac{F}{L^2} = \frac{Nm\overline{w^2}}{3V} \tag{3.12}$$

This equation is an example of the relation between the macroscopic thermodynamic property (pressure) and the microscopic properties (the number, mass, and the velocity of molecules).

Comparing the aforementioned equation with the ideal-gas law ($pV = nRT$), one obtains the following relation.

$$pV = \frac{Nm\overline{w^2}}{3} = nRT \tag{3.13}$$

Using Avogadro's number N_A ($=N/n$), the aforementioned equation is rewritten as

$$pV = \frac{2}{3}nN_A \times \frac{1}{2}m\overline{w^2} = nN_Ak_BT \tag{3.14}$$

where k_B ($=R/N_A$) is the called the Boltzmann constant. Using the microscopic properties, the temperature is expressed as

$$T = \frac{m\overline{w^2}}{3k_B} \tag{3.15}$$

From Eqs. (3.2) and (3.15), it can be seen that the pressure and the temperature are related to the microscopic kinetic energy per unit volume and that of each molecule, respectively.

3.1.2 The zeroth law of thermodynamics

3.1.2.1 The zeroth law of thermodynamics

Temperature is a familiar measure of hotness or coldness of an object. It is a common experience that when a body contacts with another one of different temperature, heat is transferred from the higher-temperature body to the lower-temperature one. The two bodies finally attain the same temperature and no heat is transferred anymore. This situation is called the thermal equilibrium. The equality of temperature is the only requirement for the thermal equilibrium. The zeroth law of thermodynamics is derived from this experience and expressed as

- *If the two bodies A and B are in thermal equilibrium with a third body C, the bodies A and B are also in thermal equilibrium with each other.*

The zeroth law is an important law serving as a basis of temperature measurement and it cannot be derived from other laws. Historically, the zeroth law was recognized explicitly more than half a century after the formulation of the first and second laws. In consequence, it was named the zeroth law of thermodynamics since it precedes the first and second laws of thermodynamics from the physical viewpoint.

3.1.2.2 Temperature scales

The zeroth law of thermodynamics suggests the presence of temperature, but it does not provide the temperature scales. The Celsius and Fahrenheit scales would be the two most famous temperature scales used in the world. On the Celsius scale, the melting and boiling points of water under the atmospheric pressure were assigned the values of 0°C and 100°C. These values correspond to 32°F and 212°F on the Fahrenheit scale.

In thermodynamics, more desirable temperature scale called a thermodynamic temperature scale whose zero point does not depend on the properties of any substances can be defined using the first and second laws of thermodynamics. The unit of the thermodynamic temperature scale is Kelvin, that is designated by K. The lowest temperature on the Kelvin scale is absolute zero (0K) and one nonzero temperature value is necessary to calculate other temperatures. The state where gas, liquid, and solid phases can coexist is called the triple point. The temperature at the triple point of water is commonly assigned the value of 273.16K. The triple point is more preferable than the melting and boiling points since it does not depend on the pressure. By this treatment, the Kelvin scale can be related to the Celsius scale simply as

$$T(\mathrm{K}) = T(^{\circ}\mathrm{C}) + 273.15 \tag{3.16}$$

In a constant volume ideal-gas thermometer, the absolute temperature T is proportional to the pressure of the gas p. Therefore, if the ideal-gas pressure at the temperature of the triple point of water is denoted by p_{triple}, the absolute temperature of arbitrary object can be calculated by

$$T(\mathrm{K}) = 273.16 \times \frac{p}{p_{triple}} \tag{3.17}$$

where p is the pressure of the ideal-gas when the thermometer is in the thermal equilibrium with the object. This is the ideal-gas scale of temperature. In the thermodynamic temperature scale, the temperature ratio is determined from the maximum work that can be extracted using the two heat reservoirs, but it is known that the thermodynamic temperature scale and the ideal-gas scale are equivalent.

3.1.2.3 Heat capacity and specific heat

When a higher-temperature substance is in contact with a lower-temperature substance, energy is transferred from the higher- to the lower-temperature substances. This form of energy transfer is called the heat. As a result, the temperature of the higher-temperature substance decreases while the temperature of the lower-temperature substance increases. It is known that different amount of heat is needed to raise the temperature of different substances by one degree. The heat capacity is defined as the heat that is required to raise the temperature of a substance by one degree. The heat capacity is proportional to the mass of the substance. Thus, the heat capacity for a unit mass of the substance is called the specific heat. The amount of heat needed to increase the temperature of a unit mass of a substance by one degree is depends on how the process is executed. The important specific heats are those when the volume or the pressure are kept constant during the process. The former is called the specific heat at constant volume c_v and the latter is called the specific heat at constant pressure c_p, respectively.

3.1.3 The first law of thermodynamics

3.1.3.1 Heat and work

Heat

It is important in the thermodynamics to recognize that there two forms for energy to cross the boundary of a closed system. The first one is the heat Q that is defined as the form of energy that is transferred by virtue of a temperature difference between a system and its surroundings. If the system is in thermal equilibrium with its surroundings, they are at the same temperature and no energy is transferred in the form of heat. The mechanism of heat transfer is commonly categorized to the heat conduction, convective heat transfer, and radiative heat transfer. The amount of heat transferred per unit area and unit time called the heat flux. The methods to calculate the heat flux are different between the three categories and discussed in the academic discipline of heat transfer engineering. The value of heat flux is dependent on the wall material surrounding the system. Even when the temperature difference between the system and its surroundings is the same, the value of heat flux is usually smaller for wooden wall than metal wall. As an ideal concept, the wall that permits no heat transfer between the system and its surroundings is called adiabatic wall. The wall that permits heat transfer is called diathermal wall.

Work (boundary work)

Energy can cross the boundary of the closed system only in the form of heat or work. This implies that if the energy crossing the boundary of a closed system is not heat, it is work. The most elementary expression of work would be given as the product of the force F acting on a body and the displacement of the body s in the direction of the force.

$$W = Fs \tag{3.18}$$

Generally, F is not constant during the displacement. In this case, the earlier equation is rewritten as

$$W = \int_1^2 F ds \tag{3.19}$$

To calculate the quantity of work by performing the aforementioned integration, variation of F with displacement should be known. The most typical energy transfer in the form of work in thermodynamics would be the boundary work in a piston-cylinder device schematically shown in Fig. 3.4. From Eq. (3.19), one can understand that there are two requirements for a work being nonzero: (1) nonzero force acts on the boundary, and (2) the boundary moves. In consequence, when force acts on the boundary but the boundary is stationary, no energy transfer in the form of work is present between the system and its surroundings. Likewise, the displacement of

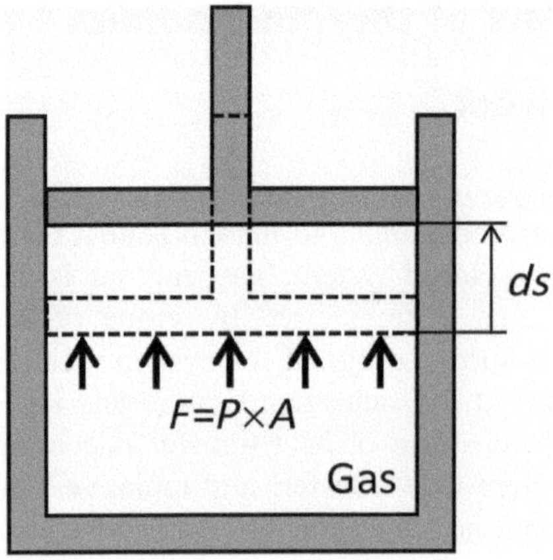

Figure 3.4 Boundary work in a piston-cylinder device.

boundary without any force cannot be a cause of energy transfer in the form of work. Gas expansion to an evacuated space is a typical example of such process.

Several other forms of work

There are several forms of work other than the boundary work. The shaft work is a common form of work and shown schematically in Fig. 3.5. The constant force F acts in the tangential direction of the rim of a disk. If the number of rotations of a disk is denoted by n, the displacement is calculated as $2\pi rn$ and the shaft work W_{shaft} is calculated by

$$W_{shaft} = 2\pi rnF = 2\pi nT \tag{3.20}$$

where T is the torque defined by $T = Fr$.

Let us consider a force acting on the one end of a linear spring. Since the force F can be written using the spring constant k and the displacement x as $F = kx$, the spring work W_{spring} is calculated by

$$W_{spring} = \int_1^2 kxdx = \frac{1}{2}k(x_2^2 - x_1^2) \tag{3.21}$$

where x_1 and x_2 are the initial and final displacements of the spring from its rest position, respectively (Fig. 3.6).

If a charged particle of q coulomb moves for a distance x in the electric field of the strength E, the force acting on the particle is given by qE. Thus, the electrical work W_e is calculated by

$$W_e = qEx = qV \tag{3.22}$$

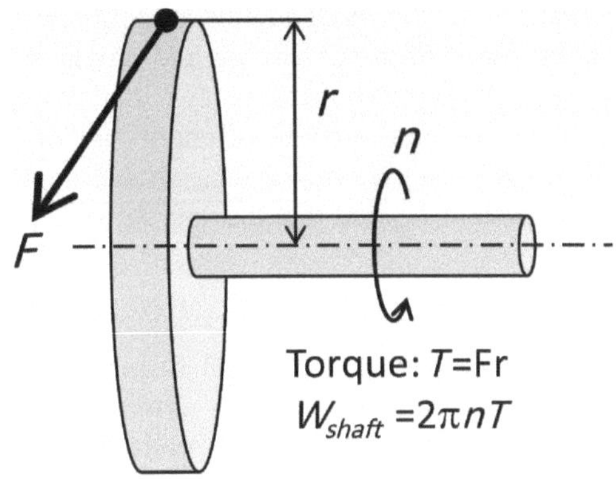

Figure 3.5 Schematic diagram of shaft work.

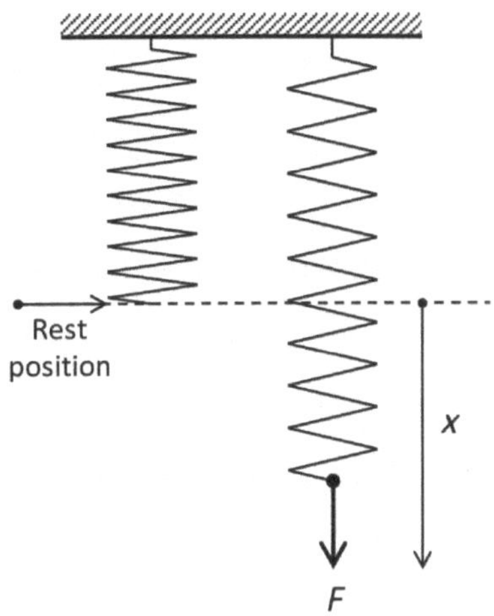

Figure 3.6 Schematic diagram of spring work.

where V (=Ex) is the electric potential difference between the initial and final particle positions. The work done per unit time is called the power. The electric current I is the number of electric charges flowing per unit time. Thus, the electrical power $\dot{W}_e$ is calculated by

$$\dot{W}_e = IV \tag{3.23}$$

3.1.3.2 *The first law of thermodynamics*

Energy conservation

Several forms of energy and energy transfer such as the kinetic energy, potential energy, internal energy, heat, and work have been discussed individually in the previous section. The first law of thermodynamics is the most basic principle expressing the relations between these forms of energy and energy transfer. Based on our experiences, the first law of thermodynamics says that the form of energy may change during a process but the total amount of energy is conserved. Therefore, the net change in the total energy of a system ΔE_{system} should be equal to the difference of the energy entering the system E_{in} and that leaving the system E_{out}.

$$\Delta E_{system} = E_{in} - E_{out} \tag{3.24}$$

The system can possess the energy in several forms such as kinetic, potential, internal, and so on. If we consider a stationary system in which the changes of kinetic and potential energies are zero and extra components such as electric and magnetic energies are assumed negligible, the net change in the total energy of a system equals to the change of the internal energy.

$$\Delta E_{system} = \Delta U \tag{3.25}$$

As for the energy transfer crossing the boundary, there are three forms: heat, work, and mass flow. When mass enters and/or leaves a system, the total energy of the system changes since the mass contains energy. In a closed system, only heat and work are the mechanisms of energy transfer to the system since no mass flow crossing the boundary is permitted. The mass flow is the additional energy transfer mechanism that is present only in open systems.

Application to closed system

Under the simplified condition discussed earlier, the first law of thermodynamics for a closed system is given by

$$\Delta U = Q - W \tag{3.26}$$

where Q and W are the energy transfer rates by heat and work, respectively. The energy transfer by mass flow is not considered since there is no mass flow crossing the boundary in the closed system. One of the most important systems in the field of mechanical engineering is the heat engine that receives energy in the form of heat and converts it to energy in the form of work. Thus, in the aforementioned equation, the sign of heat is supposed positive when the energy enters the system while the sign of work is supposed positive when the energy leaves the system.

Application to open system (steady flow system)

In an open system, mass flow crossing the boundary exists and the first law of thermodynamics is written as

$$\Delta U = Q - W + E_{mass,in} - E_{mass,out} \tag{3.27}$$

where $E_{mass,in}$ and $E_{mass,out}$ are the energies entering and leaving the system associated with mass flow, respectively. As shown schematically in Fig. 3.7, some work that is called the flow work is required to push the mass into the system. At the inlet, energy enters the system in the two ways. One is the energy contained in the mass entering the system and the other is the work that is required to push the mass into the system. If the kinetic and potential energies are neglected, the energy contained in the entering mass is internal energy and expressed by U_{in}. Denoting the volume occupied by the entering mass by V_{in}, the flow work at the inlet is $F_{in}L_{in} = p_{in}A_{in}L_{in} = p_{in}V_{in}$. Thus, $E_{mass,in}$ is written as

$$E_{mass,in} = U_{in} + p_{in}V_{in} \tag{3.28}$$

Using enthalpy H defined by $H = U + pV$, Eq. (3.27) is rewritten as

$$\Delta U = Q - W_{tech} + H_{in} - H_{out} \tag{3.29}$$

where W is replaced by W_{tech} to emphasize that W does not include the flow work at the inlet and outlet. W_{tech} is call the technical work and the sum of W_{tech} and the flow work is call the absolute work. In a steady flow system, the mass flow rate $\dot{m}$ is constant and the system properties do not change with time ($\Delta U = 0$). Thus, the first law of thermodynamics for a steady flow system is given by

$$\dot{m}(h_{out} - h_{in}) = \dot{Q} - \dot{W}_{tech} \tag{3.30}$$

where h is enthalpy per unit mass (specific enthalpy); $\dot{Q}$ and $\dot{W}_{tech}$ are the heat and the technical work per unit time, respectively.

Many components used in energy plants are regarded as steady flow systems. In pumps, compressors, and turbines, the energy transfer in the form of technical work exists but there is no heat transfer. Thus, Eq. (3.30) becomes

$$\dot{m}(h_{out} - h_{in}) = -\dot{W}_{tech} \tag{3.31}$$

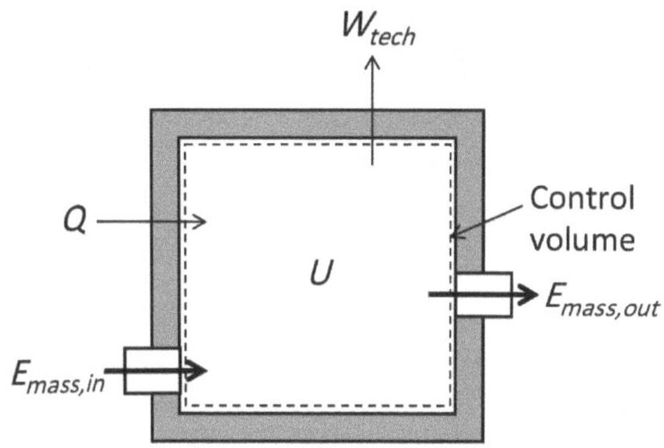

Figure 3.7 Schematic diagram of flow work.

In a boiler of thermal power plant and a reactor of nuclear power plant, energy is given to the working fluid in the form of heat. Then, Eq. (3.30) is written by

$$\dot{m}(h_{out} - h_{in}) = \dot{Q} \tag{3.32}$$

In a valve, neither heat transfer nor technical work exist. Thus, Eq. (3.30) is simplified to

$$h_{out} - h_{in} = 0 \tag{3.33}$$

When fluid passes through a valve, pressure decreases but enthalpy is conserved.

3.1.3.3 Thermodynamic process

Quasi-static or quasi-equilibrium process

As discussed earlier, the thermodynamics deals with the properties of a system in the equilibrium state where the system does not undergo a change. The change of a system from one equilibrium state to another is called a process. During the process, the system is obviously not in the equilibrium state since it undergoes a change. However, if the change of the system is slow enough, the system is expected infinitesimally close to equilibrium state throughout the process. For example, if the piston is pushed quickly in a cylinder in a piston-cylinder device, the properties such as pressure and temperature are not uniform in the system and the system properties during the process are indeterminate. If the movement of the piston is sufficiently slow, the system properties are supposed uniform. Thus, in this case, all the system properties can be determined even during the change. Such process is called a quasi-static or quasi-equilibrium process.

Reversible and irreversible processes

If the process can be reversed without leaving any trace on the surroundings, it is called a reversible process. If the process is not reversible, it is called an irreversible process. From the definition, both the system and the surroundings are returned to their initial states after a reversible process and its reverse process are undergone. It should be noted that the system can be returned to its initial state even after an irreversible process. In this case, however, some trace is always left on the surroundings.

Let us consider the isothermal expansion of gas enclosed in a piston-cylinder device as shown in Fig. 3.8A. The space surrounded by the inner walls of the piston and cylinder is supposed to be the system. It is assumed that the gas expands but its temperature is maintained since it receives heat from the heat reservoir. In this case, since some force acts on the piston wall and the piston moves, energy is transferred from the system to the surroundings in the form of boundary work. After the isothermal expansion, the gas temperature is unchanged but its volume is greater

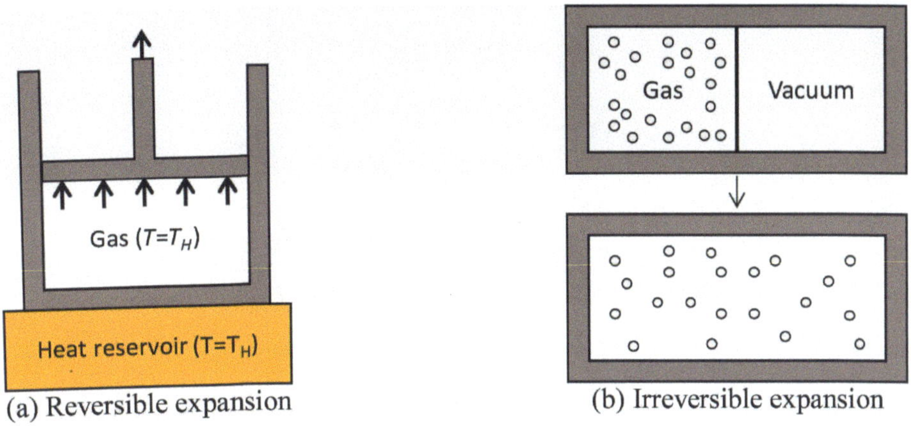

Figure 3.8 Reversible and irreversible isothermal expansion. (A) Reversible expansion. (B) Irreversible expansion.

than its initial value. If the boundary work is stored in some appropriate form of energy, it may be used to compress the gas to the initial volume. The gas temperature may be maintained during the compression process if energy is transferred from the gas to the same heat reservoir in the form of heat. It is known that both the system and its surroundings are returned to their initial states if the expansion and compression processes are quasi-static and the temperature difference between the gas and the heat reservoir is negligible throughout the process. The isothermal expansion process described earlier can hence be regarded as a reversible process.

A similar but different process is shown schematically in Fig. 3.8B. In the initial condition, a gas is enclosed in the left-half of a box. If a hole is made on the wall separating the right and lift spaces, both the spaces are filled with the gas. It is known experimentally that change in the gas temperature is small during this process. Hence, since the gas volume increases and the temperature is maintained, the variations of gas properties are similar to those in the reversible process shown in Fig. 3.8A. In this case, however, no boundary work is done to the surroundings during the expansion process. Thus, work must be prepared separately to compress the gas to the initial volume. The gas temperature increases after the compression. Therefore, energy must be removed from the gas in the form of heat to reduce the gas temperature to the initial value. After this, all the gas properties may return to the initial states. However, work is converted to heat in the surroundings. It is known that there is no way to convert heat to work without leaving any trace. It is concluded that the gas expansion process shown in Fig. 3.8B is irreversible.

In friction, two bodies in contact move relative to each other to generate heat at the interface as illustrated in Fig. 3.9. Some work is needed to move the bodies to the original positions since a friction force acts at the interface. The temperatures of the bodies once increase by the heat due to friction but may return to the initial value through heat transfer with ambient air. The surroundings, however, release

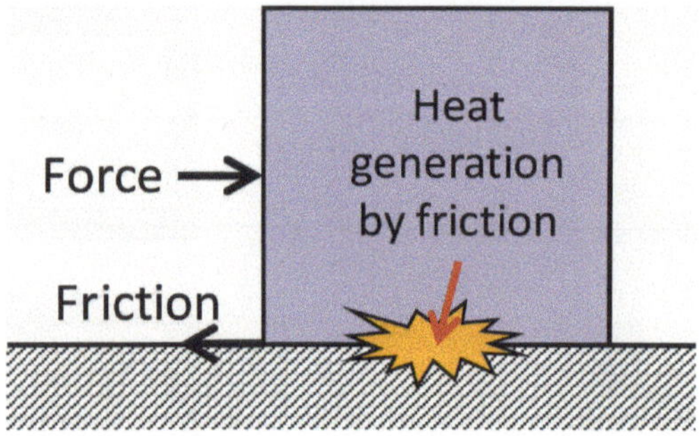

Figure 3.9 Heat generation by friction.

energy in the form of work and receive it in the form of heat. Therefore, any process involving friction is irreversible.

Heat transfer is the process in which energy moves from a higher-temperature body to a lower-temperature one. Finally, the temperatures of the two bodies become equal to reach a thermal equilibrium state. To reverse the temperatures of the two bodies to the original values, appropriate amount of energy must be removed from the originally lower-temperature body and it must be given to the originally higher-temperature body. It is known that in the surroundings, converting the energy removed from the originally lower-temperature body to that given to the originally higher-temperature body without leaving any traces is not possible. Heat transfer between bodies of different temperatures is therefore irreversible.

In the process in Fig. 3.8A, heat transfer between the gas and the heat reservoir was supposed to occur by infinitesimally small temperature difference and the resulting process was judged to be reversible. It should however be noted that no heat transfer occurs if there is no temperature difference. This implies that irreversibility of heat transfer process decreases with a decrease in the temperature difference between the two bodies but the process cannot be completely reversible. A reversible process is an ideal concept used in the thermodynamics. It is approached with a decrease in irreversibility and a completely reversible process never occurs.

3.1.4 Properties of various substances

3.1.4.1 Properties of gas

Equation of state for ideal gases

An equation of state determines the relation between the pressure p, temperature T and specific volume defined as the volume per unit mass v of a substance. The simplest and most useful equation of state would be that for an ideal-gas. That is given by

$$pv = RT \tag{3.34}$$

This ideal-gas equation of state is derived by combining the two experimentally-derived gas laws: Boyle's law (the gas pressure is inversely proportional to its volume under the constant temperature) and Charles's law (the gas temperature is proportional to its volume under the constant pressure). The proportionality factor R in the aforementioned equation is called the gas constant and its value is different for each gas. If v in Eq. (3.34) is replaced by the gas volume per unit molar mass v_m, Eq. (3.34) is rewritten as

$$pv_m = R_0T \tag{3.35}$$

The proportionality factor R_0 in the aforementioned equation has the same value of 8.31 J/mol · K for all the substances and is called the universal gas constant. Denoting the molecular weight by M, the relation between R and R_0 is

$$R_0 = MR \tag{3.36}$$

Using the mass m or the density ρ of the gas, Eq. (3.34) can be rewritten as

$$pV = mRT \tag{3.37}$$

$$p = \rho RT \tag{3.38}$$

Similarly, using the mole number n, Eq. (3.35) is rewritten as

$$pV = nR_0T \tag{3.39}$$

The ideal-gas equation of state is fairly valid for many familiar gases within ordinary ranges of the pressure and temperature as depicted in Fig. 3.10.

Equation of state for real gases

The ideal-gas equation of state (34) becomes less valid when the volume of gas molecules and/or intermolecular interactions are not negligible. Hence, its deviation from a real gas becomes significant with an increase in the pressure, decrease in the temperature, and increase in the molecular weight. Such trends can be confirmed particularly for CO_2 in Fig. 3.10. Several equations of state have been proposed so

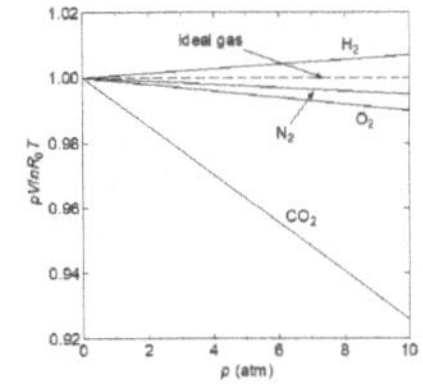

Figure 3.10 Validity of the ideal-gas equation of state for several gases at 273K.

far in order to express the relation between p, T and v of real gases more accurately. The van der Waals equation of state is one of the earliest and most famous ones and written by

$$\left(p + \frac{a}{v^2}\right)(v - b) = RT \tag{3.40}$$

It is considered that the pressure tends to decrease if attraction forces act between the gas molecules and that the volume permitted for gas molecules to move is smaller than v if the size of gas molecules is not negligible. The terms a/v^2 and b are included to compensate these effects, respectively. The two constants a and b are different for each gas. The van der Waals equation of state is not very accurate, but it is of historical importance since it is one of the earliest models for real gases and used as the base of many later models such as the Beattie-Bridgeman and Benedict-Webb-Rubin equations of state [1–3].

Another approach to express the equation of state for real gases is using the virial expansion. Fig. 3.10 suggests that pv/RT can be approximated by the power series of p as

$$\frac{pv}{RT} = 1 + b_p(T)p + c_p(T)p^2 + \cdots \tag{3.41}$$

Alternatively, the power series of $1/v$ may be used as

$$\frac{pv}{RT} = 1 + \frac{b_v(T)}{v} + \frac{c_v(T)}{v^2} + \cdots \tag{3.42}$$

These equations are call the virial equations of state. The coefficients b_p, c_p, b_v and c_v are the functions of temperature and called the virial coefficients; more specifically, b_p and b_v are called the second and c_p and c_v are called the third virial coefficients. The accuracy of the virial equation of state can be improved by increasing the number of terms. As expected in Fig. 3.10, adequate accuracy is usually achieved by taking the second virial coefficient when the pressure is not very high.

Internal energy and enthalpy of ideal gases

A facility used in a famous free-expansion experiment of air performed by Joule is shown schematically in Fig. 3.11. Two tanks connected by a tube with a valve were submerged in a thermally insulated water bath. In the initial condition, the valve was closed, and one tank contained high-pressure air and the other was evacuated. After thermal equilibrium was established, the valve was opened to let the air eventually fill the two tanks equally. It was observed that the water temperature was unchanged throughout the process, implying that there was no heat transfer between the air and water. Obviously, no work was done to the air and water in this process. Since there was no energy transfer to the air both in the forms of heat Q and work W, the internal energy of air U should be maintained

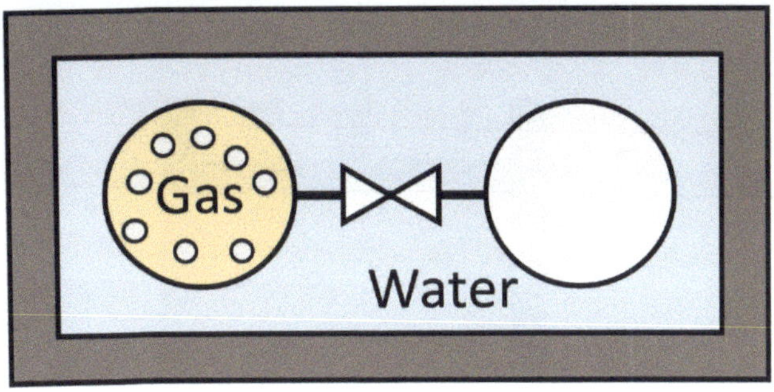

Figure 3.11 Experimental apparatus used by Joule.

from the first law of thermodynamics ($\Delta U = Q - W$). Generally, U can be expressed as a function of the temperature T and the volume V. From the experimental results described earlier, it was concluded that the internal energy of air is a function of temperature only and not a function of volume since the internal energy and temperature of air were constant but the air volume was obviously different before and after the process. Later, it was found that measurement error in Joule's experiment was too large to derive the aforementioned conclusion. It was, however, proved theoretically using the second law of thermodynamics that as suggested in Joule's experiment, the internal energy of an ideal-gas can be written as a function of temperature only.

$$u = u(T) \tag{3.43}$$

From the definition of enthalpy and the equation of state for ideal gases, the specific enthalpy of an ideal-gas is also a function of temperature only:

$$h = u(T) + pv = u(T) + RT = h(T) \tag{3.44}$$

Specific heat of ideal gases

Let us consider an ideal gas enclosed in a piston-cylinder device as shown schematically in Fig. 3.12. If heat Q is applied to the gas under the constant volume condition, all the heat is spent to increase the internal energy U of the gas.

$$Q = \Delta U = C_v \Delta T \text{ or } \Delta u = c_v \Delta T \tag{3.45}$$

The specific heat at constant volume c_v of an ideal-gas is hence written as

$$c_v = c_v(T) = \left(\frac{\partial u}{\partial T}\right)_v \tag{3.46}$$

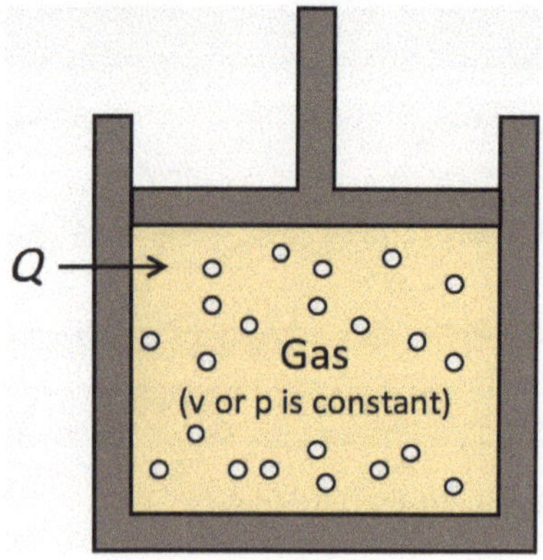

Figure 3.12 Ideal gas enclosed in a piston-cylinder device.

Next, under the constant pressure condition, the applied heat is spent to increase the internal energy of a gas as well as work done to the surroundings.

$$Q = \Delta U + W = \Delta U + p\Delta V = \Delta(U + pV) = \Delta H = C_p \Delta T \text{ or } \Delta h = c_p \Delta T \quad (3.47)$$

The specific heat at constant pressure c_p of an ideal-gas is hence written as

$$c_p = c_p(T) = \left(\frac{\partial h}{\partial T}\right)_p \quad (3.48)$$

It can be seen that since u and h of an ideal-gas depend only on temperature, the specific heats of an ideal-gas at constant volume c_v and constant pressure c_p are also functions of temperature only.

From the definition of h $(= u + pv)$ and the ideal-gas equation of state $(pv = RT)$, one obtains

$$dh = du + d(pv) = du + RdT \quad (3.49)$$

Since h and u of an ideal-gas are dependent only on temperature, the following important relation is derived.

$$c_p = c_v + R \quad (3.50)$$

This is called the Mayer relation for an ideal-gas. Here, let us define the following property called the specific heat ratio κ.

$$\kappa = \frac{c_p}{c_v} \quad (3.51)$$

If R and κ are given, c_v and c_p are calculated, respectively, by

$$c_v = \frac{R}{\kappa - 1} \tag{3.52}$$

$$c_p = \frac{\kappa R}{\kappa - 1} \tag{3.53}$$

3.1.4.2 Properties of liquids and solids

As the equation of state (34) for an ideal-gas indicates, the specific volume of gas changes noticeably with changes of other properties such as pressure and temperature. On the other hand, a change in the specific volume of liquids and solids is negligibly small in many processes. In this case, liquids and solids can be approximated as incompressible substances and the energy transfer associated with the boundary work can be eliminated. It should however be noted that the incompressible approximation is not always appropriate. For example, the compressibility of solids plays important a role in the analysis of thermal stress.

When the incompressible approximation is valid, the specific heats at constant volume and constant pressure are equal and written by

$$c_v = c_p = c \tag{3.54}$$

The specific heat of an incompressible substance is a function of temperature. A change in specific internal energy is hence written by

$$du = c(T)dT \tag{3.55}$$

When the specific heat is constant, the change of internal energy is calculated by

$$\Delta u = c\Delta T \text{ or } \Delta U = mc\Delta T \tag{3.56}$$

3.1.5 Quasi-static change of ideal-gas

A quasi-static change of an ideal-gas enclosed in a piston-cylinder device would be one of the most fundamental processes in thermodynamics. In this situation, the force acting on the piston per unit surface is equal to the gas pressure and the boundary work is written by

$$W = \int_1^2 Fds = \int_1^2 pAds = \int_1^2 pdV \text{ or } w = \frac{W}{m} = \int_1^2 pdv \tag{3.57}$$

where A is the area of the piston surface. Using the equation of state for ideal gases, the aforementioned equation is rewritten as

$$W = \int_1^2 pdV = mR\int_1^2 \frac{T}{V}dV \text{ or } w = \int_1^2 pdv = R\int_1^2 \frac{T}{v}dv \tag{3.58}$$

Similarly, the heat transfer rate is written by

$$Q = \Delta U + W = m\int_1^2 c_v dT + W \text{ or } q = \frac{Q}{m} = \Delta u + w = \int_1^2 c_v dT + w \tag{3.59}$$

The aforementioned equations indicate that the work and heat can be calculated from the quantities of state of the gas if a constraint condition satisfied during the process is given. In the following, the energy transfers are discussed for the five typical processes: isothermal process (constant temperature), isobaric process (constant pressure), isochoric process (constant volume), adiabatic process (no heat transfer), and polytropic process. Here, the specific heats are assumed constants for simplicity.

3.1.5.1 Isothermal process

Since the gas temperature is kept constant during the process, it can be assumed that $T_1 = T_2 = T$, $p_1v_1 = p_2v_2 = RT$. Hence, energy transfers per unit mass of the ideal-gas as the work and heat are calculated, respectively, by

$$w = RT\int_1^2 \frac{dv}{v} = RT\ln\frac{v_2}{v_1} = p_1v_1\ln\frac{v_2}{v_1} = p_1v_1\ln\frac{p_1}{p_2} \tag{3.60}$$

$$q = c_v\int_1^2 dT + w = 0 + w = w \tag{3.61}$$

In isothermal processes, internal energy of the gas is maintained ($\Delta u = 0$). Thus, all the heat given to the working fluid is transformed to work ($w = q > 0$) or all the work given to the working fluid is transferred to heat ($w = q < 0$).

3.1.5.2 Isobaric process

Since the gas pressure is maintained ($p_1 = p_2 = p$), the boundary work can simply be calculated by

$$w = \int_1^2 pdv = p(v_2 - v_1) = R(T_2 - T_1) \tag{3.62}$$

The heat transfer rate is calculated by

$$q = c_v(T_2 - T_1) + p(v_2 - v_1) = h_2 - h_1 = c_p(T_2 - T_1) \tag{3.63}$$

3.1.5.3 Isochoric process

Since the gas volume does not change ($v_1 = v_2$), the boundary work is zero obviously.

$$w = \int_1^2 pdv = 0 \tag{3.64}$$

The heat transfer rate is hence calculated as

$$q = c_v(T_2 - T_1) = u_2 - u_1 \tag{3.65}$$

It can be confirmed that the applied heat is spent for the increase of the internal energy of gas since there is no work. Comparing Eqs. (3.63) and (3.65), it can also be confirmed that more heat is needed in isobaric processes than in isochoric processes to raise the gas temperature from T_1 to T_2 since c_p is greater than c_v.

3.1.5.4 Adiabatic process

In adiabatic processes where there is no energy transfer in the form of heat, the following relation is derived from Eqs. (3.58) and (3.59).

$$\delta q = c_v dT + pdv = 0 \tag{3.66}$$

Using the equation of state (34) and the Mayer relation (50) for ideal gases, the aforementioned equation is transformed to

$$c_v(pdv + vdp) + Rpdv = c_v vdp + c_p pdv = 0 \tag{3.67}$$

This equation is further transformed to

$$\frac{dp}{p} + \kappa \frac{dv}{v} = 0 \tag{3.68}$$

Integrating the aforementioned equation yields

$$\ln p + \kappa \ln v = \text{const.} \tag{3.69}$$

It can hence be concluded that pv^{κ} is kept constant in adiabatic processes:

$$pv^{\kappa} = \text{const.} \tag{3.70}$$

Using the equation of state (34), this equation is rewritten as

$$Tv^{\kappa-1} = \text{const.} \tag{3.71}$$

$$\frac{T}{p^{(\kappa-1)/\kappa}} = \text{const.} \tag{3.72}$$

The boundary work done during adiabatic processes is calculated as

$$w = \int_1^2 pdv = p_1 v_1^{\kappa} \int_1^2 \frac{dv}{v^{\kappa}} = \frac{p_1 v_1}{\kappa - 1}\left[1 - \left(\frac{v_1}{v_2}\right)^{\kappa-1}\right] \tag{3.73}$$

Several other expressions of w in adiabatic processes are

$$w = \frac{p_1 v_1}{\kappa - 1}\left[1 - \left(\frac{p_2}{p_1}\right)^{(\kappa-1)/\kappa}\right] = \frac{1}{\kappa - 1}(p_1 v_1 - p_2 v_2) = \frac{R}{\kappa - 1}(T_1 - T_2) \tag{3.74}$$

Since q equals zero in adiabatic processes, w is equals to a decrease in the internal energy of the working fluid.

$$w = u_1 - u_2 \tag{3.75}$$

3.1.5.5 Polytropic process

In polytropic processes, the following relation is satisfied.

$$pv^{n} = \text{const.} \tag{3.76}$$

Various processes can be regarded as polytropic processes by changing the polytropic index n. The p-v diagram is drawn for the aforementioned described processes in Fig. 3.13. It can be seen that the polytropic process is reduced to an isothermal process when $n = 1$, isobaric process when $n = 0$, isochoric process when $n = \infty$, and adiabatic process when $n = \kappa$.

If the p-v diagrams for adiabatic and isothermal processes in Fig. 3.13 are compared, it can be seen that pressure reduction during gas expansion is more significant in the adiabatic process than in the isothermal process since no heat is transferred from the surroundings to the gas enclosed in the cylinder. Even if the piston-cylinder device is covered with thermal insulation material, completely

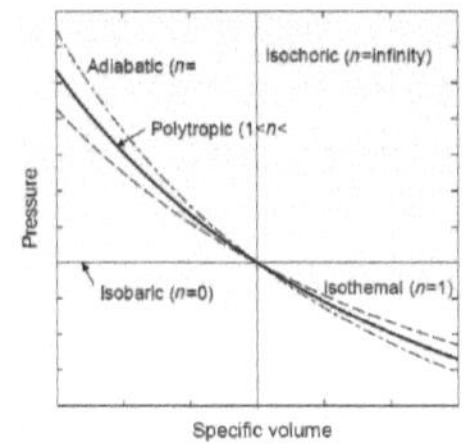

Figure 3.13 *P-v* diagram for various processes.

adiabatic process cannot be achieved in the real world and some heat would be transferred to the gas due to the temperature decrease caused by the expansion. The polytropic process can be used to approximate such process encountered in the actual device.

The boundary work during the polytropic process is expressed by the following equation where κ is replaced by n in Eqs. (3.73) and (3.74).

$$w = \frac{p_1 v_1}{n-1}\left[1 - \left(\frac{v_1}{v_2}\right)^{n-1}\right] = \frac{p_1 v_1}{n-1}\left[1 - \left(\frac{p_2}{p_1}\right)^{(n-1)/n}\right]$$

$$= \frac{1}{n-1}(p_1 v_1 - p_2 v_2) = \frac{R}{n-1}(T_1 - T_2) \tag{3.77}$$

From the first law of thermodynamics, the heat transfer rate is calculated by

$$q = c_v(T_2 - T_1) + w \tag{3.78}$$

Substituting the last expression in Eq. (3.77) to the aforementioned equation yields

$$q = c_v(T_2 - T_1) + \frac{R}{n-1}(T_1 - T_2) = \frac{n-\kappa}{n-1}c_v(T_2 - T_1) \tag{3.79}$$

3.1.6 The second law of thermodynamics

3.1.6.1 Cycle

A heat engine outputs mechanical work by expansion of the working fluid. To perform the expansion process to the surroundings, the working fluid must be heated and pressurized to be high temperature and high-pressure condition. The working fluid is heated by a heat Q_H from a heat source at the temperature T_H. After the expansion, the working fluid is compressed by an input of mechanical work $W_{\rm in}$ for a continuous operation. To make the input work smaller than the output work, the energy of the working fluid after the expansion is decreased by a heat radiation $Q_{\rm L}$ to a heat sink at the temperature $T_{\rm L}$. The energy conversion of heat to work by the heat engine is modeled as schematically shown in Fig. 3.14. The variable W represents the effective work calculated as the net work output through the operation, and is equal to the net heat input to the heat engine expressed as $Q_H - Q_L$. While mechanical work can be treated equivalently independent of the state, heat must be treated differently depending on the temperature. The performance of the heat engine is evaluated by thermal efficiency defined as the ratio of the effective work to the heat input, and is expressed as

$$\eta = \frac{W}{Q_H} = \frac{Q_H - Q_L}{Q_H} = 1 - \frac{Q_L}{Q_H} < 1 \tag{3.80}$$

In a stable operation of heat engine, the change in the state of the working fluid forms a closed curve on the state diagram as shown in Fig. 3.15. The process of the working fluid is called thermodynamic cycle. The direction of the change for thermal engines is clockwise, and the area surrounded by the closed curve is equal to the effective work.

By operating a thermodynamic cycle in reverse, heat can be transferred from low temperature to high temperature as schematically shown in Fig. 3.16. Since such operation is similar to a water pump to lift water to a higher place, the system is called heat pump system. When the system operates to the surroundings as a high temperature heat reservoir, the system operates as a refrigerating system to generate low temperature condition. The state of the working fluid changes counterclockwise as shown in Fig. 3.17. The performance of the heat pump or refrigerating cycle is evaluated by coefficient of performance (often called COP) defined as the ratio of the effective heat to the work input. The definition of coefficient of performance is different between heat pump and refrigerating cycle due to the difference in the

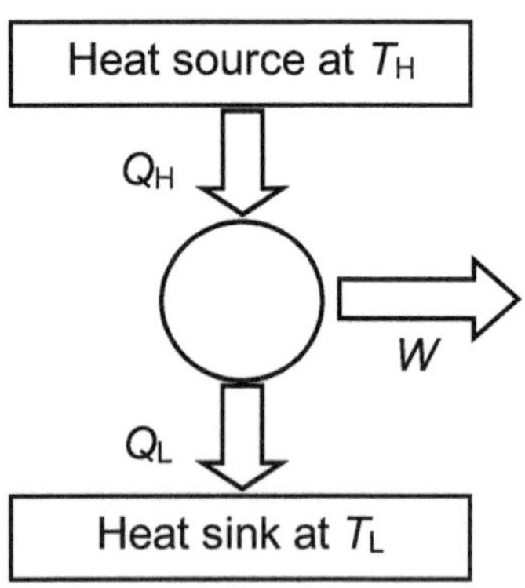

Figure 3.14 Heat engine.

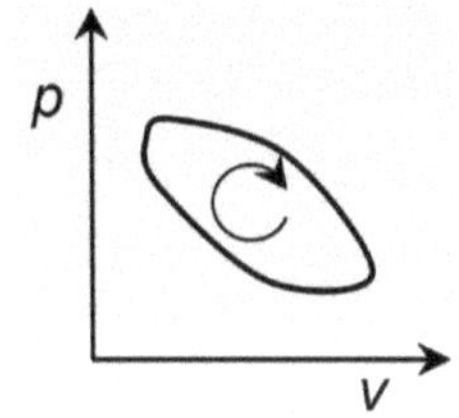

Figure 3.15 Heat engine cycle on *p-v* diagram.

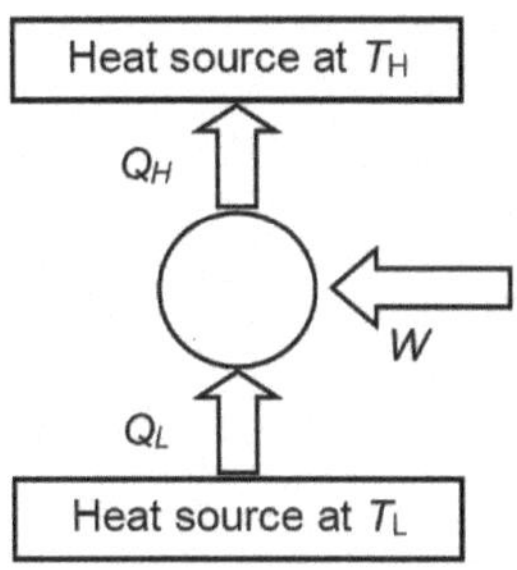

Figure 3.16 Heat pump or refrigerating system.

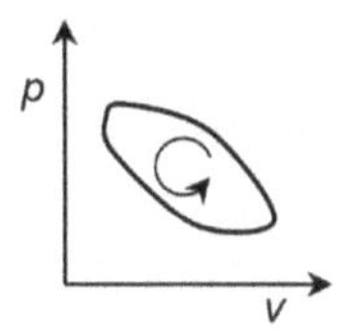

Figure 3.17 Heat pump or refrigerating cycle on *p-v* diagram.

efficient heat. The COP of the refrigerating cycle COP_R and of the heat pump cycle COP_{HP} are defined by the following equations, respectively.

$$COP_R = \frac{Q_L}{W} = \frac{Q_L}{Q_H - Q_L} \tag{3.81}$$

$$COP_{HP} = \frac{Q_H}{W} = \frac{Q_H}{Q_H - Q_L} = COP_R + 1 \tag{3.82}$$

When the heat pump and refrigerating cycle are operating for the same heat transfer amount, the COP of the heat pump cycle is higher than that of the refrigerating cycle because the work input is utilized for the heat supply in the heat pump system. However, attention must be paid that the coefficient of performance depends on the temperature conditions of heat reservoirs as described later.

3.1.6.2 Expressions of the second law of thermodynamics

The first law of thermodynamics treats heat and work as energy equivalently, and expresses energy conservation in an energy conversion process. In contrast, the second law of thermodynamics expresses the particularity of heat flow, that is, irreversibility in heat and mass transfer.

Clausius statement

Rudolf Clausius explained the second law of thermodynamics as the following statement. Heat cannot be transferred from a low temperature to a high temperature substance without leaving any trace in the surroundings. In other words, heat is transferred spontaneously only from a high temperature to a low temperature substance. Such spontaneous change is irreversible change. The device shown in Fig. 3.18 is defied by the Clausius principle.

Kelvin-Plank statement

The Kelvin-Plank statement states that it is impossible to construct a device which operates by the heat input from a single heat source and produces the equivalent work. The device is known as the perpetual motion of the second kind as shown in Fig. 3.19. If the device is realized, a system in which a heat pump cycle is driven by the output power of the device can be constructed as shown in Fig. 3.20. The system is equivalent to the device shown in Fig. 3.18, and is contrary to the Clausius statement.

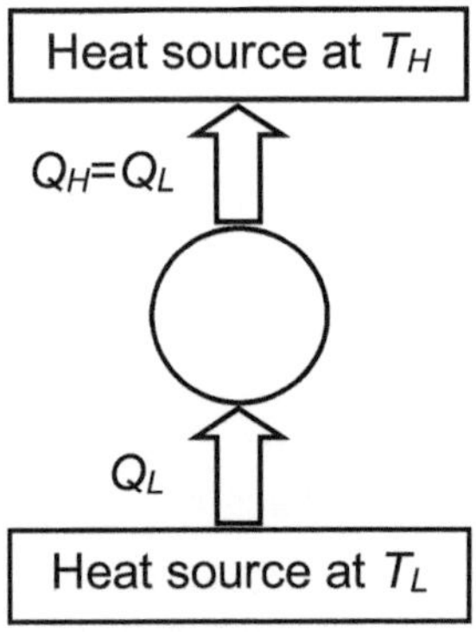

Figure 3.18 Device defied by the Clausius statement.

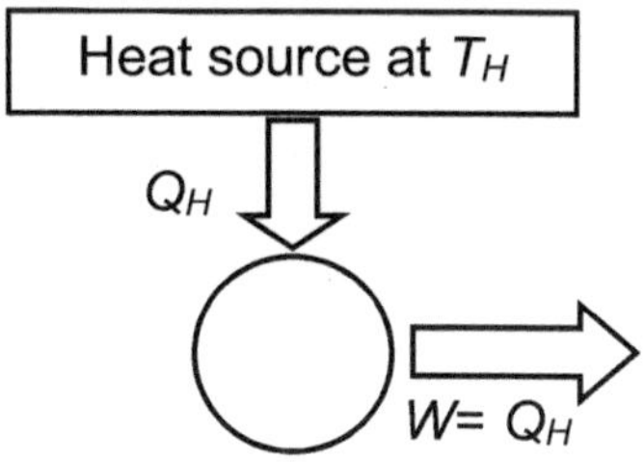

Figure 3.19 Devise defied by the Kelvin-Plank statement.

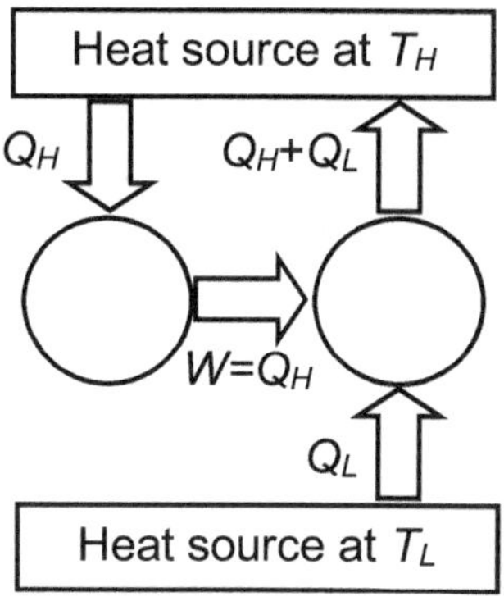

Figure 3.20 System defying the Clausius statement.

3.1.6.3 Carnot theorem

Carnot cycle proposed by Sadi Carnot is a ideal reversible cycle operating between two heat reservoirs. No cycle produces higher thermal efficiency than Carnot cycle for given temperatures of the heat reservoirs. The cycle consists of four reversible processes, and those are two isothermal processes and two adiabatic processes. Heat input and heat release are made during the isothermal processes in which the temperature of the working fluid is the same with the temperature of each heat reservoir. A *p-v* diagram of Carnot cycle is shown in Fig. 3.21. Let us derive the thermal efficiency of the Carnot cycle operating between two temperatures at T_H and T_L using an ideal-gas with unit mass as the working fluid.

Adiabatic compression process from State 1 to 2

The mechanical work w_c is equal to the difference in the internal energy, and is expressed by the following equation from Eq. (3.74).

$$w_{12} = -(u_2 - u_1) = c_v(T_L - T_H) \tag{3.83}$$

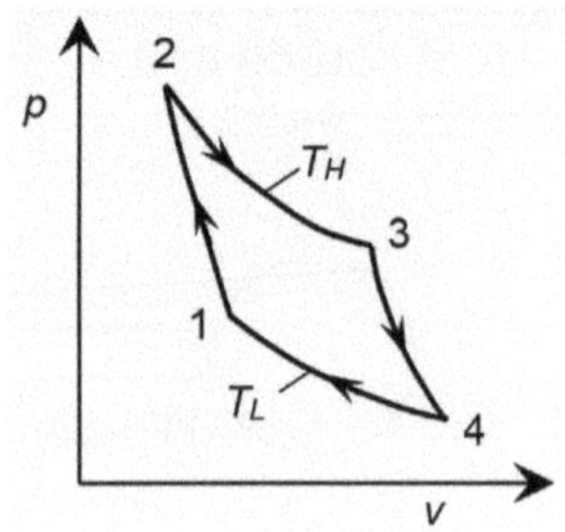

Figure 3.21 Carnot cycle.

Isothermal heating process from State 2 to 3

The heat input q_H is equal to the expansion work w_{23}, and is expressed by the following equation from Eq. (3.60).

$$q_H = w_{23} = RT_H \ln \frac{v_3}{v_2} \tag{3.84}$$

Adiabatic expansion process from State 3 to 4

The mechanical work w_{34} is expressed by the following equation.

$$w_{34} = -(u_4 - u_3) = c_v(T_H - T_L) \tag{3.85}$$

Isothermal cooling process from State 4 to 1

The heat input q_L is equal to the expansion work w_{41}, and is expressed by the following equation.

$$q_L = w_{41} = RT_L \ln \frac{v_1}{v_4} \tag{3.86}$$

Since the exhaust heat to the low temperature heat reservoir is $(-q_L)$, the thermal efficiency is expressed by

$$\eta_{Carnot} = 1 - \frac{(-q_L)}{q_H} = 1 - \frac{RT_L(v_4/v_1)}{RT_H(v_3/v_2)} \tag{3.87}$$

The following relations of the quantity of state are obtained for each adiabatic process.

$$T_L {v_1}^{\kappa-1} = T_H {v_2}^{\kappa-1} \tag{3.88}$$

$$T_H {v_3}^{\kappa-1} = T_L {v_4}^{\kappa-1} \tag{3.89}$$

From Eqs. (3.88) and (3.89),

$$\frac{v_4}{v_1} = \frac{v_3}{v_2} \tag{3.90}$$

By substituting Eq. (3.90) into Eq. (3.87), the thermal efficiency is expressed by

$$\eta_{Carnot} = 1 - \frac{T_L}{T_H} \tag{3.91}$$

The thermal efficiency of the reversible cycle can be expressed by the temperatures of the heat reservoirs only. Carnot theorem is summarized later.

Reversible heat engines operating between two heat reservoirs have the same thermal efficiency that depends only on the temperatures of the two heat reservoirs regardless of the working fluids, and no heat engine can exceed this thermal efficiency.

You can find from Carnot efficiency in Eq. (3.91) that higher T_{H} and lower T_{L} leads to higher thermal efficiency.

Because the thermal efficiency of actual heat engines must be lower than Carnot efficiency, the following inequality is obtained.

$$\eta = 1 - \frac{Q_L}{Q_H} \le \eta_{Carnot} = 1 - \frac{T_L}{T_H} \tag{3.92}$$

Separating the terms for the interaction with each heat reservoir, the inequality is rewritten as

$$\frac{Q_H}{T_H} - \frac{Q_L}{T_L} \le 0 \tag{3.93}$$

Treating heat flows from reservoirs as positive values, the inequality rewritten as

$$\frac{Q_H}{T_H} + \frac{Q_L}{T_L} \le 0 \tag{3.94}$$

where the equal sign is true for reversible heat engines. When a heat engine operates with thermal connections to n heat reservoirs, the relation is expressed by

$$\frac{Q_1}{T_1} + \frac{Q_2}{T_2} + \cdots + \frac{Q_n}{T_n} = \sum_{i=1}^{n} \frac{Q_i}{T_i} \le 0 \quad (3.95)$$

Because the temperature of heat reservoir changes continuously for a practical heat engine, the left term is expressed by a contour integral.

$$\oint \frac{\delta Q}{T} \le 0 \quad (3.96)$$

This inequality is Clausius inequality. Remind that T is temperature of the heat reservoirs.

3.1.6.4 Entropy

Consider a reversible cycle passing the states at A and B as shown on P-v diagram in Fig. 3.22.

For reversible cycle, Eq. (3.96) has equality.

$$\oint \frac{\delta Q}{T} = 0 \quad (3.97)$$

Separating the process C1 and C2 between the states at A and B,

$$\int_{A,C1}^{B} \frac{\delta Q}{T} + \int_{B,C2}^{A} \frac{\delta Q}{T} = 0 \quad (3.98)$$

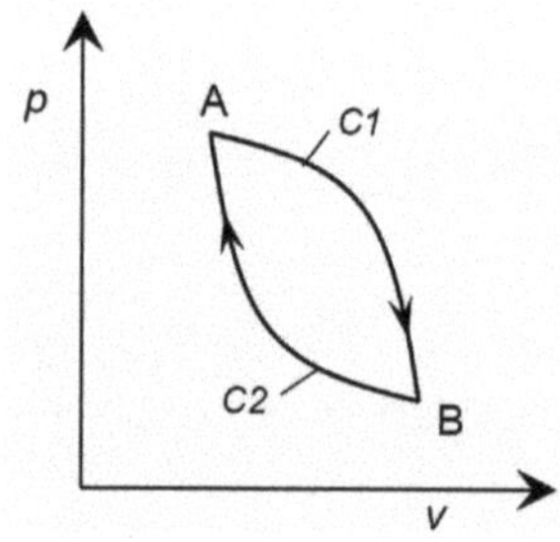

Figure 3.22 Reversible cycle.

The equation can be rewritten for the change from the state at A to B as

$$\int_{A,C1}^{B} \frac{\delta Q}{T} = \int_{A,C2}^{B} \frac{\delta Q}{T} \tag{3.99}$$

This equation means that $\delta Q/T$ is independent of the pass of the change, if the processes are reversible. Therefore, the function of $\delta Q/T$ is a quantity of state, and is defined as entropy by

$$dS = \frac{\delta Q}{T} \tag{3.100}$$

Entropy is extensive quantity. In the calculation of the change in entropy of a substance, the temperature of the substance is applied to T. For example, the change in entropy of an ideal-gas in a closed system from the state A to B is expressed from Eqs. (3.58) and (3.59) by

$$\begin{aligned} \Delta S_{AB} = \mathrm{S}_B - \mathrm{S}_A &= \int_A^B \frac{\delta Q}{T} = \int_A^B \frac{du + pdv}{T} = \int_A^B \frac{c_v dT}{T} + \int_A^B \frac{Rdv}{v} \\ &= c_v ln\left(\frac{T_B}{T_A}\right) + Rln\left(\frac{v_B}{v_A}\right) \end{aligned} \tag{3.101}$$

3.1.6.5 The principle of entropy increase

Actual processes always include irreversible processes, such as heat diffusion due to temperature difference, mass diffusion due to pressure and concentration difference, and so on. Here, consider a cycle consisting of a reversible process along C1 and an irreversible process along C2 as shown in Fig. 3.23. From Clausius inequality,

$$\oint \frac{\delta Q}{T} = \int_{A,C1}^{B} \frac{\delta Q}{T} + \int_{B,C2}^{A} \frac{\delta Q}{T} \leq 0 \tag{3.102}$$

For the reversible process,

$$\int_{A,C1}^{B} \frac{\delta Q}{T} = S_B - S_A \leq 0 \tag{3.103}$$

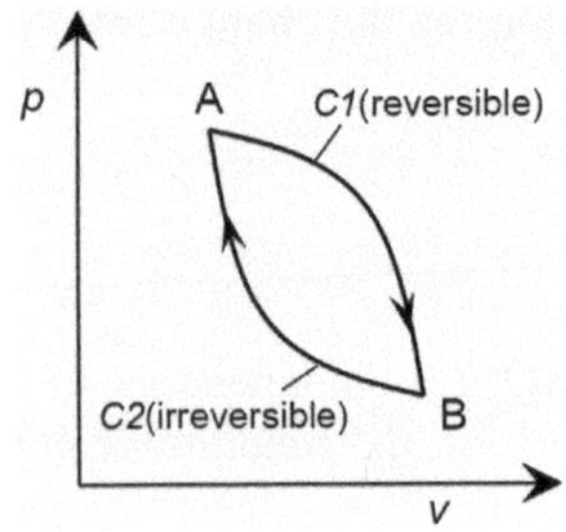

Figure 3.23 Cycle with irreversible process.

Finally, the following equation is derived.

$$S_A - S_B \geq \int_{B,C2}^{A} \frac{\delta Q}{T} \tag{3.104}$$

For an infinitesimal change,

$$dS \geq \frac{\delta Q}{T} \tag{3.105}$$

This equation expresses the change in entropy caused by a certain change. This equation is one of the expression of the second law of thermodynamics. When the change is occurred in an isolated system where no energy and mass is transferred through the boundary, because $\delta Q = 0$, the change in entropy is expressed as

$$dS \geq 0 \tag{3.106}$$

This equation expresses that when any change is occurred in an isolated system, the entropy in the isolated system must increase due to the existence of irreversibility. If the change is reversible, the entropy in the isolated system is kept constant. In other words, the direction of the change is decided by Eq. (3.106). The equation on the change in entropy expresses the principle of entropy increase.

3.1.7 Analysis of heat engine using entropy

Let us consider the change in entropy in an operation of a heat engine shown in Fig. 3.14. The changes in entropy of two heat reservoirs at T_H and T_L due to the heat flow are expressed by the following equations based on the definition of entropy, respectively.

$$S_H = -\frac{Q_H}{T_H} \tag{3.107}$$

$$S_L = \frac{Q_L}{T_L} \tag{3.108}$$

Because the change in entropy of the working fluid is zero in one cycle, the entropy change caused by the operation of the heat engine is expressed by

$$\Delta S = S_H + S_L = -\frac{Q_H}{T_H} + \frac{Q_L}{T_L} \tag{3.109}$$

For a reversible engine,

$$\Delta S = -\frac{Q_H}{T_H} + \frac{Q_L}{T_L} = 0 \tag{3.110}$$

From the relation, Carnot coefficient can be derived as follows.

$$\eta_{Carnot} = 1 - \frac{Q_L}{Q_H} = 1 - \frac{T_L}{T_H} \tag{3.111}$$

Let us reconsider the Carnot cycle using entropy. Figure 3.24 is *T-S* diagram of the Carnot cycle. The area surrounded by the cycle shows the net heat input, and is equal to the net work. It is clear that the maximum work obtained for two given temperatures of heat reservoirs can be obtained by a cycle with isothermal changes

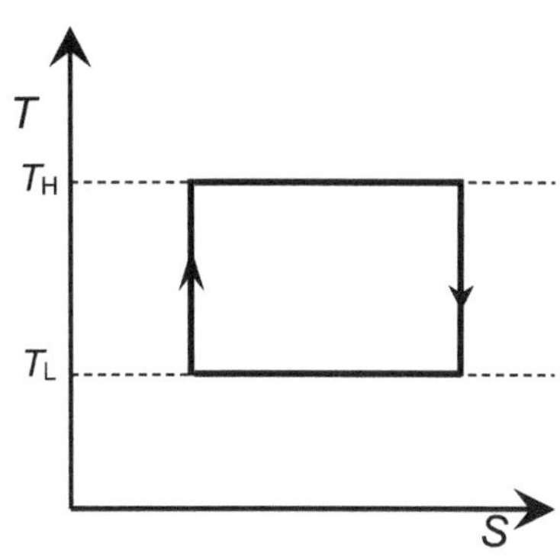

Figure 3.24 *T-S* diagram of Carnot cycle.

at the same temperature with each heat reservoir, because the temperature of the working fluid cannot exceed T_H and go blow T_L.

On the other hand, for a actual heat engine, the processes include irreversibility, and the thermal efficiency is lower than the Carnot efficiency. The effective work is expressed using the Carnot efficiency by

$$W = \left(1 - \frac{T_L}{T_H}\right) Q_H - LW \tag{3.112}$$

where LW is loss of available work due to irreversibility in the processes. From the energy conservation,

$$Q_L = \frac{T_L}{T_H} Q_H + LW \tag{3.113}$$

Applying the Clausius inequality, LW is expressed by

$$\frac{Q_H}{T_H} - \frac{Q_L}{T_L} = -\frac{LW}{T_L} \le 0 \tag{3.114}$$

$$LW = \left(\frac{Q_L}{T_L} - \frac{Q_H}{T_H}\right) T_L = T_L \Delta S \ge 0 \tag{3.115}$$

The loss of available work can be calculated from the increase in entropy during the system operation. The equation is applicable to the general process, and is called Gouy-Stodola theorem. The theorem is quite important in the design and analysis of energy conversion systems.

3.1.8 Direction of spontaneous change and free energy

3.1.8.1 Heat flow in adiabatic system

Consider a heat flow between two substances at different temperatures in an adiabatic system as shown in Fig. 3.25. The entropy change due to the heat flow, δQ, from the substance 1–2 is expressed by

$$\Delta S = \left(\frac{1}{T_2} - \frac{1}{T_1}\right) \delta Q = \frac{T_1 - T_2}{T_1 T_2} \delta Q \tag{3.116}$$

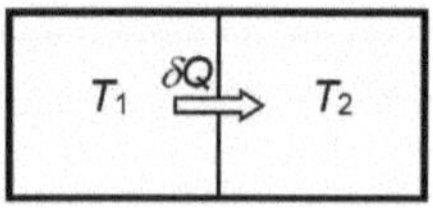

Figure 3.25 Heat flow in an adiabatic system.

To satisfy $\Delta S > 0$ in Eq. (3.106), T_1 must be higher than T_2 for the direction of heat flow.

3.1.8.2 Energy conversion in isothermal and isochoric system

Consider an energy conversion in an isothermal and isochoric system. From the first law of thermodynamics and the second law of thermodynamics of Eq. (3.105), because no expansion work is extracted in an isochoric system,

$$TdS \geq \delta Q = dU + PdV = dU \tag{3.117}$$

Applying Helmholtz free energy F defined by Eq. (3.118), Eq. (3.117) is expressed as Eq. (3.119).

$$F = U - TS \tag{3.118}$$

$$dF \leq 0 \tag{3.119}$$

Eq. (3.119) decides the direction of spontaneous change in the isothermal and isochoric system. Although no expansion work is output from an isochoric system, other type of work can be extracted if phase change and chemical reaction exist. In this case, Eq. (3.117) is rewritten as

$$TdS \geq \delta Q = dU + PdV + \delta L = dU + \delta L \tag{3.120}$$

where δL is an extracted work except for the expansion work. From Eq. (3.120), it is proved that δL cannot be exceed the amount of the decrease in the Helmholtz free energy.

$$\delta L \leq -dF \tag{3.121}$$

3.1.8.3 Energy conversion in isothermal and isobaric system

Similarly, Gibbs free energy is introduced to the change in an isothermal and isobaric system. From the first law of thermodynamics and the second law of thermodynamics of Eq. (3.105),

$$TdS \geq \delta Q = dH - VdP = dH \tag{3.122}$$

Applying Gibbs free energy G defined by Eq. (3.123), Eq. (3.122) is expressed as Eq. (3.124).

$$G = H - TS \tag{3.123}$$

$$dG \leq 0 \tag{3.124}$$

Eq. (3.124) decides the direction of spontaneous change in the isothermal and isobaric system. For the change with phase change and chemical reaction, Eq. (3.117) is rewritten as

$$TdS \geq \delta Q = dH - VdP + \delta L = dH + \delta L \tag{3.125}$$

And, it is proved that δL cannot be exceed the amount of the decrease in the Gibbs free energy.

$$\delta L \leq -dG \tag{3.126}$$

3.1.9 Phase equilibrium

In thermal and nuclear power plants, water is used as the working fluid to operate steam-power generation system, that is, Rankine cycle. In the steam-power generation system, energy transfer density of the working fluid can be made quite larger by using the latent heat for the phase change of liquid to vapor or vapor to liquid than gas cycles. In this section, state, steam diagram, and latent heat of vaporization are outlined.

The number of degrees of freedom of the state N_F under thermodynamic equilibrium condition is given by the Gibbs phase rule expressed as

$$N_F = N_C - N_P + 2 \tag{3.127}$$

where N_C and N_P are the number of components and phases, respectively. For an ideal-gas, subcooled liquid, and superheated vapor, the N_F is equal to 2 from $N_C = 1$ and $N_P = 1$. Therefore, to specify the state, two kinds of the quantity of state are required, for example, pressure and volume, pressure and temperature, pressure and enthalpy, temperature and entropy, and so on. Changes of state of the working fluid is generally indicated on two-dimensional diagrams, such as *p-v*, *T-s*, *p-h*, *h-s* diagrams, where p is pressure, T is temperature, and v, h, s are specific values of volume, enthalpy and entropy, respectively. The N_F of supercritical water is also equal to 2. On the other hand, for wet vapor, because two phases exist in thermodynamic equilibrium, the N_F becomes 1 from $N_C = 1$, $N_P = 2$. So, the phase equilibrium condition can be specified by one kind of the quantity of state. Note that the quantity of state is limited to the intensive property of p and T, because the total amount of extensive property depends on the mass fraction of phases. For the saturated liquid or vapor, the state can be decided from one kind of quantity of state. Steam diagrams on *p-v* and *T-s* diagrams are shown in Fig. 3.26A and B, respectively.

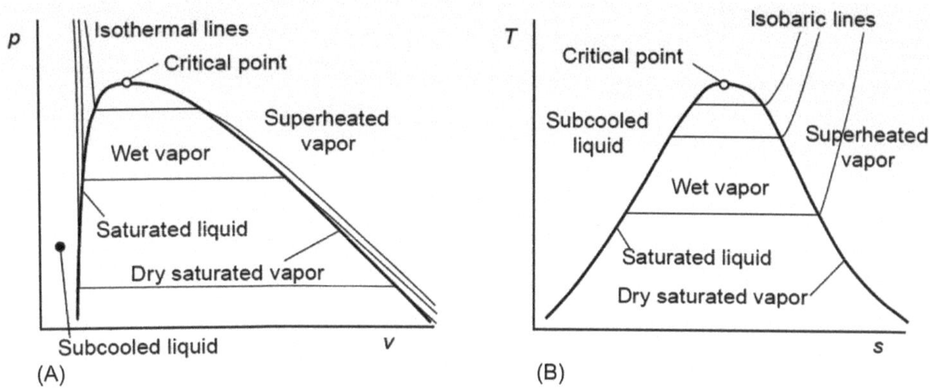

Figure 3.26 Steam diagrams of water. (A) *p-v* diagram. (B) *T-s* diagram.

Extensive properties of wet vapor are calculated by the following equations using the mass fraction of vapor to the total mass defined as vapor quality x.

$$v = (1 - x)v' + xv'' = v' + (v'' - v')x \tag{3.128}$$

$$u = (1 - x)u' + xu'' = u' + (u'' - u')x \tag{3.129}$$

$$h = (1 - x)h' + xh'' = h' + (h'' - h')x \tag{3.130}$$

$$s = (1 - x)s' + xs'' = s' + (s'' - s')x \tag{3.131}$$

where v', u', h', s' are specific values of volume, internal energy, enthalpy and entropy of the saturated liquid, and v'', u'', h'', s'' are those of the saturated vapor.

Under the phase equilibrium condition, specific Gibbs free energy of each phase of the substance is in equilibrium. For example, for wet vapor,

$$g' = h' - Ts' = g'' = h'' - Ts'' \tag{3.132}$$

where g' is specific Gibbs free energy of the saturated liquid, and g'' is that for the saturated vapor. Eq. (3.122) is transformed to

$$T(s'' - s') = h'' - h' \tag{3.133}$$

Because the left term shows the heat for vaporization, the latent heat of vaporization can be expressed by the enthalpy difference between saturated vapor and liquid.

3.1.10 Exergy

Exergy is defined as the maximum work extracted from a system in nonequilibrium with the surroundings until the system becomes equilibrium with the surroundings.

3.1.10.1 Heat Q *from heat source at the temperature* T

The maximum effective work can be obtained by a reversible cycle, and can be calculated by Carnot efficiency. Therefore, the exergy of the heat is expressed as

$$E = \left(1 - \frac{T_0}{T}\right)Q \tag{3.134}$$

where T_0 is the temperature of the surroundings.

In the case that the temperature of the heat source is lower than T_0, the effective work can be also obtained by a heat engine as shown in Fig. 3.27. Similarly, the exergy of the heat, Q, is expressed as

$$E = \left(1 - \frac{T}{T_0}\right)Q_0 = \left(1 - \frac{T}{T_0}\right)\frac{T_0}{T}Q = \left(\frac{T_0}{T} - 1\right)Q \tag{3.135}$$

Treating the heat flow from the heat source as positive value, the exergy is expressed by the same equation with Eq. (3.134).

3.1.10.2 Enthalpy H *of working fluid*

Exergy of the working fluid is important in an exergy analysis of energy systems to estimate losses of available work due to irreversible processes. The released energy from the working fluid by being in equilibrium with the surroundings is $H - H_0$, where H_0 is enthalpy of the working fluid under the condition in equilibrium with

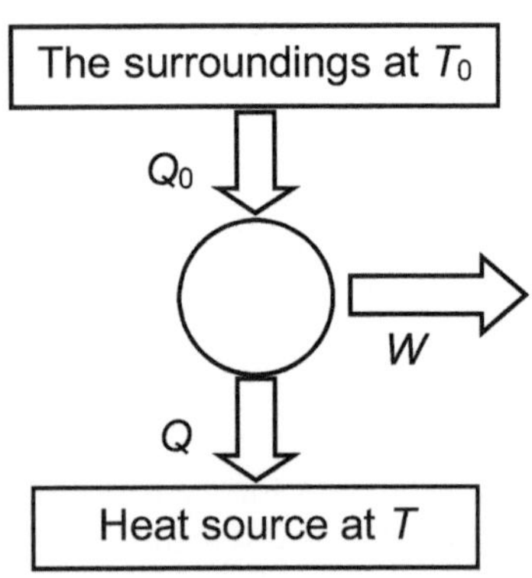

Figure 3.27 Heat engine driven by a cold heat source.

the surroundings. The maximum work can be obtained by a reversible heat engine with the heat input of $H - H_0$, and the unavailable energy is exhausted as heat of Q_0 to the surroundings. For reversible change, the change in entropy caused by the energy conversion expressed as the following equation is equal to 0.

$$\Delta S = (S_0 - S) + \frac{Q_0}{T_0} = 0 \tag{3.136}$$

where the first term is the change in entropy of the working fluid, and the second term is that of the surroundings caused by the exhaust heat. Unavailable energy can be expressed as

$$Q_0 = T_0(S - S_0) \tag{3.137}$$

Therefore, exergy of the working fluid with the enthalpy of H is expressed as

$$E = (H - H_0) - T_0(S - S_0) \tag{3.138}$$

Generally, exergy per unit mass, i.e. specific exergy, as shown in the following equation is used in the exergy analysis.

$$e = (h - h_0) - T_0(s - s_0) \tag{3.139}$$

Note that the amount of exergy depends on the condition of the surroundings.

3.1.10.3 Example of exergy analysis of equipment

Heat exchanger

Exergy of each flow is calculated by Eq. (3.138). Then, loss of available work can be calculated from the difference of the output to the input of exergy expressed as

$$LW = (E_{in} - E_{out})_H + (E_{in} - E_{out})_L \tag{3.140}$$

where subscripts H and L shows the substance at higher and lower temperature, respectively. Note that the values of H_0 and S_0 used in the calculation of exergy must be given for each substance. The causes of irreversible process are heat transfer with the temperature difference and frictional pressure loss of each flow through the heat exchanger.

Turbine

The process in a turbine is shown on an enthalpy-entropy diagram with isobaric lines in Fig. 3.28. In a reversible heat engine, the thermodynamic process of the working fluid in a turbine can be treated as a reversible adiabatic expansion

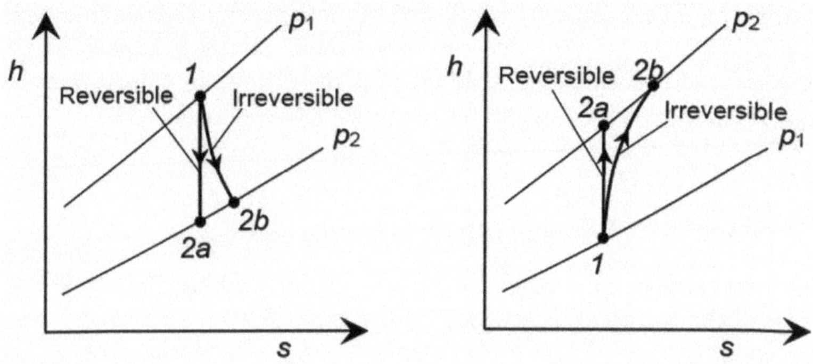

Figure 3.28 (A)Enthalpy-entropy diagram of adiabatic process. (B) Adiabatic expansion in a turbine.

process. From Eq. (3.100), the process is isentropic process. However, in actual turbines, the process has irreversibility caused by the flow structure, temperature field, and so on, even if the process is perfectly adiabatic. So, the entropy of the working fluid should increase to the state 2b as shown in Fig. 3.28. The increase in entropy leads to the increase in enthalpy at the outlet of the turbine. Therefore, to estimate or evaluate the performance of turbine, adiabatic efficiency or isentropic efficiency is introduced. The adiabatic efficiency is defined by the ratio of the actual work to the maximum work by the isentropic process. Since the work can be calculated as the enthalpy difference as described in Section 3.1.3.2, the adiabatic efficiency is expressed as

$$\eta_{ad} = \frac{h_1 - h_{2b}}{h_1 - h_{2a}} \tag{3.141}$$

Compressor

The process in a compressor is shown with isobaric lines on enthalpy-entropy diagram as shown in Fig. 3.29. For a compressor, the actual compression work becomes larger due to irreversibility, and the increase amount of the work leads to the increase in the outlet enthalpy and temperature. The adiabatic efficiency is defined by the ratio of the minimum work by the isentropic process to the actual work, and is expressed as

$$\eta_{ad} = \frac{h_{2a} - h_1}{h_{2b} - h_1} \tag{3.142}$$

Note that the relationship between the numerator and denominator is opposite of that for turbine in Eq. (3.141) in order to make the adiabatic efficiency less than unity.

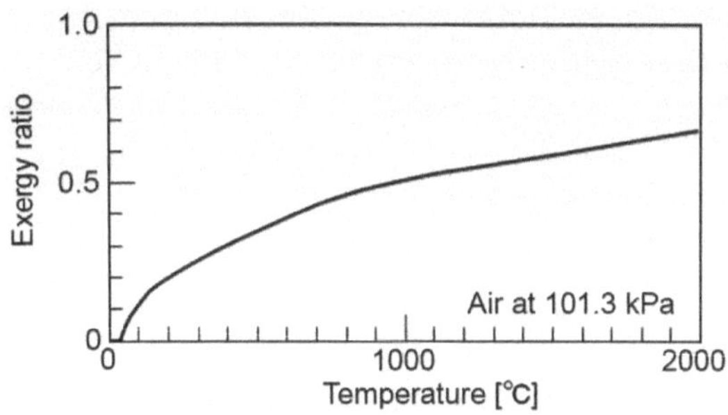

Figure 3.29 Exergy of dry air for varied temperature.

Combustion

Combustion process producing high temperature gas is composed of chemical reaction and heat diffusion. The heat diffusion causes strong irreversibility. Figure 3.29 shows the exergy of dry air for varied temperature. The vertical axis shows exergy ratio defined as the ratio of the exergy to the enthalpy difference from the value under the equilibrium condition with the surroundings. The exergy of the gas strongly depends on the temperature and increases with increasing the temperature. For example, because methane which is the main component of natural gas has the exergy ratio of about 0.92, it can be seen that the exergy loss in the combustion process is quite large.

3.2 Fundamentals of fluid dynamics of single-phase flow

3.2.1 *Introduction*

Fluid is an essential component in most power and energy machineries. For instance, water is employed as the coolant in light-water reactors. Therefore, knowledge of the fluid dynamics is necessary to design the power and energy machineries. In this section, the fundamentals of single-phase flows, for example, basic equations, are introduced. In Section 3.2.2, the concepts of fluid viscosity and compressibility is described. The basic equations, that is, continuity equation, momentum conservation equation (Navier-Stokes equation), and energy conservation equation, are derived in Section 3.2.3. To gain in-depth knowledge, the characteristics of some simple laminar and turbulent flows are discussed briefly in Section 3.2.4.

3.2.2 *Ideal fluid and viscous-compressive real fluid*

In the early days of fluid dynamics, major objects for researchers, for example, Leonhard Euler, were the ideal fluids that do not have fluid viscosity and compressibility. Then, in the 19th century, the study on viscous fluids was conducted by Claude Louis Marie Henri Navier, Sir George Gabriel Stokes, and other researchers. In addition, the compressible fluids are also explored in 19th century by, for example, Ernst Waldfried Josef Wenzel Mach.

In this subsection, the concepts of fluid viscosity and compressibility are described briefly in comparison with inviscid incompressible ideal fluids.

3.2.2.1 *Viscosity and shear stress*

Fluid viscosity takes an important role in flow-filed formation. In the (viscous) fluid dynamics, the viscosity effects can be defined as the relationship between internal strain and stress, similar to the mechanics of materials.

Ideal fluid and viscous fluid

Except for special fluids like superfluid helium, real fluids definitely have the viscosity. In some cases, however, when the inertia force is extremely larger than the viscous force, the viscous effect is negligible (see the definition of Reynolds number in Section 3.2.4.1). For example, in the flow of water or air, the viscosity effect is relatively small compared with oil or glycerin, and therefore, the viscosity can be neglected in the flow apart from an obstacle surface (see the explanation about the boundary layer in "Boundary layer equation" section).

The imaginary inviscid fluid is called "ideal fluid." Due to the lack of the viscosity, the modeling of the ideal fluid is much easier than the viscous fluid. In fact, some useful equations have been obtained with the assumption of ideal fluid (see the derivation of Bernoulli equation in "Kinetic energy conservation (Bernoulli's theorem)" section).

In contrast, the modeling of the viscosity effects is necessary to investigate the flows of real fluids. It is well known that real fluids are classified roughly into two types: one is Newtonian fluid which satisfies Newton's law of viscosity (see "Newton's law of viscosity" section) and the other is non-Newtonian fluid in which the viscosity coefficient varies with the flow regime, that is, the shear rate. The modeling of various non-Newtonian fluids, for example, Bingham fluid, dilatant fluid or pseudoplastic fluid, is beyond the scope of this book, and only the modeling of Newtonian fluid is described.

Deformation and velocity gradient

Here, we consider the small fluid parcel moving from point $A(x,y,z)$ to neighboring point $P(x+\delta x, y+\delta y, z+\delta z)$. The fluid velocity, indicated as u, at point P can be written as Eq. 3.143, by applying the Taylor expansion at point A:

$$\delta u = u(x+\delta x, y+\delta y, z+\delta z) - u(x,y,z)$$

$$= \left(\frac{\partial u}{\partial x}\right)\delta x + \left(\frac{\partial u}{\partial y}\right)\delta y + \left(\frac{\partial u}{\partial z}\right)\delta z + O\left(\delta^2\right), \tag{3.143}$$

where the fourth term in R.H.S. is the higher-order (second and higher-order) term which can be negligible. Along with the displacement, the fluid parcel can deform in nonuniform flow field. The movement of the fluid parcel can be classified into four basic factors:

1. translation,
2. expansion and contraction,
3. shearing deformation, and
4. rotation.

Among these four factors, the translation and rotation merely change the position and angle of the fluid parcel, and the expansion (or contraction) and shearing deformation lead the change in the shape of the fluid parcel.

Expansion and contraction As shown in Fig. 3.30, we consider the fluid parcel with the original length of δx. It should be noted the one-dimensional (x-direction) flow is considered here for simplicity. When the fluid parcel moves with the flow, the left edge of the fluid parcel moves with the velocity of $u(x_0)$ and the right edge moves with the velocity of $u(x_0) + \frac{\partial u}{\partial x}\delta x$, that is, after short period (δt), the length of the fluid parcel changes to $\delta x + \frac{\partial u}{\partial x}\delta t$. In the fluid dynamics, the normal strain is defined as the expansion (or contraction) per unit length and unit time, and therefore, the normal strain is calculated as:

$$\varepsilon_x = \frac{\partial u}{\partial x}. \tag{3.144}$$

Similarly, the normal strain in y and z-direction is obtained as:

$$\varepsilon_y = \frac{\partial v}{\partial y}, \tag{3.145}$$

$$\varepsilon_z = \frac{\partial w}{\partial z}. \tag{3.146}$$

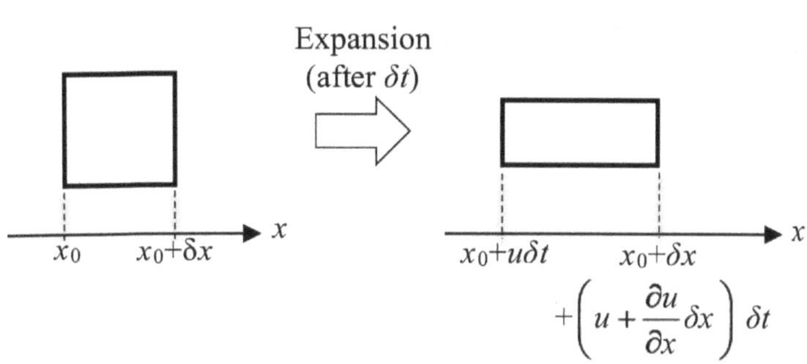

Figure 3.30 Schematic view of expansion.

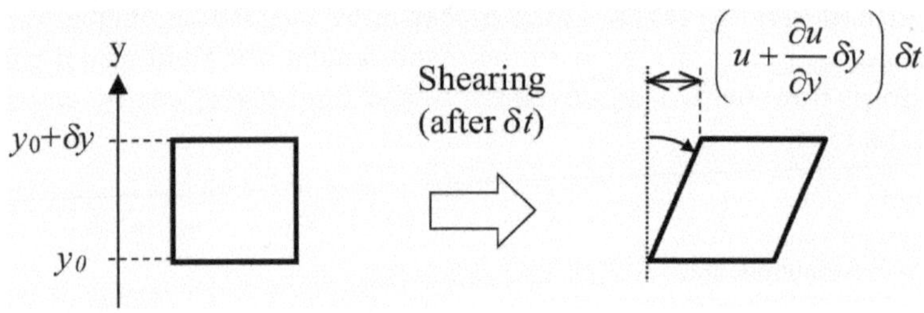

Figure 3.31 Schematic view of shearing deformation.

Shearing deformation The shearing deformation can be considered in the similar manner to the expansion (or contraction), As shown in Fig. 3.31, when the fluid parcel moves in shear flow in which the flow velocity changes in the direction normal to the flow direction, the originally square-shape fluid parcel changes its shape to a parallelogram. Namely, due to the velocity gradient in y-direction, the upper edge moves with the velocity of $u(y_0) + \frac{\partial u}{\partial y}\delta y$ which is different form the velocity of lower edge, $u(y_0)$, and the difference between these two velocity results in the shearing deformation. In the fluid dynamics, the shear rate is defined as the shearing deformation per unit length and time, and therefore, the shea rate is calculated as $\partial u/\partial y$. Here, it should be noted that the gradient of y-velocity in x-direction, $\partial v/\partial x$, leads the shearing deformation essentially the same with that by $\partial u/\partial y$. Finally, the shear rate is written as:

$$\gamma_{xy} = \gamma_{yx} = \frac{\partial v}{\partial x} + \frac{\partial u}{\partial y}. \tag{3.147}$$

Similarly, the shearing rates in y and z-direction is obtained as:

$$\gamma_{yz} = \gamma_{zy} = \frac{\partial w}{\partial y} + \frac{\partial v}{\partial z}, \tag{3.148}$$

$$\gamma_{zx} = \gamma_{xz} = \frac{\partial u}{\partial z} + \frac{\partial w}{\partial x}. \tag{3.149}$$

Rotation In a similar manner to the calculation of the shearing rate, the rotation speed of the fluid parcel, which is defined as the change in the angle of the fluid parcel per unit time, is calculated as

$$\Omega_z = \frac{1}{2}\left(\frac{\partial v}{\partial x} - \frac{\partial u}{\partial y}\right), \tag{3.150}$$

Similarly, the rotation speeds in y- and x-direction are obtained as:

$$\Omega_y = \frac{1}{2}\left(\frac{\partial u}{\partial z} - \frac{\partial w}{\partial x}\right), \tag{3.151}$$

$$\Omega_x = \frac{1}{2}\left(\frac{\partial w}{\partial y} - \frac{\partial v}{\partial z}\right), \tag{3.152}$$

By substituting Eqs. (3.144, 3.147 and 3.150) into Eq. (3.143), we obtain:

$$u(x+\delta x, y+\delta y, z+\delta z) = u(x,y,z) + \varepsilon_x \delta x + \left(\frac{1}{2}\gamma_{yx} - \Omega_z\right)\delta y + \left(\frac{1}{2}\gamma_{zx} + \Omega_y\right)\delta z. \tag{3.153}$$

In three-dimensional Cartesian coordinates, the equation is written as:

$$\begin{pmatrix} u(x+\delta x, y+\delta y, z+\delta z) \\ v(x+\delta x, y+\delta y, z+\delta z) \\ w(x+\delta x, y+\delta y, z+\delta z) \end{pmatrix} = \begin{pmatrix} u(x,y,z) \\ v(x,y,z) \\ w(x,y,z) \end{pmatrix} + \begin{pmatrix} \varepsilon_x & \frac{1}{2}\gamma_{yx} & \frac{1}{2}\gamma_{zx} \\ \frac{1}{2}\gamma_{xy} & \varepsilon_y & \frac{1}{2}\gamma_{zy} \\ \frac{1}{2}\gamma_{xz} & \frac{1}{2}\gamma_{yz} & \varepsilon_z \end{pmatrix} \begin{pmatrix} \delta x \\ \delta y \\ \delta z \end{pmatrix}$$

$$+ \begin{pmatrix} 0 & -\Omega_z & \Omega_y \\ \Omega_z & 0 & -\Omega_x \\ -\Omega_y & \Omega_x & 0 \end{pmatrix} \begin{pmatrix} \delta x \\ \delta y \\ \delta z \end{pmatrix}. \tag{3.154}$$

Newton's law of viscosity

As described in "Ideal fluid and viscous fluid" section, Newtonian fluids satisfy Newton's law of viscosity. In this law, the shear stress, τ, is proportional to the shear rate. Namely, when we consider only x-velocity, u, the shear stress in x-direction on y-plane is written as:

$$\tau_{yx} = \mu \frac{\partial u}{\partial y}, \tag{3.155}$$

where the proportionality coefficient μ is the fluid viscosity coefficient. In simple shear flow, that is, Couette flow, shown in Fig. 3.32, a fluid parcel is accelerated by the shear stress from the fluid parcel adjacent to the upper edge of the fluid parcel and decelerated by the shear stress from the fluid parcel adjacent to the lower edge. As a result, the rectilinear velocity profile is formed.

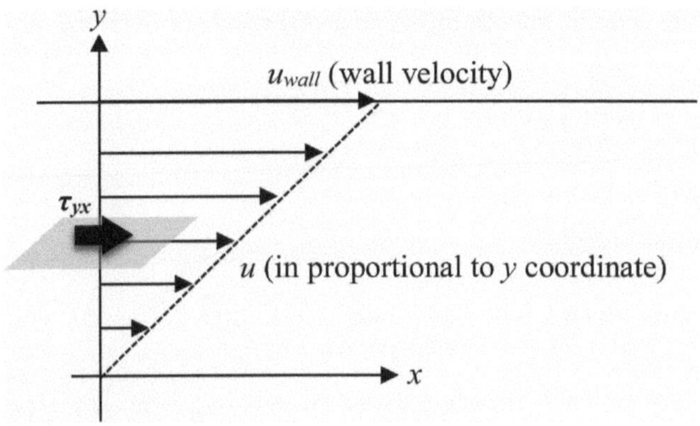

Figure 3.32 Schematic view of Couette flow (simple shear flow).

In three-dimensional Cartesian coordinates, the shear stresses are written as:

$$\tau_{xy} = \tau_{yx} = \mu\gamma_{xy} = \mu\left(\frac{\partial v}{\partial x} + \frac{\partial u}{\partial y}\right), \tag{3.156}$$

$$\tau_{yz} = \tau_{zy} = \mu\gamma_{yz} = \mu\left(\frac{\partial w}{\partial y} + \frac{\partial v}{\partial z}\right), \tag{3.157}$$

$$\tau_{zx} = \tau_{xz} = \mu\gamma_{zx} = \mu\left(\frac{\partial u}{\partial z} + \frac{\partial w}{\partial x}\right), \tag{3.158}$$

In contrast, the normal stress is defined in consideration of the normal strain and the volumetric change as:

$$\sigma_x = 2\mu\varepsilon_x - \lambda\Theta = 2\mu\frac{\partial u}{\partial x} - \lambda\Theta, \tag{3.159}$$

$$\sigma_y = 2\mu\varepsilon_y - \lambda\Theta = 2\mu\frac{\partial v}{\partial y} - \lambda\Theta, \tag{3.160}$$

$$\sigma_z = 2\mu\varepsilon_z - \lambda\Theta = 2\mu\frac{\partial w}{\partial z} - \lambda\Theta, \tag{3.161}$$

where λ is the second viscosity coefficient. Θ stands for the volumetric change in fluid, which is represented as the velocity divergence written as:

$$\Theta = \frac{\partial u}{\partial x} + \frac{\partial v}{\partial y} + \frac{\partial w}{\partial z}. \tag{3.162}$$

Here, we consider the summation of the normal stresses which is calculated as:

$$\sigma_x + \sigma_y + \sigma_z = \left(\frac{2}{3}\mu - \lambda\right)\Theta = \mu_v\Theta, \tag{3.163}$$

where μ_v is the bulk viscosity which indicate the viscosity for net volumetric change. With Stokes's hypothesis, the bulk viscosity is negligible except for special cases, for example, the attenuation of sound wave. Therefore, the second viscosity coefficient is determined as:

$$\lambda = \frac{2}{3}\mu. \tag{3.164}$$

Finally, the shear and normal stresses are summarized and the stress tensor is defined as:

$$\mathfrak{I} = \begin{pmatrix} \sigma_x & \tau_{yx} & \tau_{zx} \\ \tau_{xy} & \sigma_y & \tau_{zy} \\ \tau_{xz} & \tau_{yz} & \sigma_z \end{pmatrix}. \tag{3.165}$$

3.2.2.2 *Compressibility*

Similar to the viscosity, real fluids have the compressibility and the density of real fluid changes with pressure. The compressibility is negligible for most liquids and the fluid is treated as an incompressible fluid in such a case. However, in high-speed gas flows, the compressibility effects are essential, for example, the generation of shockwave.

Density change (equation of state)

The density change of gas is represented by the equation of state (EOS). In case of ideal gas (not ideal fluid), the ideal gas law is applicable as:

$$pV = nR_0T, \tag{3.166}$$

where R_0 is the molar gas constant. The Eq. (3.166) can be converted to:

$$pV = mRT, \tag{3.167}$$

where m is the mass of (ideal) gas and the gas constant R is defined as:

$$R = \frac{R_0}{M}, \tag{3.168}$$

where M is the molecular weight.

For nonideal gases, van der Waals equation is frequently employed, in which the real gas effects are considered as:

$$\left(p + \frac{c_1}{V_M^2}\right)(V_M - c_2) = R_0 T, \tag{3.169}$$

where c_1 and c_2 are van der Waals constants and V_M is the molar volume.

Compressible fluid and incompressible fluid

As described in "Density change (equation of state)" section, the density change should be calculated by the EOS for compressible fluids. For incompressible fluids, however, the density change is neglected and the volumetric change becomes zero. In the fluid dynamics, the volumetric change rate, that is, the volumetric change per unit volume and unit time, is represented by the velocity divergence, and therefore, for incompressible fluids, the divergence-free condition is satisfied as:

$$\Theta = \frac{\partial u}{\partial x} + \frac{\partial v}{\partial y} + \frac{\partial w}{\partial z} = 0. \tag{3.170}$$

From Eqs. (3.159–3.161), the normal stresses of incompressible fluids are written as:

$$\sigma_x = 2\mu \frac{\partial u}{\partial x}, \tag{3.171}$$

$$\sigma_y = 2\mu \frac{\partial v}{\partial y}, \tag{3.172}$$

$$\sigma_z = 2\mu \frac{\partial w}{\partial z}. \tag{3.173}$$

Mach number

The influence of the compressibility is represented by Mach number, Ma, defined as:

$$Ma = \frac{U}{U_s}, \tag{3.174}$$

where U is the representative flow velocity and U_s is the sound speed. As described in Eq. (3.174), Mach number is a nondimensional number which indicates the ratio of flow velocity to sound speed. In general, the flow with Mach number less than 0.3 can be considered as an incompressible flow and the flow with Mach number larger than 0.3 should be treated as a compressible flow in which the density change with pressure is no longer negligible. Furthermore, when Mach number is over unity, the flow is called supersonic, and the flow characteristics changes significantly, that is, the generation of shockwave. In subsonic flow with Mach number less than unity, fluid pressure increases in diverging flow. One example is the pressure recovery in a diffuser. On the other hand, in supersonic flows, the phenomenon changes drastically and fluid pressure decreases in diverging flow (see the explanation of the flow in Laval nozzle in "Critical flow" section). Most importantly, in supersonic flows, the pressure wave cannot propagate upstream because the pressure wave propagates with the sound speed. Therefore, the information at downstream region is never transferred to upstream.

Critical flow

The change in the flow characteristics from subsonic to supersonic flows lead an interesting phenomenon in Laval nozzle, also called as convergent-divergent nozzle. Fig. 3.33 shows Laval nozzle and the pressure profile in the nozzle. When the back pressure is high enough, the flow is subsonic throughout the nozzle flow path and the pressure decreases in the converging part and increases in the diverging part. As a result, the minimum pressure arises at the throat of the nozzle. In contrast, when the back pressure is low enough, the flow is supersonic throughout the

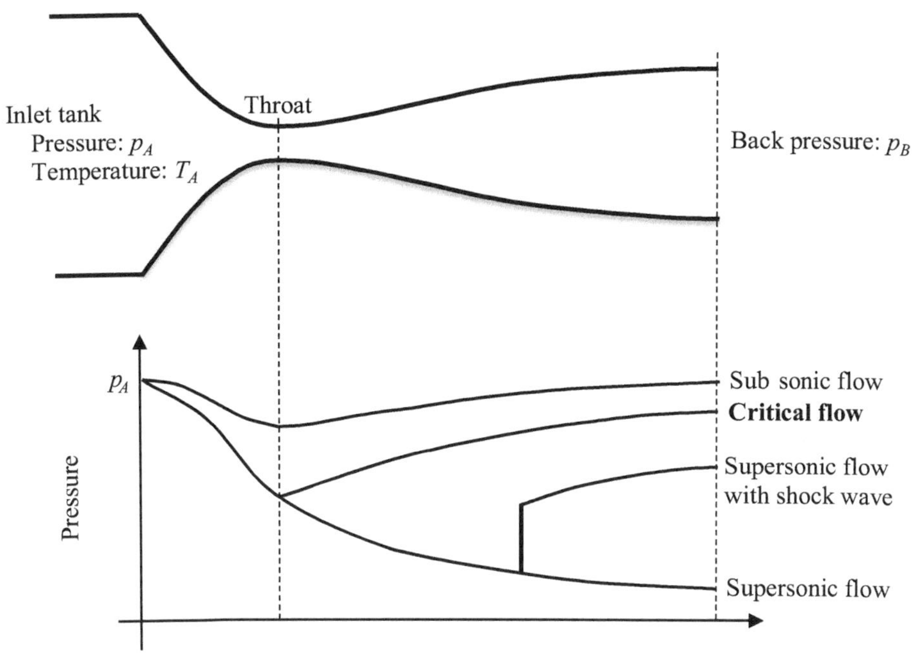

Figure 3.33 Laval nozzle and pressure profile.

diverging part of the nozzle flow path and the pressure decreases monotonically from the inlet to outlet. The back pressure in between those two conditions leads the flow with a (normal) shockwave which makes discontinuous pressure change in the diverging part.

When the back pressure decreases from the subsonic flow condition, the flow velocity at the throat increases with the decrease in the back pressure and the flow velocity at the throat reaches the sound speed at the certain state, that is, *Ma* is unity. In such a state, the flow is called critical flow. Under the critical flow condition, the pressure between the inlet tank to the throat is calculated in consideration of the isentropic pressure change as:

$$\frac{p_A}{p} = \left(1 + \frac{\chi - 1}{2} Ma^2\right)^{\chi/\chi-1}, \tag{3.175}$$

where χ is the heat capacity ratio. Since Mach number is unity at the throat, the pressure at the throat is calculated as:

$$\frac{p_T}{p_A} = \left(\frac{2}{\chi + 1}\right)^{\chi/\chi-1}. \tag{3.176}$$

It is interesting that the pressure profile in the converging part, that is, from the inlet tank to the throat, does not change with the decrease in the back pressure after the critical flow condition is satisfied. As described in "Mach number" section, the pressure wave cannot propagate upstream in supersonic flows, and therefore, the decrease in the back pressure is not detected by the flow in the converging part.

3.2.3 Basic equation

As described in the following parts, the basic equations of the fluid dynamics are derived from the conservation laws of mass, momentum and energy. Compared to the particle mechanics, the difficulty in the fluid dynamics is originated in the nonlinearity of the momentum conservation equation (see the explanation of nonlinearity in "Nonlinearity" section).

3.2.3.1 Conservation of mass

Unlike the particle mechanics, the fluid dynamics treats continua, and therefore, inflow and outflow from/to adjacent region must be considered to construct

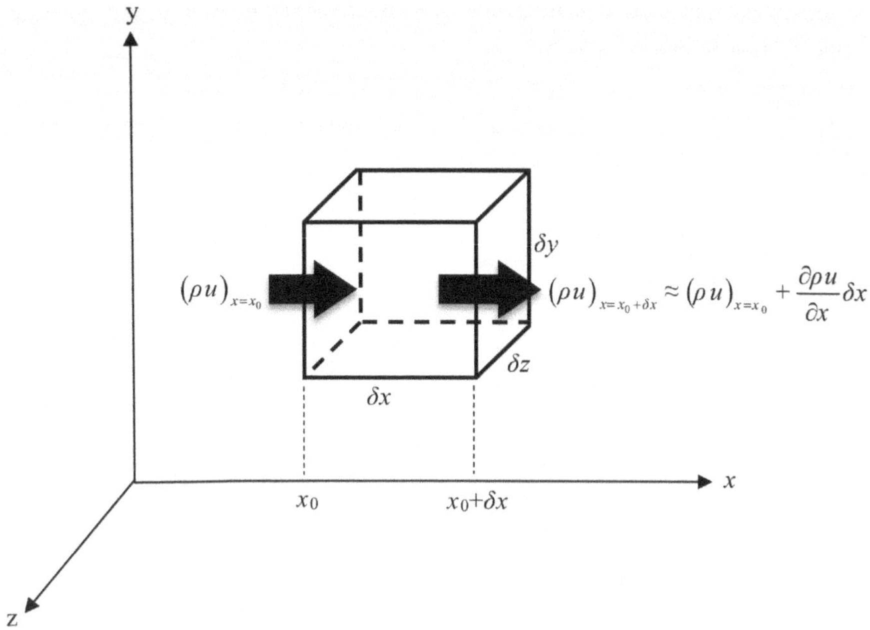

Figure 3.34 Control volume and mass flux.

the conservation laws. In such a case, the discussion based on a control volume is useful.

Control volume

As shown in Fig. 3.34, we consider a control volume, that is, a small fluid parcel, with the size of $\delta x \times \delta y \times \delta z$, and consider the inlet and outlet of the fluid mass flux. The inlet mass flux at $x = x_0$ is $(\rho u)_{x=x_0}\delta y\delta z$, and the outlet mass flux at $x = x_0 + \delta x$ is $(\rho u)_{x=x_0+\delta x}\delta y\delta z$. Similarly, in y-direction, the inlet mass flux at $y = y_0$ is $(\rho v)_{y=y_0}\delta z\delta x$, and the outlet mass flux at $y = y_0 + \delta y$ is $(\rho v)_{y=y_0+\delta y}\delta z\delta x$, and in z-direction, the inlet mass flux at $z = z_0$ is $(\rho w)_{z=z_0}\delta x\delta y$, and the outlet mass flux at $z = z_0 + \delta z$ is $(\rho w)_{z=z_0+\delta z}\delta x\delta y$. In addition, the change in fluid mass during δt (from time level t_0 to $t_0 + \delta t$) is $\left\{(\rho)_{t=t_0+\delta t} - (\rho)_{t=t_0}\right\}\delta x\delta y\delta z$.

Continuity equation

In the classical mechanics including the fluid dynamics, the mass is strictly conserved and the generation and disappearance of mass are not allowed. Therefore, in the control volume, the increase in fluid mass must be equivalent to the net inlet fluid mass flux. In consideration of the discussion in "Control volume" section, the continuity equation is written as:

$$\delta x\delta y\delta z\left[(\rho)_{t=t_0+\delta t}-(\rho)_{t=t_0}\right]=\left[(\rho u)_{x=x_0}-(\rho u)_{x=x_0+\delta x}\right]\delta y\delta z\delta t+\left[(\rho v)_{y=y_0}-(\rho v)_{y=y_0+\delta y}\right]\delta z\delta x\delta t+\left[(\rho w)_{z=z_0}-(\rho w)_{z=z_0+\delta z}\right]\delta x\delta y\delta t. \tag{3.177}$$

Here, we consider Taylor expansion as:

$$(\rho)_{t=t_0+\delta t}=(\rho)_{t=t_0}+\frac{\partial\rho}{\partial t}\delta t+\frac{1}{2}\frac{\partial^2\rho}{\partial t^2}\delta t^2+\sum_{k=3}^{\infty}\frac{1}{k!}\frac{\partial^k\rho}{\partial t^k}\delta t^k, \tag{3.178}$$

$$(\rho u)_{x=x_0+\delta x}=(\rho u)_{x=x_0}+\frac{\partial\rho u}{\partial x}\delta x+\frac{1}{2}\frac{\partial^2\rho u}{\partial x^2}\delta x^2+\sum_{k=3}^{\infty}\frac{1}{k!}\frac{\partial^k\rho u}{\partial x^k}\delta x^k, \tag{3.179}$$

$$(\rho v)_{y=y_0+\delta y}=(\rho v)_{y=y_0}+\frac{\partial\rho v}{\partial y}\delta y+\frac{1}{2}\frac{\partial^2\rho v}{\partial y^2}\delta y^2+\sum_{k=3}^{\infty}\frac{1}{k!}\frac{\partial^k\rho v}{\partial y^k}\delta y^k, \tag{3.180}$$

$$(\rho w)_{z=z_0+\delta z}=(\rho w)_{z=z_0}+\frac{\partial\rho w}{\partial z}\delta z+\frac{1}{2}\frac{\partial^2\rho w}{\partial z^2}\delta z^2+\sum_{k=3}^{\infty}\frac{1}{k!}\frac{\partial^k\rho w}{\partial z^k}\delta z^k. \tag{3.181}$$

By substituting Eqs. (3.178–3.181) into Eq. (3.177), the mass conservation is expressed by the temporal and spatial derivatives as:

$$\frac{\partial\rho}{\partial t}+\frac{1}{2}\frac{\partial^2\rho}{\partial t^2}\delta t^2+\cdots=\frac{-\frac{\partial\rho u}{\partial x}\delta x-\frac{1}{2}\frac{\partial^2\rho u}{\partial x^2}\delta x^2-\cdots}{\delta x}+\frac{-\frac{\partial\rho v}{\partial y}\delta y-\frac{1}{2}\frac{\partial^2\rho v}{\partial y^2}\delta y^2-\cdots}{\delta y}+\frac{-\frac{\partial\rho w}{\partial z}\delta z-\frac{1}{2}\frac{\partial^2\rho w}{\partial z^2}\delta z^2-\cdots}{\delta z}. \tag{3.182}$$

Finally, by neglecting the second and high-order terms, the continuity equation is obtained as:

$$\frac{\partial\rho}{\partial t}+\frac{\partial\rho u}{\partial x}+\frac{\partial\rho v}{\partial y}+\frac{\partial\rho w}{\partial z}=0. \tag{3.183}$$

For incompressible flows, the density is constant in time and space, and the continuity equation becomes simpler from as:

$$\frac{\partial u}{\partial x} + \frac{\partial v}{\partial y} + \frac{\partial w}{\partial z} = 0. \tag{3.184}$$

This is so-called the divergence-free condition which indicates no volume change in incompressible flows.

3.2.3.2 Conservation of momentum

The derivation of the momentum conservation equation is a little bit more complicated than that of the mass conservation equation because not only the inlet and outlet momentum flux from/to the control volume, but also the actions of stress, pressure, and external forces must be considered.

Momentum advection

Here, we consider the advection of x-momentum. Similar to the mass flux discussed in "Control volume" section, the inlet momentum flux at $x = x_0$ is $(\rho uu)_{x=x_0}\delta y\delta z$, and the outlet momentum flux at $x = x_0 + \delta x$ is $(\rho uu)_{x=x_0+\delta x}\delta y\delta z$. As for the advection in y-direction, the inlet mass flux at $y = y_0$ is $(\rho v)_{y=y_0}\delta z\delta x$, and the outlet mass flux at $y = y_0 + \delta y$ is $(\rho uv)_{y=y_0+\delta y}\delta z\delta x$, and in z-direction, the inlet mass flux at $z = z_0$ is $(\rho uw)_{z=z_0}\delta x\delta y$, and the outlet mass flux at $z = z_0 + \delta z$ is $(\rho uw)_{z=z_0+\delta z}\delta x\delta y$. The change in x-momentum due to the momentum advection, defined as δmf_x, is calculated as the summation of these fluxes and the change during δt is written as:

$$\begin{aligned}\delta mf_x\delta x\delta y\delta z = {} & \left[(\rho uu)_{x=x_0} - (\rho uu)_{x=x_0+\delta x}\right]\delta y\delta z\delta t + \left[(\rho uv)_{y=y_0} - (\rho uv)_{y=y_0+\delta y}\right]\delta z\delta x\delta t \\ & + \left[(\rho uw)_{z=z_0} - (\rho uw)_{z=z_0+\delta z}\right]\delta x\delta y\delta t.\end{aligned} \tag{3.185}$$

Applying Taylor expansion and neglecting the second and high-order terms gives the equation of the x-momentum change per unit volume in the derivative form as:

$$\frac{\delta mf_x}{\delta t} = -\frac{\partial \rho uu}{\partial x} - \frac{\partial \rho uv}{\partial y} - \frac{\partial \rho uw}{\partial z}, \tag{3.186}$$

For the y and z-momentums, the equations are obtained in the same manner as:

$$\frac{\delta mf_y}{\delta t} = -\frac{\partial \rho vu}{\partial x} - \frac{\partial \rho vv}{\partial y} - \frac{\partial \rho vw}{\partial z}, \tag{3.187}$$

$$\frac{\delta mf_z}{\delta t} = -\frac{\partial \rho wu}{\partial x} - \frac{\partial \rho wv}{\partial y} - \frac{\partial \rho ww}{\partial z}. \tag{3.188}$$

Viscous stress and pressure

The change in momentum is caused also by the stress and pressure. When we consider the change in x-momentum, the normal stress on x-surface and the shear stresses on y and z-surfaces affects the momentum. The normal stress at $x = x_0$ is $(\sigma_x)_{x=x_0}\delta y\delta z$, and that at $x = x_0 + \delta x$ is $(\sigma_x)_{x=x_0+\delta x}\delta y\delta z$. The shear stress at $y = y_0$ is $(\tau_{yx})_{y=y_0}\delta z\delta x$, and that at $y = y_0 + \delta y$ is $(\tau_{yx})_{y=y_0+\delta y}\delta z\delta x$. Similarly, the shear stress at $z = z_0$ is $(\tau_{zx})_{z=z_0}\delta x\delta y$, and that at $z = z_0 + \delta z$ is $(\tau_{zx})_{z=z_0+\delta z}\delta x\delta y$. The change in x-momentum due to the stress and pressure, defined as δms_x, is calculated as the summation of these stresses and pressure and the momentum change during δt is written as:

$$\begin{aligned}\delta ms_x\delta x\delta y\delta z = & \left[(\sigma_x - p)_{x=x_0+\delta x} - (\sigma_x - p)_{x=x_0}\right]\delta y\delta z\delta t + \left[(\tau_{yx})_{y=y_0+\delta y} - (\tau_{yx})_{y=y_0}\right]\delta z\delta x\delta t \\ & + \left[(\tau_{zx})_{z=z_0+\delta z} - (\tau_{zx})_{z=z_0}\right]\delta x\delta y\delta t.\end{aligned} \tag{3.189}$$

By applying Taylor expansion with neglecting the second and high-order terms, the change in x-momentum per unit volume is written in the derivative form as:

$$\frac{\delta ms_x}{\delta t} = \frac{\partial(\sigma_x - p)}{\partial x} + \frac{\partial \tau_{yx}}{\partial y} + \frac{\partial \tau_{zx}}{\partial z}, \tag{3.190}$$

For y and z-momentums, the equations are obtained in the same manner as:

$$\frac{\delta ms_y}{\delta t} = \frac{\partial \tau_{xy}}{\partial x} + \frac{\partial(\sigma_y - p)}{\partial y} + \frac{\partial \tau_{zy}}{\partial z}, \tag{3.191}$$

$$\frac{\delta ms_z}{\delta t} = \frac{\partial \tau_{xz}}{\partial x} + \frac{\partial \tau_{zy}}{\partial y} + \frac{\partial(\sigma_z - p)}{\partial z}. \tag{3.192}$$

External force

In addition to the stress and pressure, which are classified into surface forces, volumetric forces, that is, external forces, works on fluid flows to cause the change in momentum. Depending on the flow situation, the gravity, electromagnetic force, Coriolis force, etc., are considered as external forces. Here, we express the external force per unit volume as F_x, F_y and F_z for x, y, and z-direction, respectively.

Momentum change in control volume

In consideration of Taylor expansion with neglecting the second and higher-order terms, the temporal change rate in momentum per unit volume can be written in the derivative form as $\partial\rho u/\partial t$, $\partial\rho v/\partial t$ and $\partial\rho w/\partial t$ for x, y, and z-momentum, respectively. Each temporal derivative is equivalent to the changes in corresponding momentum by the summation of advection ("Momentum advection" section), stress and pressure ("Viscous stress and pressure" section), and external force ("External force" section). Therefore, the momentum conservation equation is written as:

$$\frac{\partial \rho u}{\partial t} = \frac{\delta m f_x}{\delta t} + \frac{\delta m s_x}{\delta t} + F_x, \tag{3.193}$$

$$\frac{\partial \rho v}{\partial t} = \frac{\delta m f_y}{\delta t} + \frac{\delta m s_y}{\delta t} + F_y, \tag{3.194}$$

$$\frac{\partial \rho w}{\partial t} = \frac{\delta m f_z}{\delta t} + \frac{\delta m s_z}{\delta t} + F_z. \tag{3.195}$$

3.2.3.3 Navier-Stokes equation

The basic equation of the fluid dynamics is called Navier-Stokes equation named after two great researchers in the viscous fluid dynamics. Navier-Stokes equation is written as the set of the mass and momentum conservation equations.

Conservative form and nonconservative form

In accordance with the discussion in Section 3.2.3.2, the momentum conservation equations for x, y, and z-direction are written as:

$$\frac{\partial \rho u}{\partial t} + \frac{\partial \rho uu}{\partial x} + \frac{\partial \rho uv}{\partial y} + \frac{\partial \rho uw}{\partial z} = -\frac{\partial p}{\partial x} + \frac{\partial \sigma_x}{\partial x} + \frac{\partial \tau_{yx}}{\partial y} + \frac{\partial \tau_{zx}}{\partial z} + F_x, \tag{3.196}$$

$$\frac{\partial \rho v}{\partial t} + \frac{\partial \rho vu}{\partial x} + \frac{\partial \rho vv}{\partial y} + \frac{\partial \rho vw}{\partial z} = -\frac{\partial p}{\partial y} + \frac{\partial \tau_{xy}}{\partial x} + \frac{\partial \sigma_y}{\partial y} + \frac{\partial \tau_{zy}}{\partial z} + F_y, \tag{3.197}$$

$$\frac{\partial \rho w}{\partial t} + \frac{\partial \rho wu}{\partial x} + \frac{\partial \rho wv}{\partial y} + \frac{\partial \rho ww}{\partial z} = -\frac{\partial p}{\partial z} + \frac{\partial \tau_{xz}}{\partial x} + \frac{\partial \tau_{yz}}{\partial y} + \frac{\partial \sigma_z}{\partial z} + F_z. \tag{3.198}$$

The basic equation for viscous fluid flow, called Navier-Stokes equation, are the aforementioned equations coupled with the continuity equation.

$$\frac{\partial \rho}{\partial t} + \frac{\partial \rho u}{\partial x} + \frac{\partial \rho v}{\partial y} + \frac{\partial \rho w}{\partial z} = 0. \tag{3.183}$$

This type of description is called the conservative form of Navier-Stokes equation. Such a naming is reasonable in consideration of the fact that the conservative form of Nacier-Stokes equation is obtained directly from the mass and momentum conservation laws.

By introducing the velocity vector $\vec{u} = (u, v, w)$, stress tensor $\mathfrak{J}$ and external force vector $\vec{F} = (F_x, F_y, F_z)$, the conservative form of Navier-Stokes equation can be written in the vector form as:

$$\frac{\partial \rho}{\partial t} + \nabla \times (\rho \vec{u}) = 0, \tag{3.199}$$

$$\frac{\partial \rho \vec{u}}{\partial t} + \nabla \times (\rho \vec{u}\vec{u}) = -\nabla p + \nabla \times \mathfrak{J} + \vec{F}, \tag{3.200}$$

where $\vec{u}\vec{u}$ is the direct product of the velocity vector defined as:

$$\vec{u}\vec{u} = \begin{pmatrix} uu & uv & uw \\ vu & vv & vw \\ wu & wv & ww \end{pmatrix}. \tag{3.201}$$

In addition, the divergence of the stress tensor is written as:

$$\nabla \times \mathfrak{J} = \begin{pmatrix} \dfrac{\partial \sigma_x}{\partial x} + \dfrac{\partial \tau_{yx}}{\partial y} + \dfrac{\partial \tau_{zx}}{\partial z} \\ \dfrac{\partial \tau_{xy}}{\partial x} + \dfrac{\partial \sigma_y}{\partial y} + \dfrac{\partial \tau_{zy}}{\partial z} \\ \dfrac{\partial \tau_{xz}}{\partial x} + \dfrac{\partial \tau_{yz}}{\partial y} + \dfrac{\partial \sigma_z}{\partial z} \end{pmatrix}. \tag{3.202}$$

The L.H.S. of Eq. (3.196) can be rewritten by applying partial derivation as:

$$\frac{\partial \rho u}{\partial t} + \frac{\partial \rho uu}{\partial x} + \frac{\partial \rho uv}{\partial y} + \frac{\partial \rho uw}{\partial z} = u\left(\frac{\partial \rho}{\partial t} + \frac{\partial \rho u}{\partial x} + \frac{\partial \rho v}{\partial y} + \frac{\partial \rho w}{\partial z}\right)$$
$$+ \rho\frac{\partial u}{\partial t} + \rho u\frac{\partial u}{\partial x} + \rho v\frac{\partial u}{\partial y} + \rho w\frac{\partial u}{\partial z}. \tag{3.203}$$

Then, in consideration of the continuity equation, Navier-Stokes equation is written in the following form:

$$\rho\left(\frac{\partial u}{\partial t} + u\frac{\partial u}{\partial x} + v\frac{\partial u}{\partial y} + w\frac{\partial u}{\partial z}\right) = -\frac{\partial p}{\partial x} + \frac{\partial \sigma_x}{\partial x} + \frac{\partial \tau_{yx}}{\partial y} + \frac{\partial \tau_{zx}}{\partial z} + F_x, \tag{3.204}$$

$$\rho\left(\frac{\partial v}{\partial t} + u\frac{\partial v}{\partial x} + v\frac{\partial v}{\partial y} + w\frac{\partial v}{\partial z}\right) = -\frac{\partial p}{\partial y} + \frac{\partial \tau_{xy}}{\partial x} + \frac{\partial \sigma_y}{\partial y} + \frac{\partial \tau_{zy}}{\partial z} + F_y, \tag{3.205}$$

$$\rho\left(\frac{\partial w}{\partial t} + u\frac{\partial w}{\partial x} + v\frac{\partial w}{\partial y} + w\frac{\partial w}{\partial z}\right) = -\frac{\partial p}{\partial z} + \frac{\partial \tau_{xz}}{\partial x} + \frac{\partial \tau_{yz}}{\partial y} + \frac{\partial \sigma_z}{\partial z} + F_z. \tag{3.206}$$

This type of description is called the nonconservative form of Navier-Stokes equation. In the vector form, the nonconservative form of Navier-Stokes equation is written as:

$$\rho\left(\frac{\partial \vec{u}}{\partial t} + (\vec{u} \times \nabla)\vec{u}\right) = -\nabla p + \nabla \times \mathfrak{I} + \vec{F}, \tag{3.207}$$

Here, we define the substantial derivative as:

$$\frac{D}{Dt} \equiv \frac{\partial}{\partial t} + (\vec{u} \times \nabla), \tag{3.208}$$

By using the substantial derivative, the nonconservative form of Navier-Stokes equation can be written simply as:

$$\rho \frac{D\vec{u}}{Dt} = -\nabla p + \nabla \times \mathfrak{I} + \vec{F}, \tag{3.209}$$

Nonlinearity

The main difficulty in the study of Navier-Stokes equations is the existence of the advection term, that is, the change in the momentum by advection. Mathematically, the advection term has strong nonlinearity and therefore, the rigorous solutions of Navier-Stokes equation are known for only several simple flow cases. This is why the fluid dynamics is much more difficult than the particle mechanics. It should be noted here that the following part of this subsection is focused on the mathematical aspect of Navier-Stokes equation. Therefore, it is better to skip this subsection for the readers without the interest in such a discussion.

Here, we consider a simplified equation, that is, one-dimensional equation in consideration of only the temporal and advection terms. Such an equation is written as:

$$\frac{\partial u}{\partial t} + u\frac{\partial u}{\partial x} = 0, \tag{3.210}$$

Then, we assume that two rigorous solutions, u_1 and u_2, are given for Eq. 3.210. Therefore, u_1 and u_2 satisfy the following equation as:

$$\frac{\partial u_1}{\partial t} + u_1\frac{\partial u_1}{\partial x} = \frac{\partial u_2}{\partial t} + u_2\frac{\partial u_2}{\partial x} = 0, \tag{3.211}$$

In linear systems, it is well known that the superposition principal is established and the linear combination of u_1 and u_2 also becomes the solution of the equation. However, by substituting the linear combination of u_1 and u_2 into Eq. (3.210), we can realize that the linear combination no longer becomes the solution. The actual calculation is written as:

$$\begin{aligned}
&\frac{\partial \alpha u_1 + \beta u_2}{\partial t} + (\alpha u_1 + \beta u_2)\frac{\partial \alpha u_1 + \beta u_2}{\partial x} \\
&= \alpha\left(\frac{\partial u_1}{\partial t} + u_1\frac{\partial u_1}{\partial x}\right) + \beta\left(\frac{\partial u_2}{\partial t} + u_2\frac{\partial u_2}{\partial x}\right) + \alpha\beta\left(u_1\frac{\partial u_2}{\partial x} + u_2\frac{\partial u_1}{\partial x}\right) \\
&= \alpha\beta\left(u_1\frac{\partial u_2}{\partial x} + u_2\frac{\partial u_1}{\partial x}\right) \\
&\neq 0,
\end{aligned} \tag{3.212}$$

where α and β are the constants for linear combination. Therefore, it is evident that the superposition principal is not established for Navier-Stokes equation with nonlinear advection terms. As described later in "Basic theory" section, the nonlinear advection term becomes the source of the turbulent viscosity effect, that is, the nonlinearity effect is also important to understand complicated turbulent flows.

The lack of the superposition principal makes it extremely difficult to construct a generalized solution of Navier-Stokes equation. In fact, three-dimensional Navier-Stokes existence and smoothness has been listed as one of the millennium prize problems by Clay Mathematics Institute.

Equation of incompressible flow

For incompressible flow in which the density is assumed to be constant, Navier-Stokes equation can be simplified. Based on the discussions of "Newton's law of viscosity" in Section 3.2.2.1, the viscosity terms in Navier-Stokes equation can be expressed as:

$$\frac{\partial \sigma_x}{\partial x} + \frac{\partial \tau_{yx}}{\partial y} + \frac{\partial \tau_{zx}}{\partial z} = \mu\left(\frac{\partial^2 u}{\partial x^2} + \frac{\partial^2 u}{\partial y^2} + \frac{\partial^2 u}{\partial z^2}\right) + \mu\frac{\partial \Theta}{\partial x}, \tag{3.213}$$

where it is assumed that the viscosity coefficient is constant. Here, we define the dynamic viscosity, υ, as $\upsilon \equiv \mu/\rho$. Then, from Eqs. (3.204–3.205), coupled with the continuity equation, Eq. (3.184), Navier-Stokes equation for incompressible flows are obtained as:

$$\frac{\partial u}{\partial x} + \frac{\partial v}{\partial y} + \frac{\partial w}{\partial z} = 0. \tag{3.184}$$

$$\frac{\partial u}{\partial t} + u\frac{\partial u}{\partial x} + v\frac{\partial u}{\partial y} + w\frac{\partial u}{\partial z} = -\frac{\partial \tilde{p}}{\partial x} + \upsilon\left(\frac{\partial^2 u}{\partial x^2} + \frac{\partial^2 u}{\partial y^2} + \frac{\partial^2 u}{\partial z^2}\right) + f_x, \tag{3.214}$$

$$\frac{\partial v}{\partial t}+u\frac{\partial v}{\partial x}+v\frac{\partial v}{\partial y}+w\frac{\partial v}{\partial z}=-\frac{\partial \tilde{p}}{\partial y}+\upsilon\left(\frac{\partial^2 v}{\partial x^2}+\frac{\partial^2 v}{\partial y^2}+\frac{\partial^2 v}{\partial z^2}\right)+f_y, \tag{3.215}$$

$$\frac{\partial w}{\partial t}+u\frac{\partial w}{\partial x}+v\frac{\partial w}{\partial y}+w\frac{\partial w}{\partial z}=-\frac{\partial \tilde{p}}{\partial z}+\upsilon\left(\frac{\partial^2 w}{\partial x^2}+\frac{\partial^2 w}{\partial y^2}+\frac{\partial^2 w}{\partial z^2}\right)+f_z, \tag{3.216}$$

where $\tilde{p}=p/\rho$ and f is the external force per unit mass. In vector form, above equations are written as:

$$\nabla\times\vec{u}=0, \tag{3.217}$$

$$\rho\frac{D\vec{u}}{Dt}=-\nabla\tilde{p}+\upsilon\Delta\vec{u}+\vec{f}, \tag{3.218}$$

where $\vec{f}=(f_x,f_y,f_z)$.

3.2.3.4 Conservation of energy

For compressible flows, pressure changes with temperature as described by EOS in "Density change (equation of state)" section, and therefore, the energy conservation equation, which gives the change in temperature, must be considered simultaneously with the mass and momentum conservation equations. On the other hand, for incompressible flows without, for example, the buoyancy effect, the temperature is considered as an independent variable on the flow velocity, and therefore, the energy conservation equation is considered independent from the mass and momentum conservation equations.

Fourier's law

The characteristic phenomenon considered in the energy conservation equation is the heat conduction. In Fourier's law, the conductive heat flux is proportional to the spatial temperature gradient. Therefore, the conductive heat flux, $\vec{J}$, is written as:

$$\vec{J}=-k\nabla T, \tag{3.219}$$

where the proportionality coefficient k is the thermal conductivity.

Equation of energy conservation

The energy conservation is examined on a control volume with the volume of $\delta x\delta y\delta z$ as well as the mass and momentum conservation. First, the total energy is defined as:

$$E=e+\frac{|\vec{u}|^2}{2}=h-\frac{p}{\rho}+\frac{|\vec{u}|^2}{2}, \tag{3.220}$$

where e is the internal energy per unit mass and h is the enthalpy per unit mass. Then, when we consider the energy fluxes advected through x-surfaces, the inlet energy flux at

$x = x_0$ is $(\rho \text{Eu})_{x=x_0}\delta y \delta z$, and that at $x = x_0 + \delta x$ is $(\rho \text{Eu})_{x=x_0+\delta x}\delta y \delta z$. As for y-direcion, the inlet energy flux at $y = y_0$ is $(\rho \text{Ev})_{y=y_0}\delta z \delta x$, and that at $y = y_0 + \delta y$ is $(\rho \text{Ev})_{y=y_0+\delta y}\delta z \delta x$. Similarly, the inlet energy flux at $z = z_0$ is $(\rho \text{Ew})_{z=z_0}\delta x \delta y$, and that at $z = z_0 + \delta z$ is $(\rho \text{Ew})_{z=z_0+\delta z}\delta x \delta y$. The works by the surface forces, that is, the stress and pressure, is considered on each surfaces of the control volume. The contributions of surface force is $(\sigma_x u + \tau_{yx} v + \tau_{zx} w - pu)_{x=x_0}\delta y \delta z$ at $x = x_0$, $(\sigma_x u + \tau_{yx} v + \tau_{zx} w - pu)_{x=x_0+\delta x}\delta y \delta z$ at $x = x_0 + \delta x$, $(\tau_{xy} u + \sigma_y v + \tau_{zy} w - pv)_{y=y_0}\delta z \delta x$ at $y = y_0$, $(\tau_{xy} u + \sigma_y v + \tau_{zy} w - pv)_{y=y_0+\delta y}\delta z \delta x$ at $y = y_0 + \delta y$, $(\tau_{xz} u + \tau_{yz} v + \sigma_z w - pw)_{z=z_0}\delta x \delta y$ at $z = z_0$, $(\tau_{xz} u + \tau_{yz} v + \sigma_z w - pw)_{z=z_0+\delta z}\delta x \delta y$ at $z = z_0 + \delta z$. The works by the external forces in x, y and z-direction are represented as $F_x u$, $F_y v$ and $F_z w$, respectively. The change in energy by the hear conduction is formulated on the basis of Fourier's law and the inlet heat flux at $x = x_0$ is $\left(-k\frac{\partial T}{\partial x}\right)_{x=x_0}\delta y \delta z$, the outlet heat flux at $x = x_0 + \delta x$ is $\left(-k\frac{\partial T}{\partial x}\right)_{x=x_0+\delta x}\delta y \delta z$, the inlet heat flux at $y = y_0$ is $\left(-k\frac{\partial T}{\partial y}\right)_{y=y_0}\delta z \delta x$, the outlet heat flux at $y = y_0 + \delta y$ is $\left(-k\frac{\partial T}{\partial y}\right)_{y=y_0+\delta y}\delta z \delta x$, the inlet heat flux at $z = z_0$ is $\left(-k\frac{\partial T}{\partial z}\right)_{z=z_0}\delta x \delta y$, and the outlet heat flux at $z = z_0 + \delta z$ is $\left(-k\frac{\partial T}{\partial z}\right)_{z=z_0+\delta z}\delta x \delta y$. The heat source, that is, heat energy generation, should be considered in the energy conservation equation and we use Q to indicate the heat source per unit volume. In consideration of Taylor expansion with neglecting the second and higher-order terms, the temporal change rate in energy per unit volume is written as $\partial \rho E / \partial t$, which is equivalent to the summation of the change in energy by advection, stress and pressure, external forces and heat conduction. Therefore, with the similar manner to the derivations of the mass and momentum conservation equations, the energy conservation equation is obtained in the temporal and spatial derivative form:

$$\begin{aligned}\frac{\partial \rho E}{\partial t} + \frac{\partial \rho E u}{\partial x} + \frac{\partial \rho E v}{\partial y} + \frac{\partial \rho E w}{\partial z} = & -\frac{\partial pu}{\partial x} - \frac{\partial pv}{\partial y} - \frac{\partial pw}{\partial z} + \frac{\partial \sigma_x u}{\partial x} + \frac{\partial \tau_{yx} v}{\partial x} + \frac{\partial \tau_{zx} w}{\partial x} \\ & + \frac{\partial \tau_{xy} u}{\partial y} + \frac{\partial \sigma_y v}{\partial y} + \frac{\partial \tau_{zy} w}{\partial y} + \frac{\partial \tau_{xz} u}{\partial z} + \frac{\partial \tau_{yz} v}{\partial z} + \frac{\partial \sigma_z w}{\partial z} \\ & + \frac{\partial}{\partial x}\left(k\frac{\partial T}{\partial x}\right) + \frac{\partial}{\partial y}\left(k\frac{\partial T}{\partial y}\right) + \frac{\partial}{\partial z}\left(k\frac{\partial T}{\partial z}\right) \\ & + F_x u + F_y v + F_z w + Q.\end{aligned} \tag{3.221}$$

When we use the vector form, the energy conservation equation is written as:

$$\frac{\partial \rho E}{\partial t} + \nabla \times (\rho E \vec{u}) = -\nabla \times (p\vec{u}) + \nabla \times (\Im \vec{u}) + k\Delta T + \vec{F} \times \vec{u} + Q, \tag{3.222}$$

where the thermal conductivity is assumed to be constant.

Kinetic energy conservation (Bernoulli's theorem)

The kinetic energy conservation equation can be obtained as Eq. (3.200) multiplied by $\vec{u}$:

$$\frac{\partial \rho \frac{|\vec{u}|^2}{2}}{\partial t} + \nabla \times \left(\rho \frac{|\vec{u}|^2}{2} \vec{u} \right) = -\vec{u} \times \nabla p + \vec{u} \times (\nabla \times \Im) + \vec{F} \times \vec{u}, \tag{3.223}$$

Here, we consider one-dimensional incompressible steady inviscid flow. In such a case, the kinetic energy conservation equation is simplified as:

$$u\frac{\partial u}{\partial x} = -\frac{\partial}{\partial x}\left(\frac{p}{\rho}\right) + f_x, \tag{3.224}$$

This is the so-called Euler equation. As for the external force, it is assumed that only the gravity works and the gravitational acceleration in negative x-direction is indicated as g_x. Then Eq. (3.224) is rewritten as:

$$\frac{\partial}{\partial x}\left(\frac{u^2}{2} + \frac{p}{\rho} + \rho g_x x\right) = 0, \tag{3.225}$$

which can be integrated and the algebraic equation, that is, Bernoulli equation, is obtained as:

$$\frac{u^2}{2} + \frac{p}{\rho} + \rho g_x x = const. \tag{3.226}$$

This simple equation declares that the summation of kinetic energy, pressure potential and gravity potential is constant in incompressible and inviscid flows. Of course, the assumption of inviscid fluid is not satisfied in whole flow region, especially near obstacle surfaces. However, Bernoulli's principle is useful to estimate the flow velocity and pressure in various simple flow paths with relatively slow flow velocity.

As shown in Fig. 3.35, Bernoulli's principle gives the relationship of physical variables between point 1 and point 2 in a flow path. Namely, the Bernoulli equation can be written as:

$$\frac{u_1^2}{2} + \frac{p_1}{\rho} + \rho g H_1 = \frac{u_2^2}{2} + \frac{p_2}{\rho} + \rho g H_2, \tag{3.227}$$

where H is the hydrostatic head. It should be noted that u_1 and u_2 stand for velocities in flow direction, not the x-component of velocity. In addition, the mass conservation law requires that the flow velocity changes in accordance with the

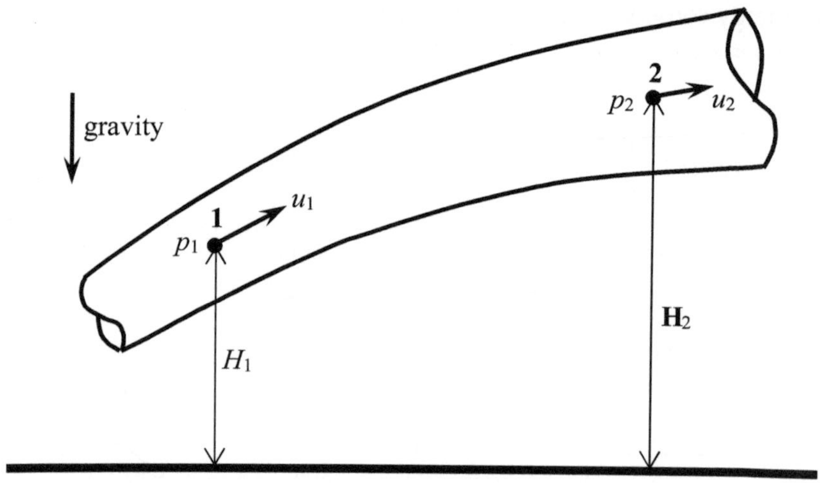

Figure 3.35 Schematic explanation of Bernoulli's principle.

cross-sectional area of the flow path, indicated with A. Therefore, the relationship between u_1 and u_2 is written as:

$$A_1 u_1 = A_2 u_2. \tag{3.228}$$

By substituting Eq. (3.228) into Eq. (3.226) gives the equation of the pressure at point 2:

$$\begin{aligned} p_2 &= p_1 + \rho\left(\frac{u_1^2}{2} - \frac{u_2^2}{2}\right) + g(H_1 - H_2) \\ &= p_1 + \rho\left\{\frac{u_1^2}{2} - \frac{1}{2}\left(\frac{A_1 u_1}{A_2}\right)^2\right\} + g(H_1 - H_2) \\ &= p_1 + \frac{\rho u_1^2}{2}\left\{1 - \left(\frac{A_1}{A_2}\right)^2\right\} + g(H_1 - H_2). \end{aligned} \tag{3.229}$$

This equation states that the pressure at any point in a flow path can be estimated with given velocity and pressure at certain point in the flow path when we know the cross-sectional areas of both points and the difference in the hydrostatic heads between two points. In fact, Bernoulli's principle is used for the measurement of flow velocity by a pitot tube. In such a case, the difference between the total pressure, that is, the pressure at a stagnation point and the static pressure, is measured by a pitot tube and the flow velocity, u, is calculated as:

$$u = \sqrt{\frac{2(p_t - p_s)}{\rho}}, \tag{3.230}$$

where p_t is the total pressure and p_s is the static pressure.

Enthalpy conservation

The substitution of the kinetic energy conservation equation, Eq. (3.223), from the total energy conservation equation, Eq. (3.222), gives the internal energy conservation equation:

$$\frac{\partial \rho e}{\partial t} + \nabla \times (\rho e \vec{u}) = -p(\nabla \times \vec{u}) + \nabla \times (\Im \vec{u}) - \vec{u} \times (\nabla \times \Im) + k\Delta T + Q, \tag{3.231}$$

Then, in consideration of the relationship between the internal energy and the enthalpy, which is written as:

$$h = e + \frac{p}{\rho}, \tag{3.232}$$

the enthalpy conservation equation is obtained as:

$$\frac{\partial \rho h}{\partial t} + \nabla \times (\rho h \vec{u}) = \frac{Dp}{Dt} + \nabla \times (\Im \vec{u}) - \vec{u} \times (\nabla \times \Im) + k\Delta T + Q. \tag{3.233}$$

When the enthalpy can be expressed as the product of the isobaric specific heat and temperature, the enthalpy conservation equation is expressed with temperature:

$$\rho C_p \frac{\partial T}{\partial t} + \rho C_p \nabla \times (T\vec{u}) = \frac{Dp}{Dt} + \nabla \times (\Im \vec{u}) - \vec{u} \times (\nabla \times \Im) + k\Delta T + Q, \tag{3.234}$$

where the isobaric specific heat C_p is assumed to be constant. In many cases, especially for the study on incompressible flows, Eq. (3.234) is employed as the energy conservation equation.

3.2.4 Laminar flow and turbulent flow

When you put milk in the coffee in a cup, you can observe very slow spreading of the milk without adding any treatment to the coffee. In such a case, you should stir the coffee with a spoon to enhance the spreading of the milk and achieve the well mixed state. At that time, the flow in the cup is turbulent, that is, the velocity fluctuates in time and space. This milk spreading phenomenon gives us an important insight to laminar and turbulent flows. In laminar flows, roughly classified into the flows with relatively slow velocity, the molecular diffusion is dominant and therefore, the milk spreads very slowly. On the other hand, in turbulent flows, the turbulent diffusion becomes much larger than the molecular diffusion and therefore, the milk spreads quickly.

In this subsection, the concepts of laminar and turbulent flows are discussed briefly.

3.2.4.1 Reynolds number

In the study on incompressible flows, the most important nondimensional number is Reynolds number. Reynolds number represents the ratio of inertia force and viscous force in viscous flows, which in turn indicates the stability of the flows.

Nondimensional Navier-Stokes equation

Reynolds number arises in the nondimensional Navier-Stokes equation. Here, we introduce the length scale, L, velocity scale, U, and the characteristic frequency of flow, f_q. Then, the nondimensional time and space coordinates are defined as:

$$\hat{t} = tf_q, \tag{3.235}$$

$$(\hat{x}, \hat{y}, \hat{z}) = \left(\frac{x}{L}, \frac{y}{L}, \frac{z}{L}\right). \tag{3.236}$$

The velocity, pressure and external force are also expressed in the nondimensional forms as:

$$\hat{\vec{u}} = (\hat{u}, \hat{v}, \hat{w}) = \left(\frac{u}{U}, \frac{v}{U}, \frac{w}{U}\right), \tag{3.237}$$

$$\hat{\tilde{p}} = \frac{\tilde{p}}{U^2}, \tag{3.238}$$

$$\hat{\vec{f}} = \left(\hat{f}_x, \hat{f}_y, \hat{f}_z\right) = \frac{L}{U^2}\left(f_x, f_y, f_z\right). \tag{3.239}$$

In addition, the nondimensional derivative operators are defined as:

$$\hat{\nabla} = \left(\frac{\partial}{\partial \hat{x}}, \frac{\partial}{\partial \hat{y}}, \frac{\partial}{\partial \hat{z}}\right) = L\left(\frac{\partial}{\partial x}, \frac{\partial}{\partial y}, \frac{\partial}{\partial z}\right), \tag{3.240}$$

$$\hat{\Delta} = \left(\frac{\partial}{\partial \hat{x}^2}, \frac{\partial}{\partial \hat{y}^2}, \frac{\partial}{\partial \hat{z}^2}\right) = L^2\left(\frac{\partial}{\partial x^2}, \frac{\partial}{\partial y^2}, \frac{\partial}{\partial z^2}\right). \tag{3.241}$$

By applying these transforms to Navier-Stokes equation for incompressible flows, that is, Eqs. (3.184, 3.214, 3.215, and 3.216), the nondimensional Navier-Stokes equation for incompressible flows is obtained as:

$$\frac{\partial \hat{u}}{\partial \hat{x}} + \frac{\partial \hat{v}}{\partial \hat{y}} + \frac{\partial \hat{w}}{\partial \hat{z}} = 0. \tag{3.242}$$

$$\frac{1}{St}\frac{\partial \hat{u}}{\partial \hat{t}} + \hat{u}\frac{\partial \hat{u}}{\partial \hat{x}} + \hat{v}\frac{\partial \hat{u}}{\partial \hat{y}} + \hat{w}\frac{\partial \hat{u}}{\partial \hat{z}} = -\frac{\partial \hat{\tilde{p}}}{\partial \hat{x}} + \frac{1}{Re}\left(\frac{\partial^2 \hat{u}}{\partial \hat{x}^2} + \frac{\partial^2 \hat{u}}{\partial \hat{y}^2} + \frac{\partial^2 \hat{u}}{\partial \hat{z}^2}\right) + \hat{f}_x, \tag{3.243}$$

$$\frac{1}{St}\frac{\partial \hat{v}}{\partial \hat{t}} + \hat{u}\frac{\partial \hat{v}}{\partial \hat{x}} + \hat{v}\frac{\partial \hat{v}}{\partial \hat{y}} + \hat{w}\frac{\partial \hat{v}}{\partial \hat{z}} = -\frac{\partial \hat{p}}{\partial \hat{y}} + \frac{1}{Re}\left(\frac{\partial^2 \hat{v}}{\partial \hat{x}^2} + \frac{\partial^2 \hat{v}}{\partial \hat{y}^2} + \frac{\partial^2 \hat{v}}{\partial \hat{z}^2}\right) + \hat{f}_y, \tag{3.244}$$

$$\frac{1}{St}\frac{\partial \hat{w}}{\partial \hat{t}} + \hat{u}\frac{\partial \hat{w}}{\partial \hat{x}} + \hat{v}\frac{\partial \hat{w}}{\partial \hat{y}} + \hat{w}\frac{\partial \hat{w}}{\partial \hat{z}} = -\frac{\partial \hat{p}}{\partial \hat{z}} + \frac{1}{Re}\left(\frac{\partial^2 \hat{w}}{\partial \hat{x}^2} + \frac{\partial^2 \hat{w}}{\partial \hat{y}^2} + \frac{\partial^2 \hat{w}}{\partial \hat{z}^2}\right) + \hat{f}_z, \tag{3.245}$$

where St is Strouhal number defined as:

$$St \equiv \frac{Lf_q}{U}. \tag{3.246}$$

Reynolds number is defined as:

$$Re \equiv \frac{UL}{\upsilon} = \frac{\rho \mathrm{UL}}{\mu}, \tag{3.247}$$

which arises in the nondimensional Navier-Stokes equation as the inverse of the viscous term coefficient. In the vector form, the nondimensional Navier-Stokes equation is organized as:

$$\hat{\nabla} \times \hat{\vec{u}} = 0, \tag{3.248}$$

$$\frac{1}{\mathrm{St}}\frac{D\hat{\vec{u}}}{Dt} = -\hat{\nabla}\hat{p} + \frac{1}{Re}\hat{\Delta}\hat{\vec{u}} + \hat{\vec{f}}. \tag{3.249}$$

For steady flows, the temporal derivative becomes zero and the nondimensional Navier-Stokes equation is written as:

$$\left(\hat{\vec{u}} \times \hat{\nabla}\right)\hat{\vec{u}} = -\hat{\nabla}\hat{p} + \frac{1}{Re}\hat{\Delta}\hat{\vec{u}} + \hat{\vec{f}}. \tag{3.250}$$

This equation indicates that the incompressible steady flow is represented by only one nondimensional number, that is, Reynolds number, when the viscosity effect is indispensable. Namely, Reynolds similarity rule is satisfied and the flows in the flow paths with the same shape but different size becomes similar when Reynolds number for both flows are the same. This knowledge is frequently used in the scaled-model tests for the design of large-scale power and energy machineries.

Physical meaning of Reynolds number

Reynolds number can be redefined as:

$$Re \equiv \frac{\rho U^2}{\mu \frac{U}{L}}. \tag{3.251}$$

As described in "Momentum advection" section, the product of the density and the velocity squared represents the momentum advection, that is, the numerator in Eq. (3.251) is regarded as the inertia force. On the other hand, the denominator is regarded as the product of the viscosity coefficient and the velocity gradient, which represents the viscous force. Therefore, Reynolds number represents the ratio of inertia force and viscous force in viscous flows.

Reynolds number can be defined in various shapes. In a vortical flow, for instance, Reynolds number is defined as the ratio of the circulation, that is, the closed-loop integral of velocity, and the dynamic viscosity. In this case, Reynolds number is regarded as the indicator of the stability of the vortical flow.

When Reynolds number is much smaller than unity, the inertia force can be neglected compared with the viscous force. Therefore, the advection term in Navier-Stokes equation is omitted and the equation can be linearized. Such an assumption at the low Reynolds number limit is called Stokes' approximation and the linearized equation is called Stokes equation which is written as:

$$\frac{1}{St}\frac{\partial \hat{\vec{u}}}{\partial t} = -\hat{\nabla}\hat{p} + \frac{1}{Re}\hat{\Delta}\hat{\vec{u}} + \hat{\vec{f}}. \tag{3.252}$$

By omitting the advection term, that is, the source of the nonlinearity, the main difficulty in the study on Navier-Stokes equation is removed. Therefore, rigorous solutions can be obtained easily. For example, we can calculate the very slow flow around a sphere as shown in Fig. 3.36.

In preparation for the flow calculation, we define the polar coordinates as:

$$r = \sqrt{x^2 + y^2 + z^2}. \tag{3.253}$$

$$x = rcos\theta, \tag{3.254}$$

where x-coordinate indicates the mainstream direction and the origin of the radial coordinate is at the center of the sphere. Then, the flow around the sphere is calculated in consideration of the boundary conditions:

$$\vec{u} = (u, v, w) = (0, 0, 0)\ at\ r = a, \tag{3.255}$$

$$\vec{u} = (u, v, w) = (U_\infty, 0, 0),\ p = p_\infty\ at\ r = \infty, \tag{3.256}$$

where a is the radius of the sphere, U_∞ and p_∞, are the x-velocity and pressure at far field. The solution is obtained as:

$$u = \left\{\frac{3ax^2}{4r^3}\left(\frac{a^2}{r^2} - 1\right) - \frac{a}{4r}\left(\frac{a^2}{r^2} + 3\right) + 1\right\}U_\infty, \tag{3.257}$$

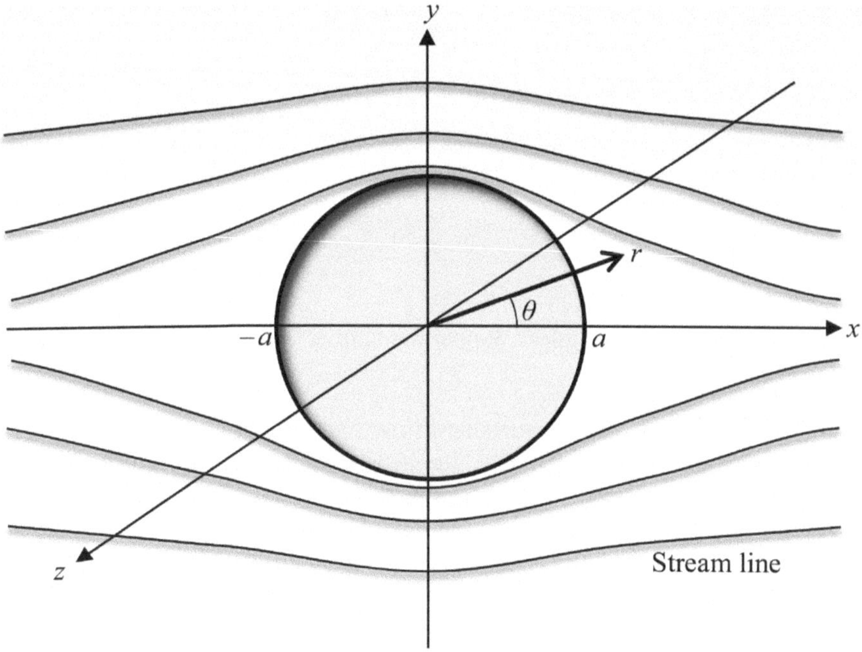

Figure 3.36 Stokes flow around sphere.

$$v = \frac{3axy}{4r^3}\left(\frac{a^2}{r^2} - 1\right)U_\infty, \tag{3.258}$$

$$w = \frac{3axz}{4r^3}\left(\frac{a^2}{r^2} - 1\right)U_\infty. \tag{3.259}$$

The pressure on the sphere surface is also calculated as:

$$(p)_{r=a} = p_\infty - \frac{3\mu U_\infty}{2a} cos\theta. \tag{3.260}$$

The shear stress in θ-direction on the sphere surface is calculated as:

$$(\tau_{r\theta})_{r=a} = -\frac{3\mu U_\infty}{2a} sin\theta. \tag{3.261}$$

It should be noted that the sphere does not receive the forces in y and z-direction, due to the flow symmetricity. Therefore, the flow around the sphere provides only x-force, that is, the drag force, to the sphere. By integrating Eqs. (3.260 and 3.261) around the sphere, the drag force, F_D, is calculated as:

$$F_D = -2\pi a^2 \int_0^\pi \left\{(\tau_{r\theta})_{r=a} sin\theta + (p)_{r=a} cos\theta\right\} d\theta = 6\pi a \mu U_\infty. \tag{3.262}$$

In general, the drag coefficient, C_D, is defined as:

$$F_D = C_D \frac{1}{2} \rho U_\infty^2 S, \tag{3.263}$$

where S is the projected area of the sphere to mainstream direction and calculated as πa^2. Finally, by the comparison between Eqs. (3.262 and 3.263), the drag coefficient is given as:

$$C_D = \frac{12}{a \rho U_\infty} = \frac{24}{Re}, \tag{3.264}$$

where the representative length in Reynolds number is the diameter of the sphere, that is, 2a.

Transition from laminar flow to turbulent flow (Orr-Sommerfeld equation)

It is well known that the flow is stable, that is, laminar, under relatively low Reynolds number conditions because the viscous force, which stabilized the flow, is dominant. With the increase in Reynolds number, the inertia force in the flow becomes dominant and the flow state changes to transition and finally turbulent. As shown in Fig. 3.37, the disturbance is generated in the flow in a channel under relatively high Reynolds number conditions, and the turbulent flow is formed in the channel. It should be noted here that the following part of this subsection contains highly-complicated mathematical discussion. Therefore, it is recommended to skip this subsection for the readers without the interest in such a mathematical discussion.

The flow transition and the generation of the turbulence have not been fully understood theoretically due to the complexity of the phenomena. However, based on a stability analysis, we can obtain the insights into the laminar-turbulent transition. Here, we consider the stability of the two-dimensional flow based on the vorticity equation. First, the vorticity, ω, is defined in the nondimensional form:

$$\hat{\omega} = \frac{\partial \hat{v}}{\partial \hat{x}} - \frac{\partial \hat{u}}{\partial \hat{y}}. \tag{3.265}$$

Then, by applying $\partial/\partial\hat{y}$ to Eq. (3.243) and $\partial/\partial\hat{x}$ to Eq. (3.244), the vorticity equation is derived as the difference between two equations:

$$\frac{\partial \hat{\omega}}{\partial \hat{t}} + \hat{u}\frac{\partial \hat{\omega}}{\partial \hat{x}} + \hat{v}\frac{\partial \hat{\omega}}{\partial \hat{y}} = \frac{1}{Re}\hat{\Delta}\hat{\omega}, \tag{3.266}$$

where external forces are not considered here. For the next step, the nondimensional stream function is introduced as:

$$\hat{u} = \frac{\partial \hat{\psi}}{\partial \hat{y}}, \tag{3.267}$$

$$\hat{v} = -\frac{\partial \hat{\psi}}{\partial \hat{x}}. \tag{3.268}$$

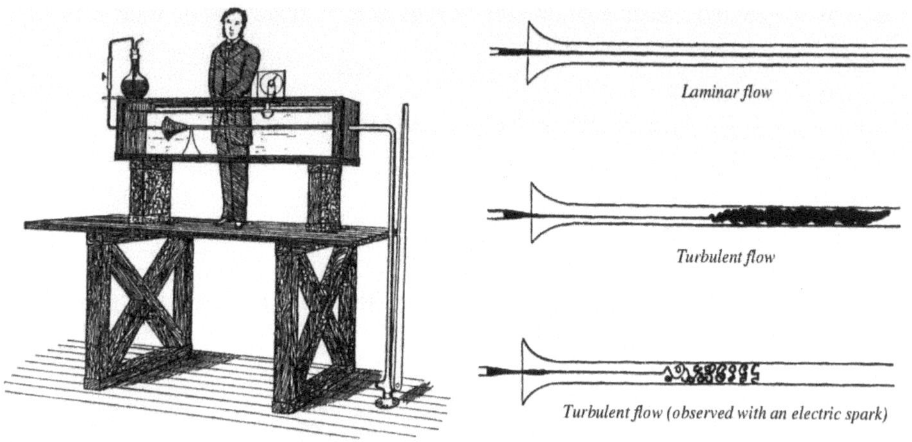

Figure 3.37 Transition from laminar flow to turbulent flow.

Since the vorticity can be expressed with the stream function:

$$\hat{\omega} = -\hat{\Delta}\hat{\psi}, \tag{3.269}$$

the vorticity equation is converted to the stream function equation:

$$\frac{\partial \hat{\Delta}\hat{\psi}}{\partial \hat{t}} + \hat{u}\frac{\partial \hat{\Delta}\hat{\psi}}{\partial \hat{x}} + \hat{v}\frac{\partial \hat{\Delta}\hat{\psi}}{\partial \hat{y}} = \frac{1}{Re}\hat{\Delta}^2\hat{\psi}. \tag{3.270}$$

We consider a parallel flow, in which the fluid flows only in x-direction on average and the averaged x-velocity is dependent only on y-coordinate. Such a flow condition is written as:

$$\hat{\bar{u}} = \hat{\bar{u}}(\hat{y}), \tag{3.271}$$

$$\hat{\bar{v}} = 0. \tag{3.272}$$

where the overbar stands for the averaged values. The parallel flow is formed in between two horizonal (x-direction) palates located at $\hat{y} = \hat{y}_1$ and $\hat{y}_2$, and the boundary conditions on the plates are set as:

$$\hat{u} = \hat{v} = 0, \tag{3.273}$$

which represents the nonslip wall condition. For the stream function, this nonslip condition is determined as:

$$\frac{\partial \hat{\psi}}{\partial \hat{x}} = \frac{\partial \hat{\psi}}{\partial \hat{y}} = 0. \tag{3.274}$$

The averaged and the fluctuation components in the stream function are defined as:

$$\hat{\psi} = \hat{\overline{\psi}} + \hat{\psi}', \tag{3.275}$$

which gives the formulation of the averaged x-velocity as:

$$\hat{\overline{u}} = \frac{\partial \hat{\overline{\psi}}}{\partial \hat{y}}. \tag{3.276}$$

Then, by introducing Eq. (3.275) into Eq. (3.270), the stream function equation can be linearized by neglecting the infinitesimal terms. The linearized stream function equation is written as:

$$\frac{\partial \hat{\Delta}\hat{\psi}'}{\partial \hat{t}} + \hat{\overline{u}}\frac{\partial \hat{\Delta}\hat{\psi}'}{\partial \hat{x}} - \frac{\partial^2 \hat{\overline{u}}}{\partial \hat{y}^2}\frac{\partial \hat{\psi}'}{\partial \hat{x}} = \frac{1}{\text{Re}}\hat{\Delta}^2\hat{\psi}', \tag{3.277}$$

Here, we assume that the solution of the following shape as:

$$\hat{\psi}'\left(\hat{x}, \hat{y}, \hat{t}\right) = \hat{\phi}(\hat{y})e^{i\hat{\alpha}(\hat{x} - \hat{c}\hat{t})}, \tag{3.278}$$

where ϕ is the spatial factor of the stream function fluctuation, α and c are constants. Finally, by substituting Eq. (3.278) into Eq. (3.277), Orr-Sommerfeld equation is obtained as:

$$\frac{\partial^4 \hat{\phi}}{\partial \hat{y}^4} - 2\hat{\alpha}^2\frac{\partial^2 \hat{\phi}}{\partial \hat{y}^2} + \hat{\alpha}^4\hat{\phi} = i\hat{\alpha}Re\left\{\left(\hat{\overline{u}} - \hat{c}\right)\left(\frac{\partial^2 \hat{\phi}}{\partial \hat{y}^2} - \hat{\alpha}^2\hat{\phi}\right) - \frac{\partial^2 \hat{\overline{u}}}{\partial \hat{y}^2}\hat{\phi}\right\}, \tag{3.279}$$

with the nonslip boundary condition at $\hat{y} = \hat{y}_1$ and $\hat{y} = \hat{y}_2$ written as:

$$\hat{\phi} = \frac{\partial \hat{\phi}}{\partial \hat{y}} = 0. \tag{3.280}$$

The stability analysis based on Orr-Sommerfeld equation gives theoretical solutions only for a few limited flow conditions, for example, stationary state. However, some numerical solutions are obtained and the stability maps, that is, the critical stability curve, are drawn for various simple flows as shown in Fig. 3.38.

It is evident that within a certain range, the increase in Reynolds number causes the flow instability which means the growth of the fluctuating components and the transition to turbulent flow.

3.2.4.2 *Laminar flow*

Here, we consider some simple laminar flows, that is, Hagen-Poiseuille flow, Rayleigh problem and flow in boundary layer. The equations of those flows are

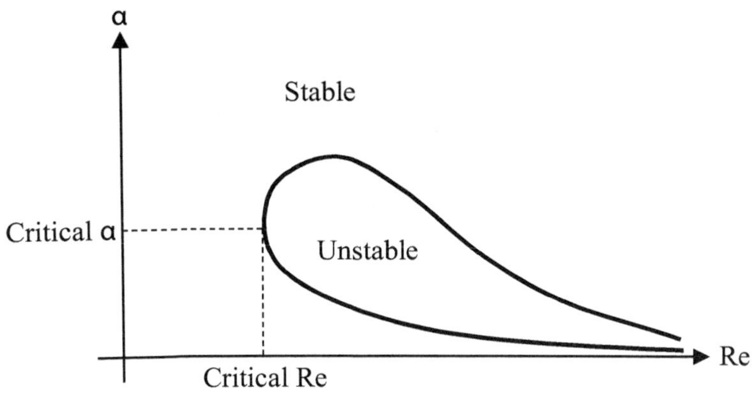

Figure 3.38 Example of stability map.

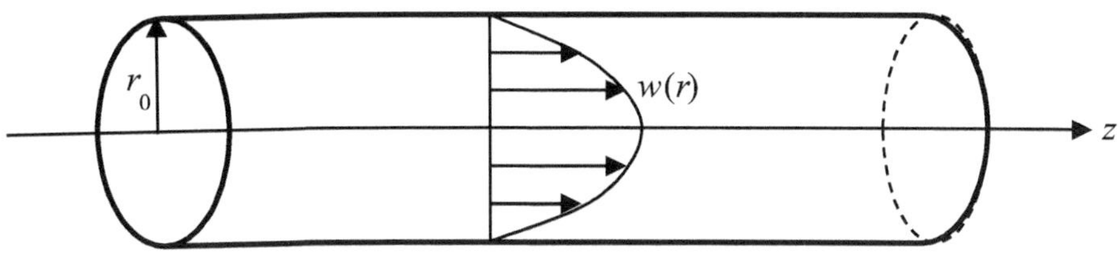

Figure 3.39 Schematic view of Hagen-Poiseuille flow.

derived from Navier-Stokes equation (see "Equation of incompressible flow" section) and gives us useful information about laminar flows.

Hagen-Poiseuille flow

Hagen-Poiseuille flow is known as one most famous laminar flow, which describes the parallel flow in a circular pipe, or the flow between two plates in two-dimensional coordinates. The equations of Hagen-Poiseuille flow is obtained as a solution of linearized Navier-Stokes equation in which the advection terms, that is, the nonlinear terms, are omitted as a result of simplified flow assumption. In spite of the relatively simplified appearance of Hagen-Poiseuille flow, the equation of Hagen-Poiseuille flow gives the important information about the flow in a pipe, especially about the wall friction (Fig. 3.39).

Here, we start from the axisymmetric two-dimensional, that is, r-z, Navier-Stokes equation for a steady incompressible flow with no external force, which is written as:

$$\frac{1}{r}\left(\frac{\partial r u_r}{\partial r}\right) + \frac{\partial w}{\partial z} = 0. \tag{3.281}$$

$$u_r \frac{\partial u_r}{\partial r} + w \frac{\partial u_r}{\partial z} = -\frac{\partial \tilde{p}}{\partial r} + \upsilon \left\{ \frac{\partial}{\partial r}\left(\frac{1}{r}\frac{\partial r u_r}{\partial r}\right) + \frac{\partial^2 u_r}{\partial z^2} \right\}, \tag{3.282}$$

$$u_r \frac{\partial w}{\partial r} + w \frac{\partial w}{\partial z} = -\frac{\partial \tilde{p}}{\partial z} + \upsilon \left\{ \frac{1}{r}\frac{\partial}{\partial r}\left(r \frac{\partial w}{\partial r}\right) + \frac{\partial^2 w}{\partial z^2} \right\}, \tag{3.283}$$

where u_r is the velocity in r-direction. For a parallel flow, the velocity components are simplified as:

$$u_r = 0, \tag{3.284}$$

$$w = w(r). \tag{3.285}$$

By substituting Eq. (3.284) into Eq. (3.281), the gradient of z-velocity is calculated as:

$$\frac{\partial w}{\partial z} = 0. \tag{3.286}$$

In addition, by substituting Eq. (3.284) into Eq. (3.282), the terms except for the pressure gradient term becomes zero, and therefore, the also the pressure gradient in radial direction becomes zero as:

$$0 = -\frac{\partial \tilde{p}}{\partial r}, \tag{3.287}$$

which indicates that the pressure is constant in radial direction. Namely, the pressure is expressed as the function of only z, which is written as:

$$\tilde{p} = \tilde{p}(z). \tag{3.288}$$

The equation for z-velocity is given by Eq. (3.283):

$$0 = -\frac{d\tilde{p}}{dz} + \frac{\upsilon}{r}\frac{d}{dr}\left(r\frac{dw}{dr}\right), \tag{3.289}$$

where it should be noted that the pressure and z-velocity are the functions of only z. By integrating Eq. (3.289) twice in radial direction, we obtain the following equation:

$$w = \frac{1}{4\upsilon}\frac{d\tilde{p}}{dz}r^2 + c_1 log r + c_2, \tag{3.290}$$

where c_1 and c_2 are the integration constants. The symmetric condition at the pipe center, $r = 0$, is written as

$$\frac{\partial w}{\partial r} = 0, \tag{3.291}$$

and therefore, c_1 is determined as:

$$c_1 = 0. \tag{3.292}$$

In addition, the nonslip boundary condition is applied on wall surface, $r = r_0$:

$$w = 0, \tag{3.293}$$

which determines c_2 as:

$$c_2 = -\frac{1}{4\upsilon}\frac{d\tilde{p}}{dz}r_0^2. \tag{3.294}$$

Finally, the equation of z-velocity is obtained as:

$$w = -\frac{r_0^2}{4\upsilon}\frac{d\tilde{p}}{dz}\left\{1 - \left(\frac{r}{r_0}\right)^2\right\}. \tag{3.295}$$

Based on Eq. (3.295), we can calculate the maximum velocity. From the flow symmetrical condition, the maximum velocity appears at the pipe center, that is, $r = 0$, which is calculated as:

$$w_{max} = -\frac{r_0^2}{4\upsilon}\frac{d\tilde{p}}{dz} = \frac{r_0^2}{4\upsilon}\left|\frac{d\tilde{p}}{dz}\right|, \tag{3.296}$$

Then, the cross-sectional averaged velocity is calculated as:

$$w_{ave} = \frac{1}{\pi r_0^2}\int_0^{r_0} w_{max}\left\{1 - \left(\frac{r}{r_0}\right)^2\right\}2\pi r dr = \frac{w_{max}}{2}. \tag{3.297}$$

Pressure drop in laminar flow

The equation of Hagen-Poiseuille flow gives the pressure drop in a circular pipe. The pressure drop in a circular pipe is caused by wall friction, which is written with Darcy friction factor, f_d, as:

$$-\rho\frac{d\tilde{p}}{dz} = \frac{f_d}{d}\frac{1}{2}\rho w_{ave}^2, \tag{3.298}$$

where d is the pipe diameter which is equal to $2r_0$. In consideration of Eqs. (3.296 and 3.297), the pressure gradient in z-direction is written as:

$$-\frac{d\tilde{p}}{dz} = \frac{8\upsilon w_{ave}}{r_0^2}. \tag{3.299}$$

Therefore, from Eqs. (3.298 and 3.299), Darcy friction factor can be defined as well-known simple form:

$$f_d = \frac{64}{\frac{w_{ave}d}{\upsilon}} = \frac{64}{Re}. \tag{3.300}$$

Rayleigh problem

Rayleigh problem is known as an example of Stokes flows. In Rayleigh problem, the flow generated by an infinitely long plate which suddenly start moving with a constant speed is considered. In this case, the flow is assumed to be parallel to the plate and the velocity changes only in the direction normal to the plate. Also, the pressure is assumed to be constant in whole region. In spite of those highly simplified flow assumption, Rayleigh problem gives the important insight for the development of a viscous layer.

Here, we define the velocity parallel to the plate as u and the coordinate normal to the plate as y. Then, the two-dimensional Navier-Stokes equation for an unsteady incompressible flow with no external force is simplified with the above assumptions:

$$\frac{\partial u}{\partial t} = \upsilon \frac{\partial^2 u}{\partial y^2}, \tag{3.301}$$

where the origin of y-coordinate is on the plate. Since the plate is moving with the constant speed, the boundary condition at $y = 0$ is written as:

$$u = u_w, \tag{3.302}$$

where u_w is the moving speed of the plate. In addition, at $y =$ infinity, the velocity becomes zero:

$$u = 0. \tag{3.303}$$

We introduce nondimensional coordinate and nondimensional velocity:

$$\zeta = \frac{y}{2\sqrt{\upsilon t}}, \tag{3.304}$$

$$u = u_w N(\zeta). \tag{3.305}$$

By using these nondimensional variables, the partial differential equation, Eq. (3.301), is transformed to an ordinary differential equation:

$$\frac{d^2 N}{d\zeta^2} + 2\zeta \frac{dN}{d\zeta} = 0. \tag{3.306}$$

The boundary conditions at $\zeta = 0$ is transformed to:

$$N = 1, \tag{3.307}$$

and also at $\zeta = \infty$, the boundary condition is transformed to:

$$N = 0. \tag{3.308}$$

The solution of the ordinary differential equation, Eq. (3.306), is known to be:

$$N(\zeta) = 1 - erf\,\zeta, \tag{3.309}$$

where the error function erf is defined as:

$$erf\,\zeta = \frac{2}{\sqrt{\pi}}\int_0^{\zeta} e^{-\xi^2}\,d\xi. \tag{3.310}$$

Therefore, the velocity profile is obtained as:

$$u = u_w(1 - erf\,\zeta). \tag{3.311}$$

The velocity profile in Eq. (3.311) is shown in Fig. 3.40. It is evident that the velocity attenuates in positive y-direction and approaches asymptotically to zero. When we define the viscous layer as the region where the velocity is larger than 1% of u_w, the viscous layer is called 99% boundary layer and the thickness of the 99% boundary layer, $\delta_{0.99}$, is calculated as:

$$\delta_{0.99} \cong 4\sqrt{\upsilon t}, \tag{3.312}$$

which indicates the viscous layer develops with the square root of time.

Boundary layer equation

As seen in Rayleigh problem, the viscous layer develops near an obstacle surface, for example, wall surface, by the effect of fluid viscosity. This viscous layer is

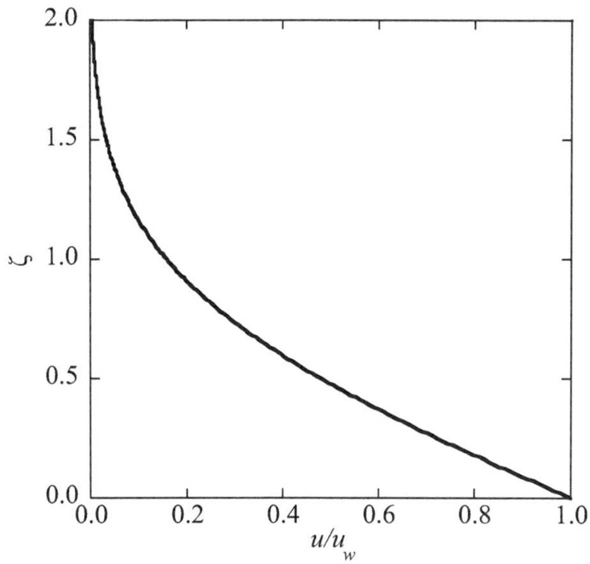

Figure 3.40 Solution of Rayleigh problem.

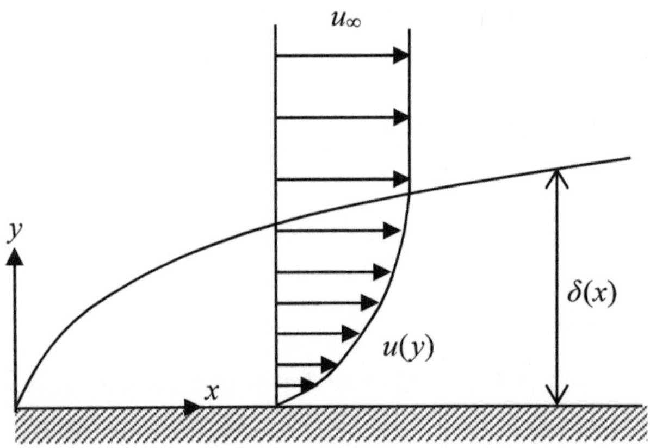

Figure 3.41 Schematic view of boundary layer.

called boundary layer and in general, the effect of fluid viscosity must be considered only in the boundary layer. In other words, the flow far from an obstacle surface can be modeled as an inviscid flow and the effect of the viscosity is not negligible only near the obstacle surface. For laminar flows, the basic equation of the laminar boundary layer, that is, the boundary layer equation, is obtained by Prandtl and known as Prandtl boundary layer theory (Fig. 3.41).

Here, we consider the two-dimensional steady flow along a plate. The two-dimensional Navier-Stokes equation with no external force is written as:

$$\frac{\partial u}{\partial x} + \frac{\partial v}{\partial y} = 0, \tag{3.313}$$

$$u\frac{\partial u}{\partial x} + v\frac{\partial u}{\partial y} = -\frac{\partial \tilde{p}}{\partial x} + \upsilon\left(\frac{\partial^2 u}{\partial x^2} + \frac{\partial^2 u}{\partial y^2}\right), \tag{3.314}$$

$$u\frac{\partial v}{\partial x} + v\frac{\partial v}{\partial y} = -\frac{\partial \tilde{p}}{\partial y} + \upsilon\left(\frac{\partial^2 v}{\partial x^2} + \frac{\partial^2 v}{\partial y^2}\right), \tag{3.315}$$

where x and y-coordinates are defined as the directions along the plate and normal to the plate, respectively. The origin of x-coordinate is at the edge of the plate where the boundary layer start developing, and the origin of y-coordinate is on the plate. Then, we introduce length scale L_1 for x-direction and L_2 for y-direction. Since the boundary layer is thin, the relationship between two length scales becomes:

$$L_1 \gg L_2. \tag{3.316}$$

Similarly, velocity scale U_1 for x-direction and U_2 for y-direction are introduced. From the result of Rayleigh problem, it can be assumed that L_2, the length scale of boundary layer thickness, is proportional to the square root of the flow time from the edge of the plate as:

$$L_2 \sim \sqrt{\upsilon t} = \sqrt{\frac{\upsilon x}{U_1}}. \tag{3.317}$$

Therefore, the scale of the dynamic viscosity is obtained as:

$$\upsilon \sim \frac{L_2^2}{L_1} U_1. \tag{3.318}$$

Then, based on the length, velocity and viscosity scales, the nondimensional form (except for pressure term) of Navier-Stokes equation is obtained as:

$$\frac{U_1}{L_1}\frac{\partial \hat{u}}{\partial \hat{x}} + \frac{U_2}{L_2}\frac{\partial \hat{v}}{\partial \hat{y}} = 0, \tag{3.319}$$

$$\frac{U_1^2}{L_1}\hat{u}\frac{\partial \hat{u}}{\partial \hat{x}} + \frac{U_1 U_2}{L_2}\hat{v}\frac{\partial \hat{u}}{\partial \hat{y}} = -\frac{\partial \tilde{p}}{\partial x} + \left(\frac{L_2^2 U_1^2}{L_1^3}\frac{\partial^2 \hat{u}}{\partial \hat{x}^2} + \frac{U_1^2}{L_1}\frac{\partial^2 \hat{u}}{\partial \hat{y}^2}\right), \tag{3.320}$$

$$\frac{U_1 U_2}{L_1}\hat{u}\frac{\partial \hat{v}}{\partial \hat{x}} + \frac{U_2^2}{L_2}\hat{v}\frac{\partial \hat{v}}{\partial \hat{y}} = -\frac{\partial \tilde{p}}{\partial y} + \left(\frac{L_1^2 U_1 U_2}{L_1^3}\frac{\partial^2 \hat{v}}{\partial \hat{x}^2} + \frac{U_1 U_2}{L_1}\frac{\partial^2 \hat{v}}{\partial \hat{y}^2}\right). \tag{3.321}$$

In the nondimensional continuity equation, Eq. (3.319), two terms should be same order to satisfy the mass continuity. Therefore, we obtain:

$$\frac{U_1}{L_1} \sim \frac{U_2}{L_2}. \tag{3.322}$$

In consideration of Eq. (3.316), Eq. (3.322) indicates:

$$U_1 \gg U_2. \tag{3.323}$$

In Eq. (3.320), two terms of the L.H.S are same order, that is, U_1^2/L_1. As for viscosity term, however, it is obvious for the orders of two terms are different:

$$\frac{L_2^2 U_1^2}{L_1^3} \ll \frac{U_1^2}{L_1}, \tag{3.324}$$

which indicates that the x-derivative viscosity term is negligible compared to the y-derivative viscosity term. Therefore, the leading order in Eq. (3.320) is U_1^2/L_1, and the pressure term is considered to be the same order as the leading terms:

$$\frac{\partial \tilde{p}}{\partial x} \sim \frac{U_1^2}{L_1}. \tag{3.325}$$

In Eq. (3.321), the order of terms is much smaller than the terms in Eq. (3.320), and the leading order in Eq. (3.321) is U_1U_2/L_1 by the y-derivative viscosity term. Therefore, the pressure term is considered to be the order of U_1U_2/L_1 as:

$$\frac{\partial \tilde{p}}{\partial y} \sim \frac{U_1U_2}{L_1}, \tag{3.326}$$

which indicates that the order of the pressure gradient in y-direction is much smaller than that in x-direction (see Eq. 3.325). Namely, the pressure change in y-direction is negligible and the pressure is the function of only x, which is determined by the outer flow, that is, the flow outside of the boundary layer, in which fluid viscosity is negligible. Based on the order comparison result, smaller order terms are omitted and Prandtl boundary layer equation is obtained as:

$$\frac{\partial u}{\partial x} + \frac{\partial v}{\partial y} = 0, \tag{3.327}$$

$$u\frac{\partial u}{\partial x} + v\frac{\partial u}{\partial y} = -\frac{\partial \tilde{p}}{\partial x} + \upsilon\frac{\partial^2 u}{\partial x^2}, \tag{3.328}$$

$$\frac{\partial \tilde{p}}{\partial y} = 0, \tag{3.329}$$

$$u_\infty\frac{\partial u_\infty}{\partial x} = -\frac{\partial \tilde{p}}{\partial x}, \tag{3.330}$$

where u_∞ is the x-velocity in the outer flow. The nonslip boundary condition is applied on the wall, $y = 0$, which is written as:

$$u = v = 0, \tag{3.331}$$

Also, the boundary condition at far field, $y = \infty$, is written as:

$$u = u_\infty, \tag{3.332}$$

$$v = 0. \tag{3.333}$$

Since the boundary layer equation is still nonlinear due to the existence of the advection terms in Eqs. (3.328 and 3.330), an algebraic rigorous solution cannot be obtained. To obtain an approximated solution, Blasius introduced nondimensional variables and transformed the partial differential equation, Eq. (3.187), to an ordinary differential equation. The nondimensional coordinate is defined as:

$$\zeta = y\sqrt{\frac{u_\infty}{\upsilon x}}, \tag{3.334}$$

and the nondimensional velocity is also defined as:

$$u = u_{\infty} \frac{dN(\zeta)}{d\zeta}. \tag{3.335}$$

Then, from Eq. (3.327), the y-velocity is calculated as:

$$v = \frac{1}{2}\sqrt{\frac{\nu u_{\infty}}{x}}\left(\zeta\frac{dN}{d\zeta} - N\right), \tag{3.336}$$

and Eq. (3.328) is transformed to:

$$N\frac{\partial^2 N}{\partial \zeta^2} + 2\frac{\partial^3 N}{\partial \zeta^3} = 0. \tag{3.337}$$

The transformed boundary condition at $\zeta = 0$ is written as:

$$N = 0, \tag{3.338}$$

$$\frac{dN}{d\zeta} = 0. \tag{3.339}$$

Also, the transformed boundary condition at $\zeta = \infty$ is written as:

$$N = 1. \tag{3.340}$$

The transformed boundary layer equation, Eq. (3.328), is solved numerically and the velocity profile in the laminar boundary layer is shown in Fig. 3.42.

3.2.4.3 Turbulent flow

As described in "Transition from laminar flow to turbulent flow (Orr-Sommerfeld equation)" section, a laminar flow changes to transient state and finally a turbulent flow with the increase in Reynolds number. In fact, most of the flows in power and energy machineries are regarded as turbulent flows. The modeling of a turbulent flow is much more difficult than that of a laminar flow, due to the complicated characteristics of the turbulent flow. The turbulence is one of the most difficult problems in the study on fluid dynamics.

Basic theory

As shown in Fig. 3.43, a turbulent flow consists of numerous eddies with various scales, which makes the flow velocity fluctuates randomly in time and space. In other words, a turbulent flow is characterized by the fluctuating flow velocity. Therefore, the modeling of a turbulent flow is highly difficult and in usual, the turbulent flow is modeled based on statistical quantities, for example, averaged velocity.

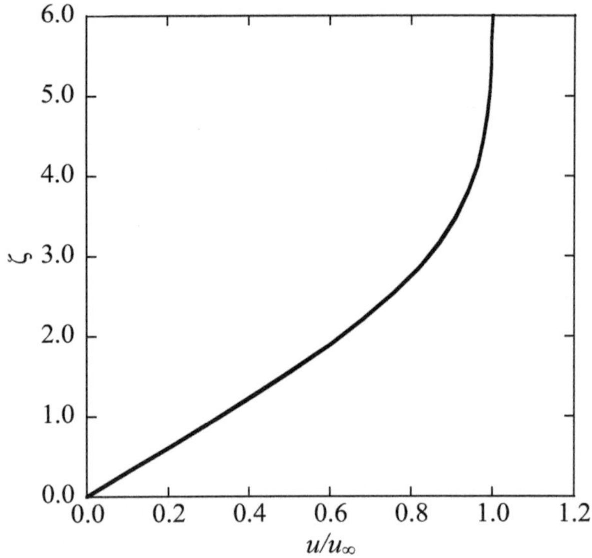

Figure 3.42 Velocity profile in laminar boundary layer.

Here, velocity and pressure are divided into averaged and fluctuation components as:

$$u = \overline{u} + u', \tag{3.341}$$

$$v = \overline{v} + v', \tag{3.342}$$

$$w = \overline{w} + w', \tag{3.343}$$

$$p = \overline{p} + p', \tag{3.344}$$

where the overbar represents the averaged values and the prime mark represents the fluctuating values. It should be noted that the average of the fluctuating value is zero as:

$$\overline{u'} = \overline{v'} = \overline{w'} = \overline{p'} = 0, \tag{3.345}$$

By substituting Eqs. (3.341, 3.342, 3.343, and 3.344) into the three-dimensional Navier-Stokes equation and applying the averaging operation, the equation for turbulent flows is obtained as:

$$\frac{\partial \overline{u}}{\partial x} + \frac{\partial \overline{v}}{\partial y} + \frac{\partial \overline{w}}{\partial y} = 0, \tag{3.346}$$

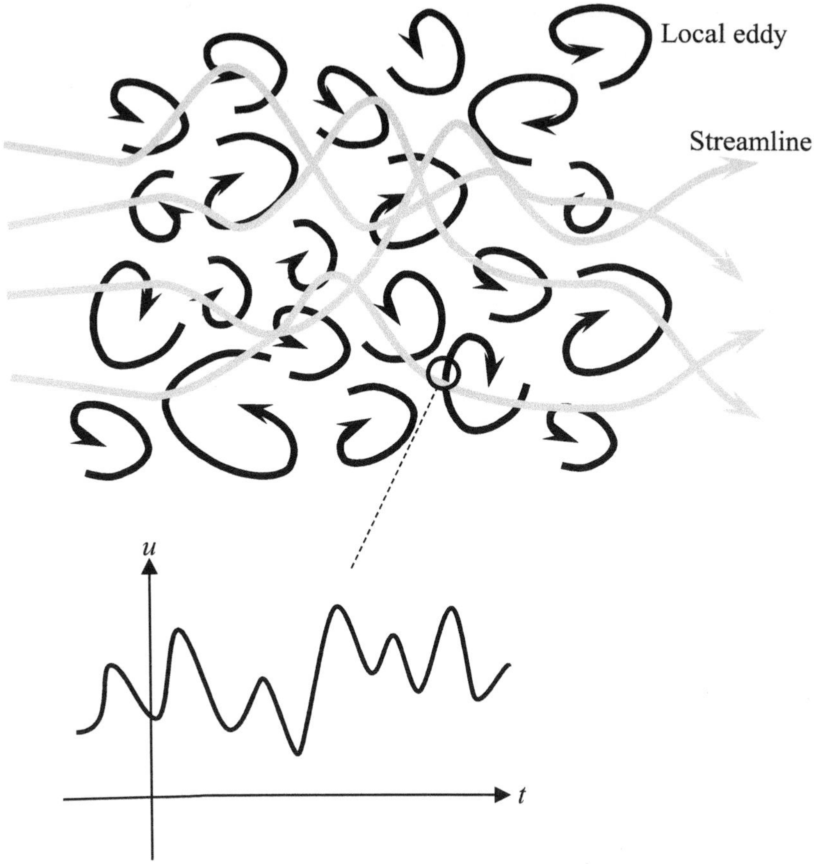

Figure 3.43 Schematic view of turbulent flow.

$$\frac{\partial \overline{u}}{\partial t} + \overline{u}\frac{\partial \overline{u}}{\partial x} + \overline{v}\frac{\partial \overline{u}}{\partial y} + \overline{w}\frac{\partial \overline{u}}{\partial z} = -\frac{\partial \overline{p}}{\partial x} + \frac{\partial}{\partial x}\left(\upsilon\frac{\partial \overline{u}}{\partial x} - \overline{u'u'}\right)$$

$$+ \frac{\partial}{\partial y}\left(\upsilon\frac{\partial \overline{u}}{\partial y} - \overline{u'v'}\right) + \frac{\partial}{\partial z}\left(\upsilon\frac{\partial \overline{u}}{\partial z} - \overline{u'w'}\right) + \overline{f}_x, \tag{3.347}$$

$$\frac{\partial \overline{v}}{\partial t} + \overline{u}\frac{\partial \overline{v}}{\partial x} + \overline{v}\frac{\partial \overline{v}}{\partial y} + \overline{w}\frac{\partial \overline{v}}{\partial z} = -\frac{\partial \overline{p}}{\partial y} + \frac{\partial}{\partial x}\left(\upsilon\frac{\partial \overline{v}}{\partial x} - \overline{v'u'}\right)$$

$$+ \frac{\partial}{\partial y}\left(\upsilon\frac{\partial \overline{v}}{\partial y} - \overline{v'v'}\right) + \frac{\partial}{\partial z}\left(\upsilon\frac{\partial \overline{v}}{\partial z} - \overline{v'w'}\right) + \overline{f}_y, \tag{3.348}$$

$$\frac{\partial \overline{w}}{\partial t} + \overline{u}\frac{\partial \overline{w}}{\partial x} + \overline{v}\frac{\partial \overline{w}}{\partial y} + \overline{w}\frac{\partial \overline{w}}{\partial z} = -\frac{\partial \overline{p}}{\partial z} + \frac{\partial}{\partial x}\left(\upsilon\frac{\partial \overline{w}}{\partial x} - \overline{w'u'}\right)$$

$$+ \frac{\partial}{\partial y}\left(\upsilon\frac{\partial \overline{w}}{\partial y} - \overline{w'v'}\right) + \frac{\partial}{\partial z}\left(\upsilon\frac{\partial \overline{w}}{\partial z} - \overline{w'w'}\right) + \overline{f}_z, \tag{3.349}$$

where the averaged values of the products of fluctuating velocity components, that is, $\overline{u'u'}$, $\overline{v'v'}$, $\overline{w'w'}$, $\overline{u'v'}$,$\overline{v'w'}$, and $\overline{w'u'}$, are called Reynolds stress which originated in the advection terms, that is, nonlinear terms. The difficulty in the turbulence modeling is summarized in the treatment of Reynolds stress. In general, Reynolds stress represents the mixing effect by the fluctuating velocity components, and therefore, Reynolds stress is modeled as an additional viscosity stress, called turbulent viscosity stress. It should be noted, however, Reynolds stress can be much larger than the molecular viscosity stress in developed turbulent flows.

In Prandtl's mixing length theory, which is known as a simple modeling of Reynolds stress, the random movement of a fluid parcel is considered based on the analogy with the mean free path of a molecular in thermodynamics. Namely, a fluid parcel travels the characteristic length, called the mixing length, on average and the momentum of the fluid parcel is transported with the movement. By applying this assumption to the two-dimensional turbulent flow on a wall, the turbulent viscosity stress, τ_t, by the dominant Reynolds stress, $\overline{u'v'}$, is modeled as:

$$\tau_t = -\rho\overline{u'v'} = \rho l^2 \left|\frac{d\overline{u}}{dy}\right| \frac{d\overline{u}}{dy}, \tag{3.350}$$

where u and v are the velocity components parallel and normal to the wall, respectively. y is the coordinate normal to the wall with its origin on the wall. Eq. (3.350) indicates the turbulent viscosity stress is proportional to the square of the mixing length, l. On the other hand, the molecular viscosity stress, τ, is written as:

$$\tau = \rho\upsilon \frac{d\overline{u}}{dy}. \tag{3.351}$$

By the comparison between Eqs. (3.350 and 3.351), the turbulent viscosity, or eddy viscosity, υ_t, is defined as:

$$\upsilon_t = l^2 \left|\frac{d\overline{u}}{dy}\right|. \tag{3.352}$$

Then, Prandtl assumed that the mixing length is proportional to the distance from the wall:

$$l = \kappa y, \tag{3.353}$$

where κ is von Karman constant. Here, it is assumed that Eq. (3.350) can be used to calculate the wall shear stress, τ_w, as:

$$\tau_w = \tau_t = \rho l^2 \left|\frac{d\overline{u}}{dy}\right| \frac{d\overline{u}}{dy}. \tag{3.354}$$

By substituting Eq. (3.353) into Eq. (3.354) and integrating in y-direction gives the averaged velocity profile:

$$\overline{u} = \frac{u^*}{\kappa} lny + const., \tag{3.355}$$

where u^* is the friction velocity defined as:

$$u^* = \sqrt{\frac{|\tau_w|}{\rho}}. \tag{3.356}$$

Energy cascade and Kolmogorov scale

In this subsection, the underlying mechanism of the turbulent flow is discussed. Therefore, it is recommended to skip this subsection for the readers without the interest in such a mechanism.

As described in "Basic theory" section, the essence of the turbulence is the numerous eddies with various scales. This eddy structure is maintained by the energy input by an external force or outer flow to large-scale eddies, that is, nearly the system size-scale eddies, and transferred sequentially to smaller-scale eddies. This mechanism is called the energy cascade, and the heterogeneous anisotropic turbulent characteristics driven by the source of the turbulence, that is, an external force or outer flow, is considered to be lost during the energy cascade. Therefore, at the terminal point of the energy cascade, that is the smallest scale eddy, it can be assumed that the turbulent characteristics is homogeneous and isotropic, which is dependent only on the dynamic viscosity and energy dissipation rate. This hypothesis is proposed by Kolmogorov and the smallest eddy scale is called Kolmogorov microscale. Since Kolmogorov microscale, L_m, is dependent only on the dynamic viscosity, υ, and the energy dissipation rate, ε_t, the Kolmogorov microscale is determined by the dimensional analysis:

$$L_m = \left(\frac{\upsilon^3}{\varepsilon_t}\right)^{1/4}, \tag{3.357}$$

which represents the range of the viscous sublayer where the turbulent kinetic energy from given by larger scale eddies is dissipated by the viscosity effect.

In between the viscous sublayer range and the large eddy scale, called energy containing range, there is an intermediate range, called inertial sublayer range. Since the eddies in the inertial sublayer range is much larger than those in the viscous sublayer range, the viscous dissipation of the turbulent kinetic energy is negligible in this range. On the other hand, the eddy scale in the inertial sublayer range is assumed to be enough small to lost the heterogeneous and anisotropic characteristics and achieve homogeneous isotropic turbulence. Therefore, it is considered that the turbulent kinetic energy is merely transferred to smaller-scale eddies in the inertial sublayer

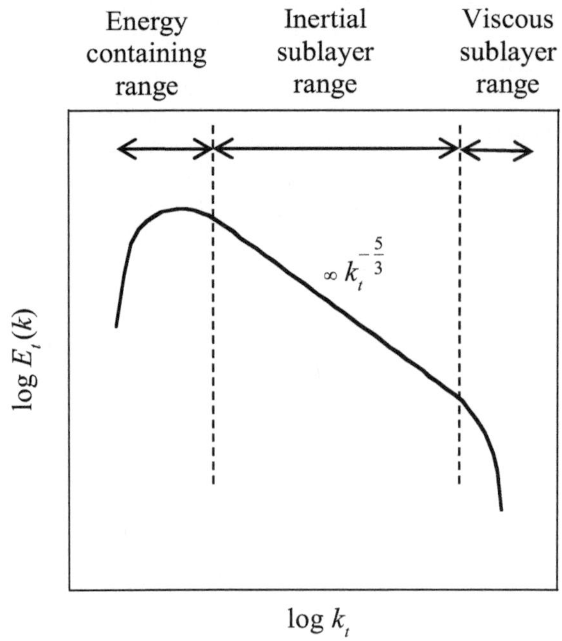

Figure 3.44 Energy cascade in turbulent flow.

range and the eddy characteristics is dependent only on the eddy length scale and the energy dissipation rate, that is, independent from the dynamic viscosity. This hypothesis is also proposed by Kolmogorov and the universal law in the inertial sublayer range is called Kolmogorov's scaling law. Here, we consider the turbulent kinetic energy in wavenumber space, E_t (k). Based on Kolmogorov's hypothesis, the turbulent kinetic energy of each eddy in the inertial sublayer range is dependent only on the eddy length scale and the energy dissipation rate. Therefore, the equation of the turbulent kinetic energy is determined by the dimensional analysis:

$$E_t(k) = c\varepsilon_t^{2/3} k_t^{-5/3}, \tag{3.358}$$

where c is a universal constant and k_t is the wavenumber of eddies. Eq. (3.358) indicates that the turbulent kinetic energy decreases in proportional to the wavenumber to minus five-thirds. In fact, experimental observations indicate that Kolmogorov's scaling law is satisfied for the small-scale eddies in turbulent flows. Based on above discussion, the turbulent kinetic energy profile in wavenumber space is drawn as Fig. 3.44.

Turbulent boundary layer

As shown in Fig. 3.35, in accordance with the development of the flow on a wall, a laminar boundary layer changes to a turbulent boundary layer which is characterized by highly disturbed flow in the boundary layer. From the statistical point of view, the

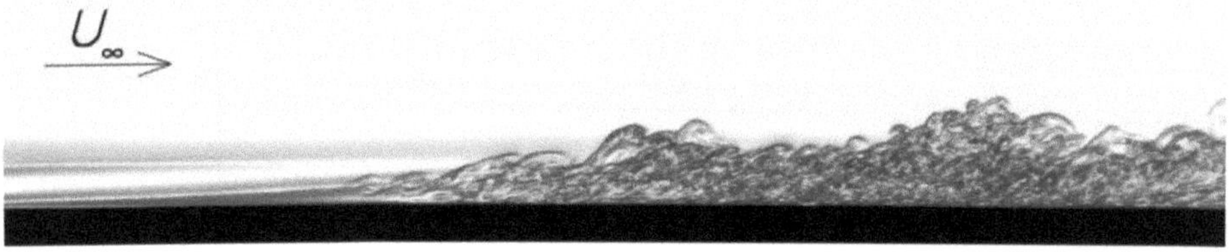

Figure 3.45 Turbulent boundary layer (Andreas Babucke Babucke, 19 September 2008. This file is licensed under the Creative Commons Attribution 3.0 Germany license).

thickness of a turbulent boundary layer is much larger than that of a laminar boundary layer. In addition, the averaged velocity profile parallel to a wall is different between laminar and turbulent boundary layers. Namely, due to the superior mixing effect in a turbulent flow, the averaged velocity tends to be uniform in the turbulent boundary layer, and therefore, the velocity gradient at the wall becomes much larger in the turbulent boundary layer than that in the laminar boundary layer (Fig. 3.45).

It is well known that the averaged velocity profile in a turbulent boundary layer is approximated by linear and logarithmic functions. Such an approximation is called the wall law which is commonly employed to investigate the characteristics of turbulent boundary layers, for example, for the calculation of the wall friction. Here, we consider Cartesian two-dimensional coordinates with x as the coordinate along a wall and y as the coordinate normal to the wall, and the velocity components u and v for each coordinate. Then, two nondimensional parameters, u^+ and y^+, are introduced as:

$$u^+ = \frac{\overline{u}}{u^*}, \tag{3.359}$$

$$y^+ = \frac{yu^*\rho}{\mu}. \tag{3.360}$$

The three layers model gives three different functions of the averaged x-velocity profile depending on y^+. The region with $y^+ < 5$ is called the viscous sublayer in which the viscosity effect is dominant and the flow is laminarized. Therefore, the averaged x-velocity profile in the viscous sublayer is written as:

$$u^+ = y^+, \tag{3.361}$$

which indicates the averaged x-velocity is proportional to the distance from the wall. On the other hand, the region away from the wall, $30 < y^+$, is known as the logarithmic velocity region in which the averaged x-velocity is approximated as a logarithmic function. In consideration of Prandtl's mixing length theory, the logarithmic velocity profile is written as:

$$u^+ = \frac{1}{\kappa} ln y^+ + const. \tag{3.362}$$

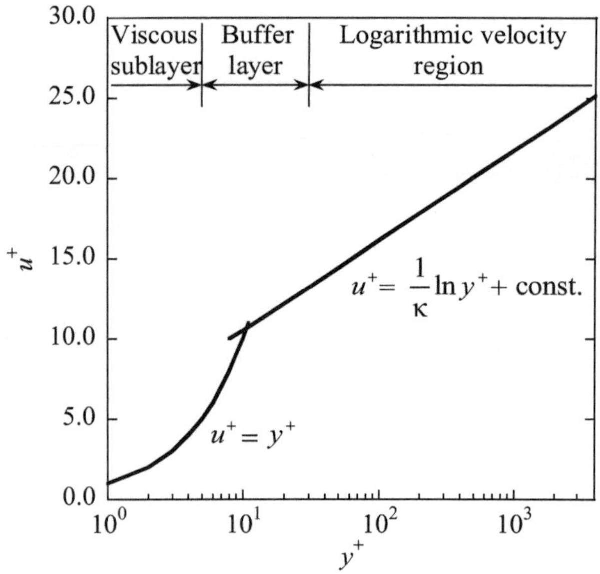

Figure 3.46 Velocity profile in turbulent boundary layer.

The region between the viscous sublayer and the logarithmic velocity region, that is, $5 < y^+ < 30$, there is the buffer layer in which the averaged x-velocity does not satisfy neither linear nor logarithmic profile. In the buffer layer, the averaged x-velocity profile is determined by the interpolation of the liner function in the viscous sublayer and the logarithmic function in the logarithmic velocity region (Fig. 3.46).

Pressure drop in turbulent flow

As described in "Turbulent boundary layer" section, the velocity gradient at the wall is much larger in a turbulent boundary layer compared to a laminar boundary layer, which in turn makes the wall fiction much larger in a turbulent boundary layer.

Here, we consider the fully developed turbulent flow in a circular pipe. In this case, it is known that the averaged velocity profile is well approximated with 1/7-power law:

$$\overline{u} = u_{max}\left(1 - \frac{r}{r_0}\right)^{1/7}, \tag{3.363}$$

where r is the radial coordinate, r_0 is the radius of the circular pipe and u_{max} is the maximum averaged velocity at the center of the pipe. From Eq. (3.361), the cross-sectional averaged velocity is calculated as:

$$u_{ave} \cong 0.82 u_{max}. \tag{3.364}$$

When the viscous sublayer exists near the pipe wall as described in "Turbulent boundary layer" section, the velocity profile in the viscous sublayer is proportional to the distance from the wall surface, and the wall shear stress can be written in the form:

$$\tau_w = c\frac{\rho u_{ave}^2}{2}\left(\frac{\upsilon}{u_{ave}d}\right)^{1/4}. \tag{3.365}$$

where c is the constant. Here, Reynolds number is defined as:

$$Re \equiv \frac{u_{ave}d}{\upsilon}. \tag{3.366}$$

Therefore, in consideration of Eq. (3.298), Darcy friction factor is calculated as:

$$f_d = cRe^{-1/4}. \tag{3.367}$$

This equation is consistent with Blasius equation written as:

$$f_d = 0.3164Re^{-1/4}, \tag{3.368}$$

which is applicable to Reynolds number range of $3\times 10^3 < Re < 10^5$. More accurately, Colebrook equation is employed to calculate Darcy friction factor. Colebrook equation is written as:

$$\frac{1}{\sqrt{f_d}} = -2log\left(\frac{\varepsilon_w}{3.71d} + \frac{2.51}{Re\sqrt{f_d}}\right), \tag{3.369}$$

where ε_w is the averaged relative roughness of the pipe wall surface. Colebrook equation is expressed in an implicit form, that is, Darcy friction factor is obtained by an iterative calculation.

Darcy friction factor of laminar and turbulent flows in a circular pipe is summarized in Moody diagram shown in Fig. 3.47.

3.3 Fundamentals of heat transfer

3.3.1 Introduction to heat transfer

3.3.1.1 Modes of thermal energy transport

Heat transfer must occur when there is a temperature difference in a medium. We consider three different modes of thermal energy transport: conduction, convection, and radiation. Conduction occurs across a stationary medium (such as a solid or fluid)

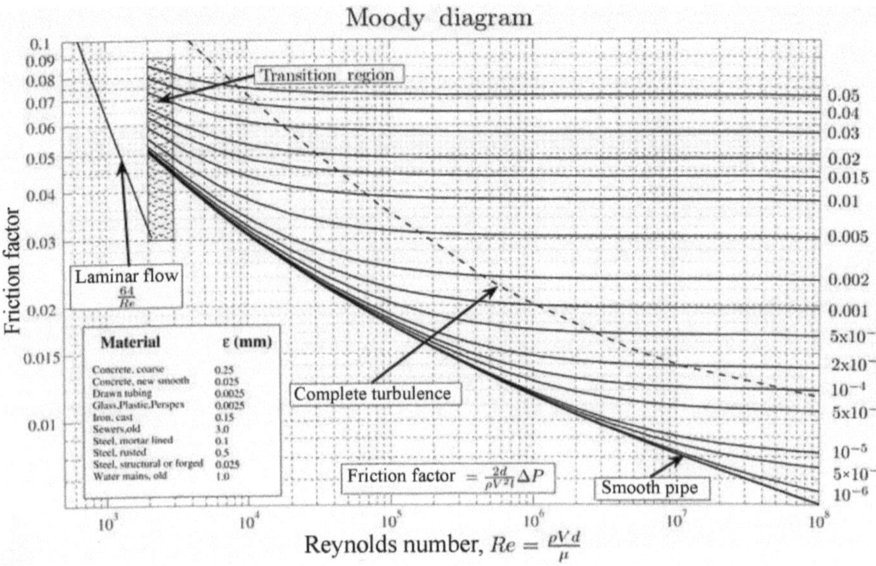

Figure 3.47 Moody diagram. (S Beck and R Collins, University of Sheffield, 24 July 2008. This file is licensed under the Creative Commons Attribution-Share Alike 3.0 Unported license).

when a temperature gradient exists, while convection occurs between a surface and a moving fluid in the case that their temperatures are different. In the case of thermal radiation, all surfaces emit energy according to temperature in the form of electromagnetic waves. Therefore, even in the absence of a medium between the surfaces, net heat transfer due to thermal radiation will occur between surfaces having different temperatures. Note that there are several good texts for heat transfer [4,5].

3.3.1.2 Conduction

It has been empirically shown that thermal energy is transferred by conduction (based on random molecular motion, otherwise referred to as the diffusion of energy) from high temperature to low temperature regions when a temperature gradient exists in an object. For one-dimensional heat transfer through a flat wall such as that presented in Fig. 3.48, for which the temperature distribution is $T(x)$, the amount of heat transfer per unit area and unit time is proportional to the temperature gradient at the location, meaning that

$$\frac{q}{A} \propto \frac{dT}{dx} \tag{3.370}$$

Introducing the proportionality coefficient k yields

$$\frac{q}{A} = -k\frac{dT}{dx} \tag{3.371}$$

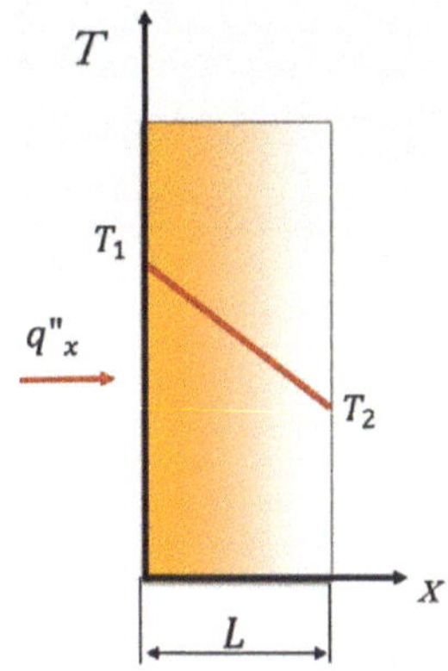

Figure 3.48 The temperature distribution in a flat wall in the case of one-dimensional heat transfer by conduction.

The parameter k is known as the thermal conductivity (W/(m·K)) and is a characteristic of the wall material. The rate equation of Eq. (3.371) is known as Fourier's law, in which the negative sign reflects the transfer of heat in the same direction as that in which the temperature decreases. Assuming steady-state conditions, as are presented in Fig. 3.48 and for which there is a linear temperature distribution, we can express the temperature gradient as

$$\frac{dT}{dx} = \frac{T_2 - T_1}{L} \tag{3.372}$$

and the heat flux is then

$$q''_x = -k\frac{T_2 - T_1}{L} = k\frac{T_1 - T_2}{L} \tag{3.373}$$

3.3.1.3 Convection

Convective heat transfer can proceed through two different mechanisms. One process involves the transfer of energy as a result of random molecular motion (i.e., energy diffusion), while energy can also be transferred through the bulk (i.e., macroscopic) motion of a fluid (otherwise known as advection). In such cases, the term convection is often employed when discussing the sum of the heat transfers for these two processes.

Consider the fluid flow over the heated surface in Fig. 3.49. The interaction between the fluid and surface generates a region in the fluid over which the velocity ranges from zero at the surface to some finite value, referred to as the *velocity boundary layer*. In the case that the fluid and surface are at different temperatures, a zone in the fluid over which there is a temperature variation from the surface appears. This is referred to as the *thermal boundary layer*. Convective heat transfer

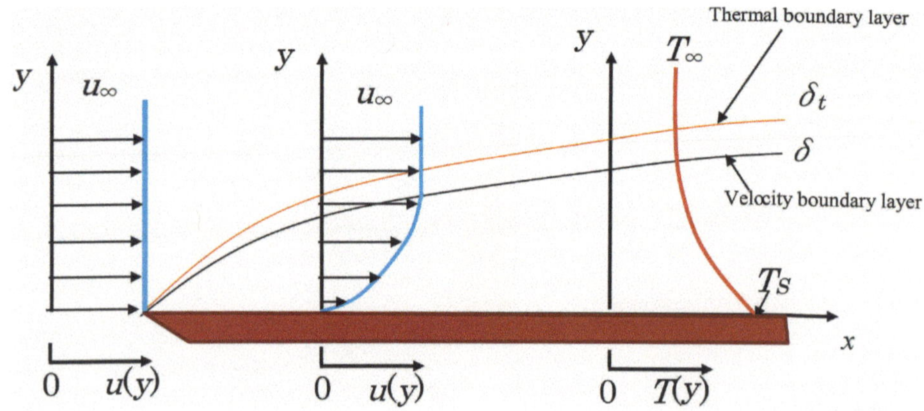

Figure 3.49 The development of velocity and thermal boundary layers via convective heat transfer.

results from the bulk fluid motion in the interior of the boundary layer and randomized molecular motion. Molecular motion (i.e., diffusion) is dominant in proximity to the surface, because the velocity of the fluid is low in this region. Typically, the velocity is zero on the surface, such that the transfer of heat proceeds exclusively by conduction (i.e., the diffusion of energy). The heat transfer is gradually dominated by advection as the fluid velocity increases moving away from the heat transfer surface, and this process is indicated by the temperature gradient in Fig. 3.49. Consequently, diffusion depends on the temperature gradient as represented by the Fourier equation, and the temperature gradient is largest at the heat transfer surface and decreases gradually as the edge of the thermal boundary layer is approached.

Convective heat transfer may be classified as forced convection, free (or natural) convection, boiling, and condensation.

Forced convection is generated by external means, such as a fan or a pump, whereas, in free (or natural) convection, the flow is induced by buoyancy forces, which are due to density differences caused by temperature variations in the fluid. The thermal energy of the fluid is transferred only by sensible heat in the case without phase change. However, in the case of certain convection processes, latent heat exchange occurs. Typically, this exchange of latent heat occurs together with a phase change in the fluid from liquid to vapor or vapor to liquid, and both boiling and condensation are of interest in this regard. Irrespective of the type of heat transfer that occurs in conjunction with convection, the associated rate equation takes the form

$$q'' = h(T_s - T_\infty) \tag{3.374}$$

where q'' is the heat flux for convection (W/m^2) and is correlated with the difference between the fluid and surface temperatures, T_∞ and T_s, respectively. This equation represents the Newtonian cooling law. The variable h (W/(m · K)) is the

Table 3.1 Typical values for the convective heat transfer coefficient.

Modes of thermal energy transport	h (W/(m^{2}K))
Natural convection	
Gases Liquids	1–20 10–500
Forced convection	
Gases Liquids	300–3000 500–8000
Convection with phase change	
Boiling and condensation	5000–100,000

heat transfer coefficient for convection. This coefficient is affected by the boundary conditions, which in turn are determined by the surface geometry, the type of fluid motion, and various transport and thermodynamic characteristics of the fluid. Table 3.1 provides some typical values for this variable.

3.3.1.4 Radiation

Radiative energy is transferred in the form of electromagnetic waves. Thus, while energy transfer due to conduction or convection requires a medium through which the heat energy is transferred, radiation does not. In fact, radiative energy transfer occurs very efficiently in a vacuum. Considering the exchange of radiative energy between a small gray surface with a temperature of Ts and a relatively large surface with a temperature of Tsur completely surrounding the smaller surface, the *net* rate of heat transfer due to radiation *from* the surface is

$$q''_{rad} = \varepsilon\sigma\left(T_S^4 - T_{Sur}^4\right) \tag{3.375}$$

Here, T_S *and* T_{Sur} are the absolute temperatures (K) of the surface and the surroundings, respectively, σ is the Stefan-Boltzmann constant ($5.67 \times 10^{-8} W/(m^{2\cdot} K^4)$), and ε is a radiative property of the surface referred to as the emissivity.

3.3.2 Boiling

3.3.2.1 Pool boiling

Boiling curve

Fig. 3.51 shows a boiling curve that demonstrates the effect of the wall superheat, ΔT_S on the wall heat flux. The superheat is defined as the difference between the liquid saturation temperature at the system pressure and the wall temperature. The associated physical mechanisms can be understood by assessing the boiling curve in this figure, which presents data for water at 1 atm, although other fluids will exhibit similar trends.

1. Free convection
   Free convection occurs when $\Delta T_s \leq \Delta T_{s,A}$. The mode of heat transfer in a gravitational field will be natural convection in the pool. As the wall superheat ΔT_s is increased, bubble inception will eventually occur at point A (referred to as the *onset of nucleate boiling*, ONB).
2. Nucleate boiling
   Nucleate boiling will occur over the range $\Delta T_{s,A} \leq \Delta T_s \leq \Delta T_{s,C}$. There are two possible flow regimes that may be distinguished in nucleate boiling. Within region *A-B, the formation of isolated bubbles* occurs at nucleation sites, and these bubbles move away from the surface. As a result, significant fluid mixing proceeds close to the surface, greatly increasing the values of h and q''_S. Increases in ΔT_s above $\Delta T_{s,B}$ will generate additional sites for nucleation and the enhanced bubble formation will lead to the interaction and coalescence of bubbles. Within region *B-C*, the vapor forms columns or jets as it leaves. The highest heat flux, $q''_{S,C} = q''_{\max}$, is normally known as the critical heat flux (CHF) and there is a CHF greater than 1 MW/m^2 at atmospheric pressure in water. At the highest heat flux, a large quantity of vapor is formed, such that the liquid does not readily wet the surface. To control the heat input rate to the heater, the heater temperature will jump to point E and subsequently stabilize. The heater will then burnout as point E reaches a temperature greater than that the heater components can melt.
   Typically, industrial devices will operate at temperatures within the nucleate boiling range, since the rate of heat transfer is extremely high in this range while the superheating is small. It has been shown that convection coefficients greater than 10^4 W/(m^2 K) are associated with this regime, which are significantly larger than the values for convection in the absence of a phase change.
3. Transition boiling
   The region corresponding to $\Delta T_{s,C} \leq \Delta T_s \leq \Delta T_{s,D}$ is associated with transition boiling. This region involves two boiling modes and shows characteristics of film and nucleate boiling. Transition boiling is highly unstable, and so it is challenging to produce a steady-state during this process, other than by controlling the heater surface temperature.
4. Film boiling
   Film boiling exists for $\Delta T_s \geq \Delta T_{s,D}$. At point D on the boiling curve, the heat flux is at a minimum, $q''_{S,D} = q''_{min}$, and the surface is completely covered by a *vapor blanket*. Heat transfer from the surface to the liquid occurs by both conduction and radiation through the vapor. As the surface temperature is increased, radiation through the vapor film becomes more significant and the heat flux increases with increases in ΔT_s. The configuration of film boiling is shown in Fig. 3.50.

Correlations for pool boiling

The characteristics of the boiling curve and the various physical mechanisms that typify the different regimes demonstrate that numerous heat transfer correlations are associated with boiling. This section therefore examines a few of the more commonly used correlations for film and nucleate boiling.

1. Nucleate pool boiling

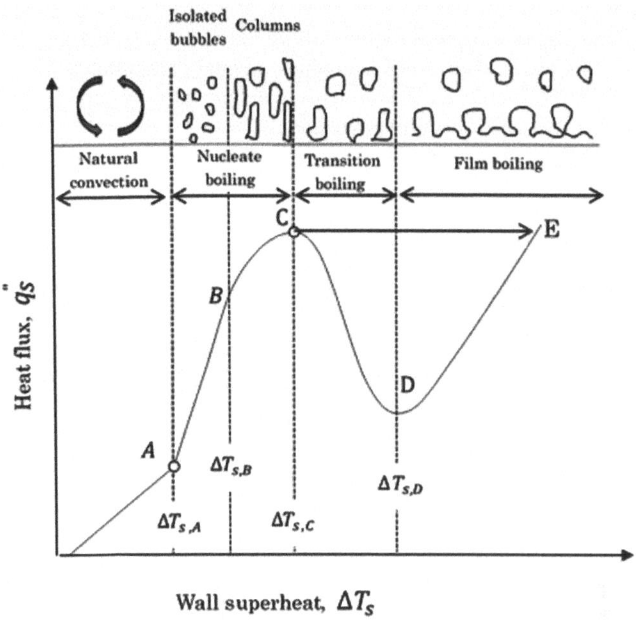

Figure 3.50 The typical boiling curve and the surface heat flux q_S'' as a function of wall superheat.

Table 3.2 Values of $C_{s,f}$ for various surface–fluid combinations.

Fluid	Solid	$C_{s,f}$
Water	Copper	0.013
Water	Brass	0.006
Water	Nickel	0.006
Water	Platinum	0.013
Benzene	Chromium	0.010
Ethanol	Chromium	0.0027
Isopropanol	Copper	0.0025
n-Butanol	Copper	0.003

Rohsenow [6] was the first to develop the following correlation, which is used widely for the nucleate boiling regime

$$q_S'' = \mu_l h_{fg}\left[\frac{g(\rho_l - \rho_v)}{\sigma}\right]^{1/2}\left(\frac{c_{p,l}\Delta T_s}{C_{s,f} h_{fg} Pr_l^n}\right)^3 \quad (3.376)$$

Here, all parameters are those for the liquid phase other than ρ_v, and all should be established at T_{sat}. $C_{s,f}$ is affected by the particular solid–fluid combination, and typical empirical values are provided in Table 3.2. The exponent, n, has a value of 1.0 for water and 1.7 for all other liquids.

2. Critical heat flux

The CHF is an important point on the boiling curve because the temperature increases dramatically at this point, potentially degrading thermal components. Kutateladze [7] and Zuber [8,9] employed dimensional and hydrodynamic stability analyses, respectively, to obtain expressions that can be approximated as

$$q''_{max} = Ch_{fg}\rho_v \left[\frac{\sigma g(\rho_l - \rho_v)}{\rho_v^2}\right]^{1/4} \tag{3.377}$$

The solution to this equation is not dependent of the properties of the surface material but is somewhat affected by the heated surface morphology via the leading constant, C. In the case of a large horizontal cylinder, a sphere, or many other large but finite heated surface morphologies, a C value of $\pi/24 = 0.131$ (the Zuber constant) provides outcomes that agree with experimental results to within 16% [10]. However, for large horizontal plates, a value of $C = 0.149$ gives improved concordance with experimental results. The parameters in Eq. (3.377) are determined at the saturation temperature and it should be noted that several proposed CHF mechanisms for pool boiling have been reviewed previously [11].

Moreover, it is noteworthy that the CHF is significantly affected by pressure, primarily as a result of the pressure dependence of the heat of vaporization and surface tension. Cichelli and Bonilla [12] experimentally established that the maximum CHF appears at one third of the critical pressure, following which the heat flux value drops to zero at the critical pressure, as can be seen in Fig. 3.51. This tendency occurs because the surface tension decreases with pressure and becomes zero at the critical pressure, while the vapor density is extremely low under low-pressure conditions.

3. Film boiling

In the superheated range beyond the temperature at the minimum heat flux point, there is no contact between the liquid phase and the surface because a continuous

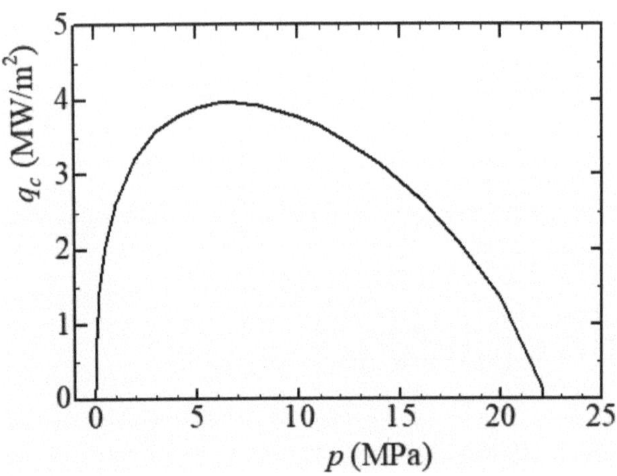

Figure 3.51 The effect of the system pressure at the critical heat flux

vapor film covers the surface. The Nusselt number for film boiling on a cylinder or sphere of diameter D can be obtained from the equation

$$\overline{Nu}_D = \frac{\overline{h}_{conv}D}{k_v} = C\left[\frac{g(\rho_l - \rho_v)h'_{fg}D^3}{\nu_v k_v(T_s - T_{sat})}\right]^{1/4} \tag{3.378}$$

$$h'_{fg} = h_{fg} + 0.8c_{p,v}(T_s - T_{sat}) \tag{3.379}$$

The constant C in this equation has a value of 0.67 for spheres and 0.62 for horizontal cylinders [13]. Here, the corrected latent heat h'_{fg} takes into account the sensible energy necessary to provide specific temperatures inside the vapor blanket at values greater than the saturation temperature. In this case, the vapor characteristics are determined at the pressure of the system and the film temperature, $T_f = (T_s + T_{sat})/2$, while ρ_l and h_{fg} are determined at the saturation temperature. The correlation of Eq. (3.378) can be obtained from laminar film condensation theory because stable vapor film is very similar to laminar film condensation. At high surface temperatures ($T_S \geq 300°C$), there is significant radiative heat transfer across the film of vapor. Because this radiation creates a thicker film, we can no longer assume that convective and radiative heat transfers can simply be summed. Bromley [13] examined film boiling from the exterior surfaces of horizontal tubes and proposed that the overall heat transfer coefficient could be calculated from a transcendental equation having the form

$$\overline{h}^{4/3} = \overline{h}_{conv}^{4/3} + \overline{h}_{rad}\overline{h}^{1/3} \tag{3.380}$$

If $\overline{h}_{rad} < \overline{h}_{conv}$, then a simpler form may be used, written as

$$\overline{h} = \overline{h}_{conv} + 3/4\overline{h}_{rad} \tag{3.381}$$

Here, the effective radiation coefficient $\overline{h}_{rad}$ is expressed as

$$\overline{h}_{rad} = \frac{\varepsilon\sigma\left(T_S^4 - T_{Sat}^4\right)}{T_S - T_{Sat}} \tag{3.382}$$

where ε is the emissivity of the solid, and σ is the Stefan–Boltzmann constant.

3.3.2.2 *Flow boiling*

Introduction to two-phase flow

To evaluate the phase change equipment, such as heat exchanger, evaporator and boiler, the fundamental knowledge of two-phase flow is quite important. Among the two-phase region, the ratio between liquid and vapor is changing, and as the results, the heat-transfer and flow characteristics are changing with the transition of the two-phase flow condition. The detailed treatments will be found in several papers and literatures [9,14], thus fundamental knowledge and general tendency to understand the two-phase flow will be explained in this chapter.

Fig. 3.52 is schematic image of heat transfer and flow characteristics of convective boiling under relatively low heat flux condition in vertical upward flow channel. The understanding of the image of this figure may be most important subject to treat the phase change equipment, and it may be main purpose of this chapter. In the channel, flow regime changes as follows; single phase liquid, bubbly flow, slug flow, churn flow, annular flow, mist flow and single vapor phase. Strictly speaking, these flow structures influence on the detail mechanisms of heat transfer and flow characteristic, but the fine treatment of those difference can be neglected in rough evaluation. In this chapter, more general image will be introduced mainly based on the simple model.

Among the several parameters, quality x should be most important parameter to treat the boiling two-phase flow, and which is the vapor flow ratio against the total flow. This parameter can be expressed by using mass flow rate M(kg/s) or mass flux G(kg/m^{2}s) as follows,

$$\mathrm{x} = \frac{M_G}{M_G + M_L} = \frac{G_G}{G_G + G_L} \tag{3.383}$$

In the actual boiling equipment, the direct measurement of amount of vapor is quite difficult, thus the thermal equilibrium quality is more widely used instead of quality. The thermal equilibrium quality can be expressed as follow by using enthalpy,

$$x_{eq} = \frac{i - i_{SAT} - (i_{in} - i_{SAT})}{H_{LG}} \tag{3.384}$$

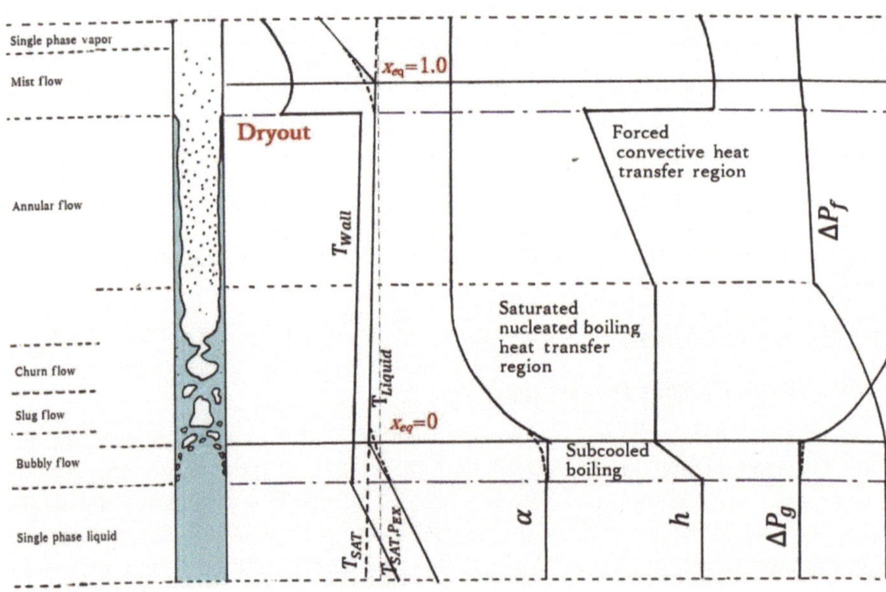

Figure 3.52 Schematic image of heat transfer and flow characteristics of convective boiling.

where i_{sat} is saturation enthalpy and i_{in} is inlet enthalpy and H_{LG} is latent heat of vaporization. This parameter is also can be calculated by using next equation, under constant heat flux condition with flow area A and perimeter length P_h,

$$x_{eq} = \frac{\frac{\int qP_h dz}{GA} - c_{PL}(T_{SAT} - T_{in})}{H_{LG}} \tag{3.385}$$

The quality x takes the value between 0 and 1, but the thermal equilibrium quality x_{eq} can take more wide range. $x_{eq} < 0$ corresponds to the subcooled condition and $x_{eq} > 1$ corresponds to the superheated condition. In Fig. 3.52, the line of $x_{eq} = 1$ and $x_{eq} = 0$ are drawn, but this condition does not agree with the transient condition to single phase flow regions. To compensate these thermal nonequilibrium tendency, the actual quality is also proposed [15,16], but it also can be neglected in the rough order estimation of the characteristics of heat transfer or pressure difference of equipment.

Another unique parameter of the two-phase flow, void fraction, is also drawn in Fig. 3.52. As the definition, void fraction expresses the vapor phase area ratio against the total flow area, thus 0 corresponds to the single phase liquid, and 1 corresponds to the single phase vapor. The measuring of void fraction in actual system is slightly difficult, and the meaning of the void fraction will be changed on the basis on the measuring procedure. These discussions will be important in the strict evaluation, but it also can be negligible in this stage.

Several models of two-phase flow have been proposed to express the relationship between liquid phase and vapor phases, and also so many empirical correlations have been used. But to understand the fundamental feature, simple homogeneous model is quite reasonable. In homogeneous model, the void fraction can be calculated by quality as follows,

$$\alpha = \frac{1}{1 + (\rho_G/\rho_L)(1 - x/x)} \tag{3.386}$$

Fig. 3.53 is the calculation results of void fraction against the quality. In the calculation, physical properties of water at 0.1, 1.0, 7, and 15 MPa are used. As shown in this figure, void fraction drastically increases during low quality region, it means the annular flow regime is dominant flow regime in the most of conventional systems. As mentioned before, the homogeneous model is simple and useful mode to rough estimation, but it must be used with the understanding of the characteristics of model. As widely known, in the case of homogeneous model, void fraction is overestimated than the actual value especially under low-pressure condition.

Heat transfer coefficient

1. Single phase flow

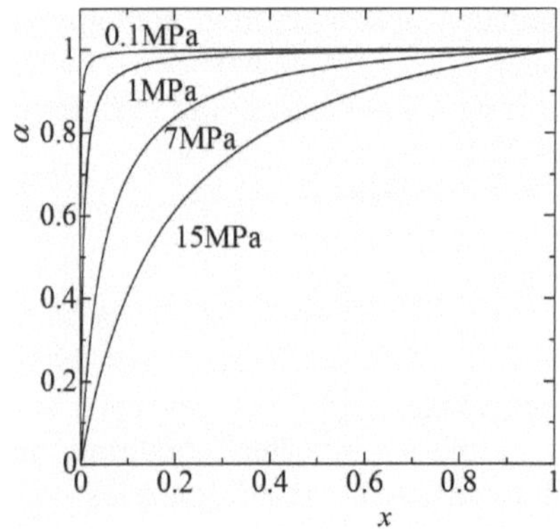

Figure 3.53 Void fraction against quality.

The subcooled liquid enters into the heated channel keeps the single phase condition until start of the ONB. The heat transfer coefficient in this region can be expressed by a simple single phase correlation [4], such as the Dittus-Boelter equation so on,

$$\mathrm{Nu} = \frac{h_L D}{\lambda_L} = 0.023 Re^{0.8} Pr^{0.4} \tag{3.387}$$

In this region heat transfer coefficient keeps the almost constant value, and the wall temperature monotonically increases with the increasing of the water temperature by heating as shown in Fig. 3.52.

2. Two-phase flow

The start of two-phase region is the ONB, and it is caused by the non-thermal equilibrium condition. Consequently, the liquid temperature takes the lower value than saturated temperature, but the wall temperature takes almost constant value by keeping the constant temperature difference with saturation temperature. At the initiation of boiling, superheated condition will be required, thus after start of ONB, the wall temperature slightly decreases owing to the increasing of the heat transfer coefficient. As the engineering usage the heat transfer coefficient during subcooled boiling region can be calculated by the interpolation between the single phase and the saturated nucleated boiling heat transfer. But for more rough estimation, this region can be negligible in the most of conventional systems.

After this subcooled boiling condition, heat transfer region enter to the saturated nucleated boiling heat transfer region. In this area heat transfer mechanism is dominated by the saturated nucleated boiling, thus heat transfer coefficient takes the constant value imaginary. With the increasing of the quality, the flow regime becomes to the annular flow regime, and the dominant heat transfer mechanism becomes the forced convection of core vapor flow. In this region the liquid film is considered as the heat restriction, and the heat through the liquid film is removed by the phase change at the interface of annular liquid and core vapor region. Thus the heat transfer coefficient slightly increases with the increase of quality. Finally, the liquid film completely evaporates, and dryout occurs. Just before the dryout condition owing to the decreasing of the liquid film thickness, the heat transfer coefficient slightly increases and the wall temperature decreases. This temperature movement is used for the detecting of dryout condition. Among these conditions, the liquid temperature keeps the saturation temperature owing to the occurrence of phase change. Strictly speaking, this saturation temperature is not constant value owing to the influence of the pressure drop of tube. The wall temperature takes the slightly higher value to keep the temperature difference with the saturation temperature. The understanding of the wall temperature characteristic will be important in the designing of the heat exchanger and to estimate the pinch point.

The typical correlation includes the expression of the transition of saturate nucleated boiling heat transfer region to forced convective heat transfer region is Schrock-Grossman [17], as follows,

$$\mathrm{h}_{TP} = h_L \times 0.739 \left[Bo \times 10^4 + 1.5 \left(\frac{1}{\chi_{tt}} \right)^{2/3} \right] \tag{3.388}$$

where boiling number $\mathrm{Bo} = q/GH_{LG}$, and they used next value on the basis on the Lockhart-Martinelli parameter as follow,

$$\chi_{tt}^2 = \left(\frac{1-x}{x} \right)^{1.8} \left(\frac{\rho_g}{\rho_L} \right) \left(\frac{\mu_g}{\mu_L} \right)^{-0.2}$$

$$\mathrm{Nu} = \frac{h_L D}{\lambda_L} = 0.023 Re^{0.8} Pr^{1/3}$$

In the Eq. (3.388), the first term corresponds to the nucleated boiling and second term corresponds to the forced convection.

These heat transfer coefficient images with the influence of the heat flux magnitude are shown in Fig. 3.54. In the region where the forced convective boiling is dominant, the heat transfer coefficient is mainly decided by the vapor velocity, thus it becomes the similar value if it has same quality. On the contrary, the region where the nucleated boiling is dominant, the influence of the heat flux appeared clearly, it is because the heat transfer coefficient in this region depend on the magnitude of the activation of nucleation site. Although these two regions show different characteristics, both regions have high heat transfer coefficients owing

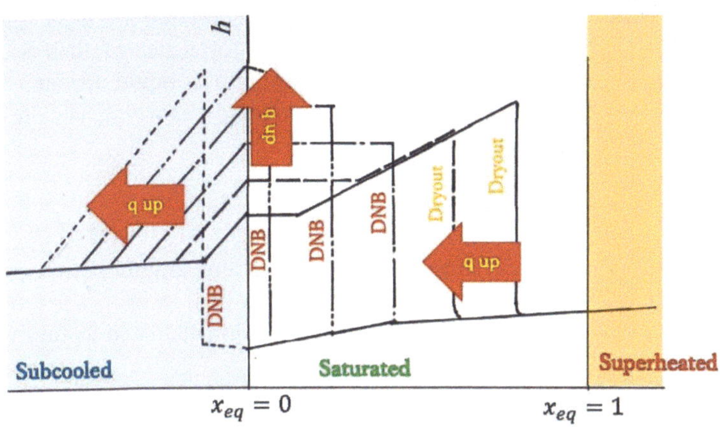

Figure 3.54 The influence of the heat flux on the heat transfer characteristics.

to the phase change. After these high heat transfer regions, the heat transfer coefficient drastically decreases. These conditions are classified into two mechanisms, that is, dryout and departure from nucleate boiling (DNB), on the basis of the heat transfer mechanism at the transition condition. In the case of the dryout, the transition is caused by the evaporation of liquid film. Thus, even after dryout, vapor phase may still keep the high heat transfer coefficient by forced convection. By increasing the heat flux, liquid film breakout will occur, then dryout condition moves to the lower quality condition. DNB mainly occurs owing to the heat transfer restriction by vapor under low quality condition. Consequently, the increasing of the vapor phase velocity is not enough, then this condition may cause the physical burnout.

3. Postdryout

Even after the dryout condition, liquid droplet still exists. In this postdryout region, heat transfer mechanism mainly consists of two steps. Firstly, heat is transferred from wall to vapor, and next, heat is transferred from vapor to liquid droplet. In this mechanism, vapor phase becomes superheated condition, thus liquid may exist even under $x_{eq} > 1$.

Several correlations of this region have also been reported [9], so far. Roko calculated the heat transfer coefficient numerically, and after that they reported the simplified correlations. In this section, Roko's simplified model is introduced briefly.

Roko [16] proposed the heat transfer coefficient of postdryout regions as follows,

$$\begin{aligned} & x_{eq} \leq 1.0 \\ & h = \frac{k h_{w,G}}{1 + \frac{h_{w,G}}{q} \frac{H_{LG}}{c_{P_e}} \frac{x_{eq} - x_a}{x_a}} \\ & x_{eq} > 1.0 \\ & h = \frac{k h_{w,G}}{1 + \frac{h_{w,G}}{q} \frac{H_{LG}}{c_{P_e}} \frac{x_{eq}(1 - x_a)}{x_a}} \end{aligned} \tag{3.389}$$

where physical properties of thermal equilibrium temperature are used, and k is the tuning factor of simplification. Roko used $k = 0.9$ for water of their experiment, Umekawa [18] used $k = 1.0$ for natural circulation loop of liquid nitrogen.

where heat transfer coefficient between the wall and vapor $h_{w,G}$ is as follows,

$$h_{w,G} = 0.0073 \frac{k_f}{D} Re_f^{0.886} Pr_f^{0.61} x_a^{0.886}$$

In this calculation, physical properties of average temperature between wall temperature and thermal equilibrium temperature is used.

x_a is the actual quality in the postdryout region, and calculated as follows,

$$x_a = \frac{(1 - x_{CHF}) x_{eq} + (x_0 - 1) x_{CHF}}{x_0 - x_{CHF}} \tag{3.390}$$

In this equation, x_0 is the quality where the entrainment mist completely evaporated.

$$x_0 = 1 + 1.09F^{-0.536}$$

$$F = 2.08 \times 10^4 \left(\frac{p}{p_{CR}}\right)^{1.6} \left(\frac{G}{1000}\right)^{2.5} x_{CHF}^2 \frac{D}{10^{-5} q}$$

where x_{CHF} is the dryout quality as follows,

$$\frac{1 - x_{CHF}}{x_{CHF}} = \left(12.01 \left(\frac{p}{p_{CR}}\right)^2 - 5.628 \left(\frac{p}{p_{CR}}\right) + 0.8623\right) \left(\frac{G}{1000}\right)^{1.2}$$

Although Roko's simplified correlation show a good agreement with experimental data, simple Dittus-Boelter's equation of single vapor phase can be used for more rough estimation to understand the general tendency of the equipment

Pressure drop

In this section the pressure drop of two-phase flow will be explained by using the fundamental components of pressure drop, i.e. the frictional term, the gravity term and acceleration term, without the treatment of detail mechanism.

1. Single phase flow

   Components of pressure drop in single phase flow are gravity term and friction term. The gravity term in vertical flow condition can be calculated as follows,

$$\Delta P_{gL} = \rho_L g dz_L \tag{3.391}$$

   And, the friction term can be expressed by using Darcy-Weisbach with Blasiuis's equation, as follows,

$$\Delta P_{fL} = \lambda_L \frac{dZ_L}{D} \frac{G^2}{2\rho_L} \tag{3.392}$$

$$\lambda_L = 0.3164 Re_L^{-0.25}$$

   As the results in single liquid-phase region, the gravity terms become dominant, and frictional term is relatively small as shown in Fig. 3.52

2. Two-phase region

   By the occurrence of the boiling, the gravity term drastically decreases as same as the tendency of void fraction. And also owing to the increasing of the vapor velocity the friction term increases as shown in Fig. 3.52. Moreover in this region the acceleration term must be considered. These characteristics can be expressed simply by using homogeneous model

$$\begin{aligned}\Delta P_{gTP} &= \int_0^{z_{TP}} \rho_{TP} g dz \\ &= \int_0^{z_{TP}} \frac{1}{v_{TP}} g dz \\ &= ln\left(\frac{v_{EX}}{v_L}\right)\frac{z_{TP}}{v_{EX}-v_L} g\end{aligned} \tag{3.393}$$

$$\begin{aligned}\Delta P_{fTP} &= \int_0^{z_{TP}} \lambda_{TP}\frac{1}{D}\frac{G^2}{2\rho_{TP}} dz \\ &= \lambda_{TP}\frac{G^2}{2D}\int_0^{z_{TP}} v_{TP} dz \\ &= \lambda_{TP}\frac{G^2}{2D}\int_0^{z_{TP}} (v_L + x v_{LG}) dz \\ &= \lambda_{TP}\frac{G^2 z_{TP}}{2D}\left(v_L + \frac{x}{2} v_{LG}\right)\end{aligned} \tag{3.394}$$

$$\begin{aligned}\Delta P_{aTP} &= \rho_{TP} u_{TP}^2 - \rho_L u_L^2 \\ &= G^2 (v_{TP} - v_L)\end{aligned} \tag{3.395}$$

In the estimation of λ_{TP}, liquid viscosity is widely used.

Critical heat flux

The drastic reduction of heat transfer coefficient in Fig. 3.54 is generally called as CHF, and it normally decides the limitation of the operating condition of boiling equipment. CHF is originally the term of the heat flux itself at the transition of heat transfer, but it is widely used as the phenomena itself. CHF includes the DNB which caused in low quality condition, and dryout which caused in high quality condition. As mentioned in former section, dryout is simple mechanism of liquid film dryout is mainly observed in annular flow regime, thus, it occurred in high quality condition. This condition can be considered as the total balance of the entrainment, deposition and evaporation with liquid film breakout condition. By increasing of the heat flux, the condition moves to the lower quality condition mainly owing to the changing of breakout condition. Several correlations of CHF have been proposed, and one of them is Katto's correlation [19,20]. As the extension of the dryout, some kind of DNB can be explained, but DNB at subcooled region is caused by the quite different mechanism. DNB in the subcooled region is traditionally explained by using two kinds of models shown in Fig. 3.55 [21,22], but both models can be considered as the restriction of the heat transfer by vapor phase. The criteria of DNB have not been cleared enough so far, but in any case,

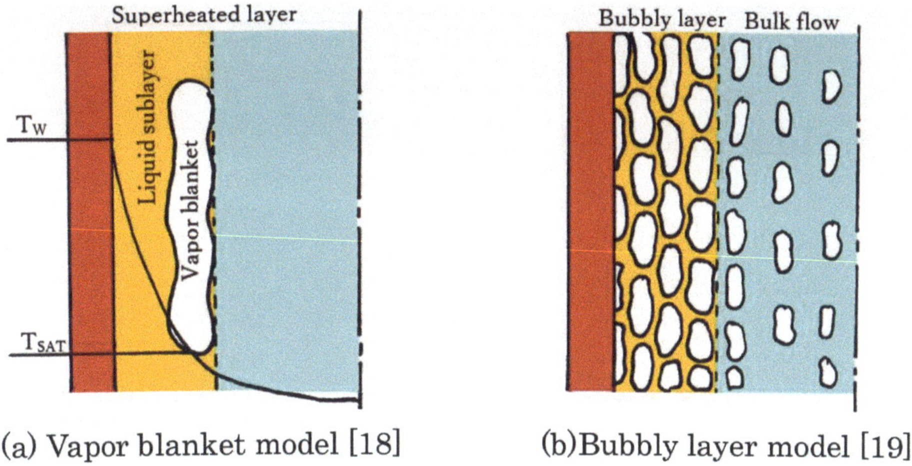

Figure 3.55 DNB models. (A) Vapor blanket model [21]. (B) Bubbly layer model [22].

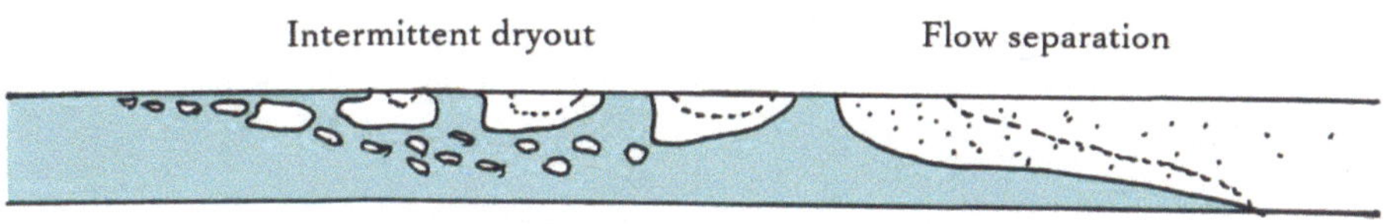

Figure 3.56 Schematic image of horizontal boiling.

DNB is occurred under high heat flux condition with low vapor velocity. Thus, DNB phenomena easily cause the physical burnout. As the technical term, CHF is also called as boiling transition, boiling crisis, and burnout, but it can be considered as the similar phenomena.

Actual phenomenon

In this section, two-phase flow is explained by using the vertical forced convection under relatively low heat flux condition. This condition is relatively easily explained, but the operating condition of actual equipment is not limited in these simple case. For example, kinds of working fluids, size of the tubes, orientation of tubes, magnitude of the gravity are widely extended. Fig. 3.56 is the schematic image of horizontal boiling channel in similar condition with Fig. 3.52. In this case the flow separation occurred owing to the gravity and it cause the non-uniform heat transfer condition, and it also caused the intermittent dryout. Also Fig. 3.57 is the example of the CHF under downward flow condition with the data of upward flow. As easily understand the quite different characteristics can be observed. As mentioned previously, the main subject of this chapter may be the understanding of the image of Fig. 3.52 to treat the boiling equipment. This image is very important to design the boiling equipment, but it must be

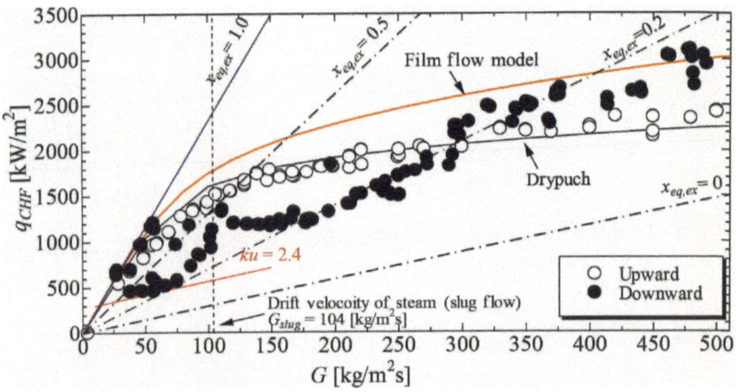

Figure 3.57 Comparison of critical heat flux under downward/upward flow.

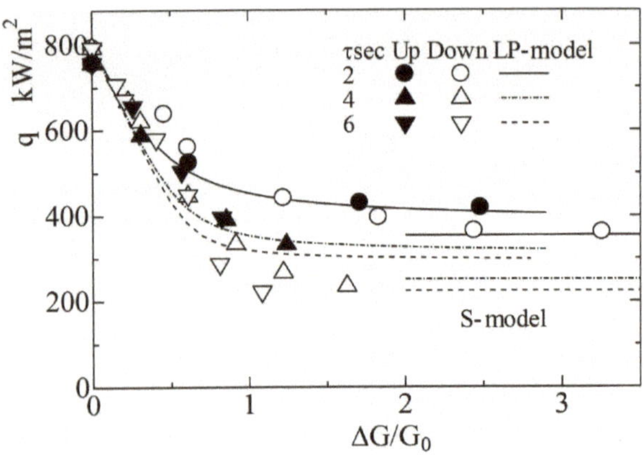

Figure 3.58 Influence of the flow oscillation on CHF.

understood that the fixed image cannot be adapted to the actual equipment. To extend the new application, with the general image, the suitable literature of knowledge must be used.

Recently, in the case of the conventional boiler, owing to the increasing of the unstable sustainable energy source, it operates under wide range including transient condition. Under these condition, even in the case of the vertical flow, flow oscillation, such as density wave oscillation, may occur. Most of the knowledge is obtained under the stable condition, and under oscillatory condition the quite different feature can occur. Fig. 3.58 expresses the influence of the flow oscillation on CHF, and drastic decreasing of CHF can be observed. Moreover under these condition periodical dryout is widely observed and it easily caused the thermal stress fatigue even under enough heat transfer condition.

3.3.3 Condensation

Condensation will proceed in the case that a vapor's temperature is lowered to a value less than its saturation temperature. In practice, this process is frequently associated with contact between a cold surface and a vapor. Upon such contact, the vapor's latent energy is transferred to the surface to produce a condensate. Two types of condensation are normally observed, based on the surface conditions. In the main type, a film of liquid coats the condensing surface and then flows off continually as a result of gravity. This co-called *film condensation* generally occurs in conjunction with an uncontaminated surface. In contrast, the presence on the surface of any substance that limits wetting can generate *dropwise condensation*, in which drops form in various cavities on the surface after which they grow and agglomerate. Typically, greater than 90% of the cold surface can be coated with such drops, with sizes from only several micrometers to visible agglomerations. Such droplets then escape from the surface in response to gravitational force. Both dropwise and film condensation of steam on a horizontal cylindrical surface are pictured in Fig. 3.59. In the case of both droplet and film condensation, liquid film generated by the condensation inhibits heat transfer from the vapor to the surface. Because this effect becomes stronger as the condensate thickness increases along the flow direction, horizontal cylinders or short vertical walls are preferred if film condensation is a possibility. Consequently, the majority of condensers comprise bundles of horizontal tubes with a liquid coolant flow through them, and circulation of the vapor being condensed around the tubes. Higher rates of condensation and heat transfer can be obtained with droplet rather than film formation. In the case of dropwise condensation, the majority of heat transfer proceeds via smaller droplets

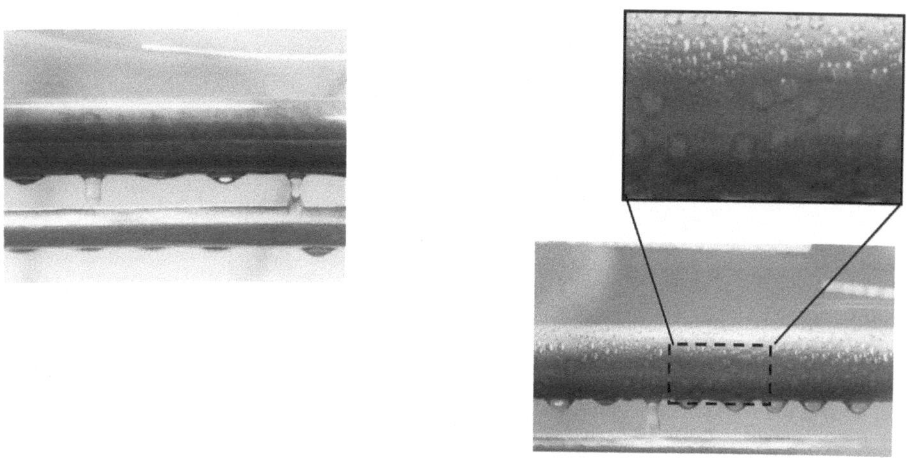

(*a*) Film condensation (*b*) Dropwise condensation

Figure 3.59 Condensation on a horizontal cylinder. (A) Film condensation. (B) Dropwise condensation.

(below 100 μm in diameter). Heat transfer rates at least ten times those possible with film condensation may be obtained. Thus, surface coatings are typically employed to decrease wetting and promote dropwise condensation. These coatings include Teflon, waxes, fatty acids and silicones. Even so, these materials are slowly degraded via fouling, physical abrasion or oxidation so that film condensation will eventually appear. Consequently, while dropwise condensation is preferable in industrial scenarios, it is challenging to maintain. As such, and also since film condensation provides smaller convective heat transfer coefficients, the design of a condenser will typically assume film condensation.

3.3.3.1 Laminar film condensation

Fig. 3.60 demonstrates the development of velocity and thermal boundary layers related to film condensation. The condensed film subsequently moves down due to gravity. Both the liquid film thickness and the mass flow rate m of the condensate are increased as x increases since condensation continually proceeds at the liquid–vapor interface, which has a temperature of T_{sat}. Thus, heat moves from the interface through the interior of the film toward the plate surface, for which $T_s < T_{sat}$. The vapor is typically in a superheated state ($T_v > T_{sat}$) and may form a

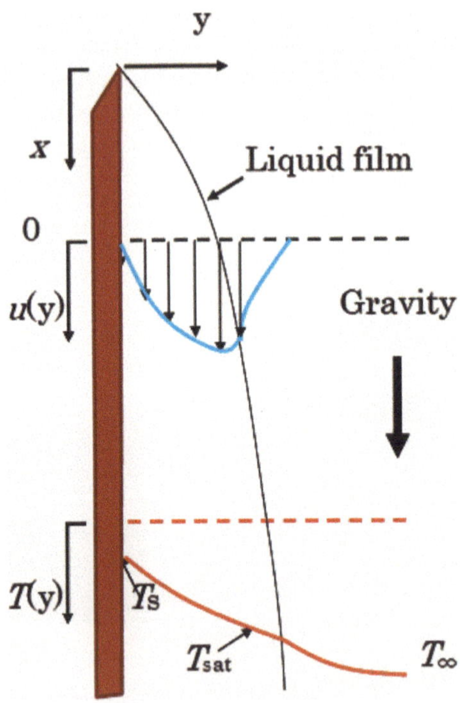

Figure 3.60 Boundary layer effects related to film condensation on a vertical surface.

component of a mixture in which several noncondensable gases are present. In addition, a finite shear stress is present at the interface between the liquid and vapor so that a velocity gradient appears in the film and in the vapor [23,24]. Useful analysis have been reported by Nusselt [25] making several assumptions.

The average Nusselt number obtained from the analysis then has the form

$$\overline{Nu}_L = \frac{\overline{h}_L L}{k_l} = 0.943\left[\frac{g(\rho_l-\rho_v)h'_{fg}L^3}{\nu_l k_l(T_{sat}-T_s)}\right]^{1/4} \quad (3.396)$$

$$h'_{fg} = h_{fg} + 0.68c_{p,l}(T_{sat} - T_s) \quad (3.397)$$

All liquid properties should be evaluated at the film temperature $T_f = (T_s + T_{sat})/2$. The vapor density ρ_v and latent heat of vaporization h_{fg} should be evaluated at T_{sat}.

3.3.3.2 Turbulent film condensation

For the vertical surface, the criterion for transition can be written based on the Reynolds number, as

$$Re_\delta \equiv \frac{4\Gamma}{\mu_l} = \frac{4g\rho_l(\rho_l-\rho_v)\delta^3}{3\mu_l{}^2} \quad (3.398)$$

$$\delta(x) = \left[\frac{4k_l\mu_l(T_{sat}-T_s)x}{g\rho_l(\rho_l-\rho_v)h_{fg}}\right]^{1/4} \quad (3.399)$$

For $Re_\delta \le 30$, the film is laminar without wave, the film is laminar with waves at $30 \le Re_\delta \le 1800$. The transition from laminar to turbulent flow occurs at $Re_\delta \approx 1800$ completely.

The average modified Nusselt number may be expressed for various ranges of Reynolds number in each regime as in the equations

$$\overline{Nu}_L = \frac{\overline{h}_L\left(\nu_l^2/g\right)^{1/3}}{k_l} = 1.47Re_\delta{}^{-1/3} \quad Re_\delta \le 30 \quad (3.400)$$

$$\overline{Nu}_L = \frac{\overline{h}_L\left(\nu_l^2/g\right)^{1/3}}{k_l} = \frac{Re_\delta}{1.08Re_\delta{}^{1.22} - 5.2} \quad 30 \le Re_\delta \le 1800 \quad (3.401)$$

$$\overline{Nu}_L = \frac{\overline{h}_L\left(\nu_l^2/g\right)^{1/3}}{k_l} = \frac{Re_\delta}{8750 + 58Pr_l{}^{-0.5}(Re_\delta{}^{0.75} - 253)} \quad Re_\delta \ge 1800, Pr_l \ge 1 \quad (3.402)$$

3.3.3.3 Dropwise condensation

The heat transfer coefficients associated with dropwise condensation are typically at least ten times those related to film condensation. Thus, in the case of a heat exchanger based on the former process, the thermal resistances attributable to condensation may be less than those for other reasons, and so precise mathematical expressions for the condensation process are not needed. Among the many surface—fluid systems that have been studied [26], the majority of the data are for steam condensation on copper surfaces that promote condensation (i.e., surfaces on which wetting is inhibited) and the results can be summarized as [27].

$$\overline{h}_{dc} = 51,104 + 2044T_{sat}(°C) \quad 22°C \leq T_{sat} \leq 100°C \tag{3.403}$$

$$\overline{h}_{dc} = 255,510 \quad 100°C \leq T_{sat} \tag{3.404}$$

3.4 Fundamentals of combustion

3.4.1 Fuel

Fossil fuels are a general term for hydrocarbon-based fuels formed by fossilization of buried ancient organisms under the geothermal pressure. For discussing the fossil fuels, it is important to distinguish between resources and reserves. Resources are the total amount of fossil fuel estimated to be underground, including the amount already produced. Resources include the discovered and undiscovered fuels, economically recoverable, or nonrecoverable. On the other hand, reserves are the amount of fossil fuels that regarded as a part of the resources, has been discovered, and are recoverable economically and technically minable at present or in the near future.

The McKelvey box proposed in 1972 shows clearly the relationship between the resources and reserves according to their geologic certainty of their occurrence in the abscissa and the economic feasibility of recovery in the ordinate. Fig. 3.61 shows the McKelvey box. The outer boundaries of the box shows the total amount of the resources. The right side of the box represents the undiscovered fuels referred to as prospective resources. Reserves are located at the economically recoverable portion of the discovered resources. Reserves is positioned as part of resources. The lower left corner of the box represents the resources that is noneconomically feasible to extract.

Reserves can be categorized into proved reserves, probable reserves, and possible reserves according to the degree of certainty. These categorization is reflected the degree of certainty of recovery and uncertainty of evaluation. Reserves-to-production ratio, obtained by dividing the proven reserves by the production rate in the last year, indicates the remaining lifespan of each fossil fuels for stable supply. Proven reserves vary from year to year due to exploration efforts and technical and

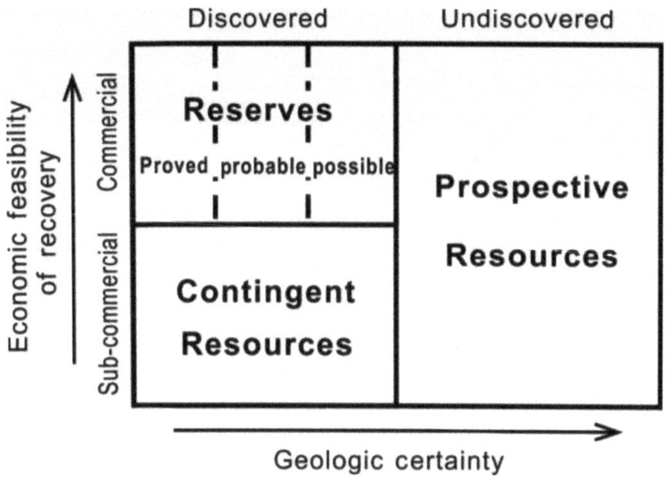

Figure 3.61 McKelvey box.

Table 3.3 Components of gaseous fuels and higher heating value.

Gaseous fuel	Main components	Higher heating value (MJ/m_N^3)
Natural gas	CH_4, C_2H_6, C_3H_8, C_4H_{10}	39–45
Liquefied petroleum gas	C_3H_8, C_4H_{10}	84–125
Coal gas	H_2, CH_4, CO	20–23
Blast furnace gas	N_2, CO, CO_2	3–4
Producer gas	N_2, CO, H_2,	5–6
Water gas	H_2, CO, CH_4	11–18
City gas	CH_4, C_2H_6, C_3H_8, C_4H_{10}	35–45

economic conditions. The definition and classification of reserves had been used to be unique standards to each country or company, but in 2007, new standards were formulated by four organizations: Society of Petroleum Engineers (SPE), World Petroleum Congress (WPC), American Association of Petroleum Geologists (AAPG), and Society of Petroleum Evaluation Engineers (SPPE).

3.4.1.1 Gaseous fuel

The fuel used in the gaseous state is called gaseous fuel. Table 3.3 shows the typical gaseous fuels. The primary fuels are natural gas and liquefied petroleum gas (LPG). Unconventional gas of natural gas is produced from sources other than conventional gas fields, including shale gas, tight sand gas, coalbed methane, and methane hydrate. Synthetic gases include coal gas, producer gas, water gas, and blast furnace gas. It is produced as byproduct gases in steel mills and refineries. Fuel gas supplied to consumers through the piping network is called city gas.

Natural gas

Natural gas is extracted directly from gas or oil fields, and its main component is methane. It is included some other hydrocarbons and incombustible gases. Table 3.4 shows typical natural gas compositions. The composition of natural gas varies depending on the gas field, however, LNG is removed impurities such as carbon dioxide CO_2 and hydrogen sulfide H_2S during liquefaction. Natural gas has a high heating value. Its higher heating value (HHV) is ranged from $39.7-45.2\,\mathrm{MJ}/m_{\mathrm{N}}^{3}$. Natural gas can be transported by pipeline for distances up to 4000 km over land. Natural gas is liquefied by cooling below $-162°C$. LNG tanker transports liquefied natural gas across oceans. The proven reserves of natural gas are 196.9 trillion cubic meters and the reserves-to-production ratio is 50.9 years [28]. Fig. 3.62 shows the proven reserves of natural gas in 2018 [28], and Fig. 3.63 shows changes in natural gas production [28].

Table 3.4 Compositions of natural gas (vol.%) [30].

Production country (Lording Port)	**Alaska (Kenai)**	**Brunei (Lumut)**	**Indonesia East Kalimantan (Bontang)**	**Indonesia North Sumatera (Arun)**	**Australia (Withnell Bay)**	**Qatar (Ras Laffan)**
CH_4	99.81	89.62	90.45	89.1	87.51	89.92
C_2H_6	0.07	5.25	6.18	8.03	8.28	6.6
C_3H_8		3.4	2.47	1.55	3.3	2.25
i-C_4H_{10}		0.71	0.45	0.34	0.39	0.41
n-C_4H_{10}		0.96	0.43	0.4	0.46	0.64
C_5H_{12}		0.03	0.01	0.03	0.01	
N_2	0.12	0.03	0.01	0.04	0.05	0.18
HHV MJ/m^3_N	39.7	45.2	44.2	44.0	45.2	44.2

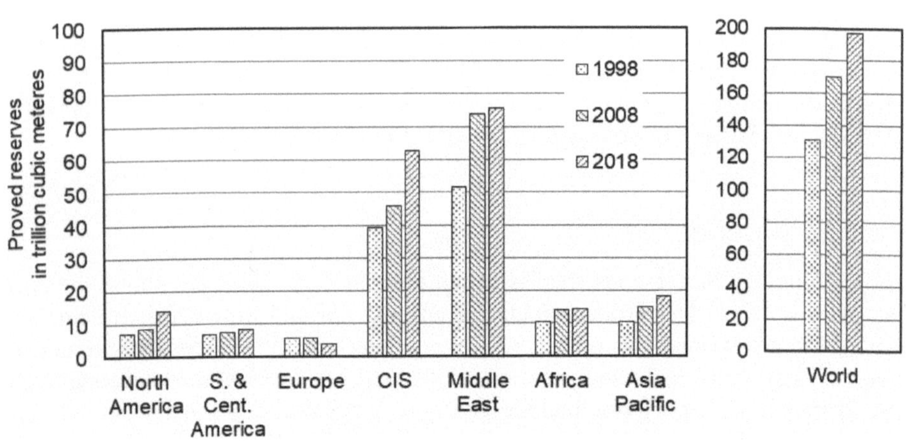

Figure 3.62 Proved reserves of natural gas from 1998 to 2018 [28].

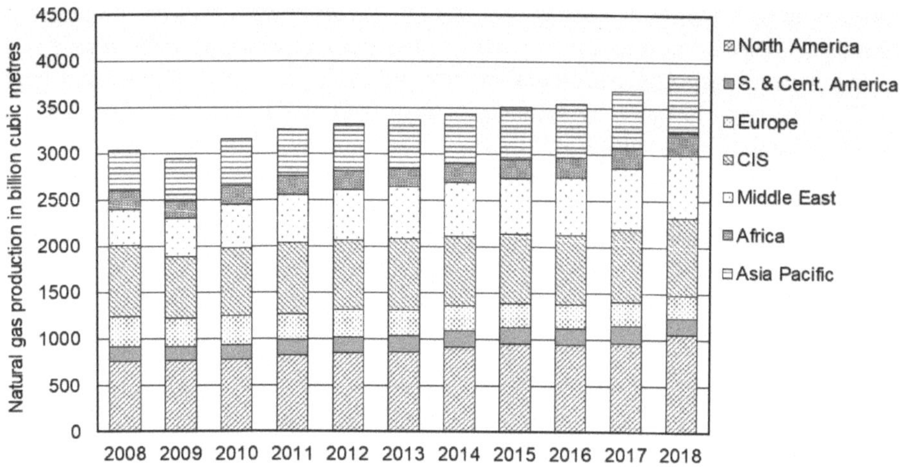

Figure 3.63 Natural gas production from 2008 to 2018 [28].

With the progress of mining technology, the production of unconventional gas has increased rapidly. Shale gas is natural gas extracted from sedimentary rocks called shale. Natural gas is collected by creating a fracture in the shale layer artificially by hydraulic fracturing. Furthermore, the technology to excavate the well horizontally according to the shale layer was established, and the shale gas production increased drastically. The technically recoverable resource of the shale gas reported by the U.S. Energy Information Administration (EIA) was 206.7 trillion cubic meters [29]. 32% of the world's total natural gas resources are in shale gas formations. The term technically recoverable resources represents the volumes of natural gas that could be produced with current technology, regardless of natural gas prices and production costs. Thus, it shows a wider range of resources than the proven reserves.

Tight gas is natural gas stored in a denser sandstone layer. Development has progressed in U.S. since the 1980s. Coalbed methane is methane gas generated in the process of coal production and stored in the coal layer. Methane hydrate is a crystalline solid in which methane molecules are trapped in a cage-like lattice structure formed by water molecules with hydrogen bonds. One cubic meters of methane hydrate produces about 165 m^3 of methane gas. However, methane hydrate exists in a low-temperature and high-pressure environment of $-80°C$ at 1 atm or 23 atm at 0°C. The technological development for commercialization is under way.

Liquefied petroleum gas

Hydrocarbons containing one or two carbon atoms in a molecule are gaseous state at normal temperature and pressure, and those containing five or more are liquids state. Petroleum gas is the hydrocarbon fuel containing 3 or 4 carbon atoms, which are a gas state under the atmospheric pressure, but can be easily liquefied by pressurizing to 1.5 MPa or cooling to −49°C LPG has both excellent properties of usability in gaseous fuels and transportation and storage performance in liquid fuels. The main components of petroleum gas are propane, propylene, butane and butylene. LPG is manufactured by separating from natural gas or recovering from oil refining processes, and is used for industrial fuel, household fuel, city gas, etc.

Coal gas

Coal gas is obtained by coal distillation at a temperature of about 1200°C in a coke oven. The components are about 50% of hydrogen, about 20% of methane, and several percent of carbon monoxide.

Producer gas

Producer gas is a fuel gas obtained by incomplete combustion of coal or coke in the presence of steam. Since air is used as the gasifying agent, it contains a lot of nitrogen and has a low heating value. Combustible components are about 25% of carbon monoxide and about 15% of hydrogen.

Water gas

Water gas is a fuel gas composed of hydrogen and carbon monoxide obtained by reacting coal or coke with steam. A gas rich in hydrogen is generated by the steam reforming reaction ($C + H_2O \rightarrow CO + H_2$). This process requires to proceed at a high temperature than 900°C due to the endothermic reaction.

Blast furnace gas

Blast furnace gas is produced from the blast furnace in a steel mill. This gas has 23% of carbon monoxide, 60% of nitrogen, and 14% of carbon dioxide. The heating value is low.

City gas

City gas is a mixture gas of various gaseous fuels to supply through a piping network. The main source is LNG. A small amount of odorant is added to detect gas leaks. In Japan, Korea, and Taiwan, high calorie gas with high heating value 43–45 $\mathrm{MJ}/m_\mathrm{N}^3$ is distributed. In the United States, low-calorie gases of 35.8–41 $\mathrm{MJ}/m_\mathrm{N}^3$ are commonly distributed. In Europe, different calorie gases are supplied, and the pipeline networks are divided in each gas.

3.4.1.2 *Liquid fuel*

The fuel used in the liquid state is called liquid fuel. Most liquid fuels are produced from petroleum. Petroleum fuels are refined into gasoline, naphtha, kerosene, light oil, and heavy oil in the order of boiling point at a stage of crude oil distillation. Table 3.5 shows the properties of petroleum fuels.

Gasoline is a light petroleum fuel, in which octane number has been adjusted to be suitable for spark ignition engines. Boiling point is 30°C–180°C. The octane number is a measure of the difficulty of self-ignition to avoid the knocking in a gasoline engine.

Table 3.5 Properties of petroleum fuels.

	Boiling point (°C)	Higher heating value (MJ/kg)
Gasoline	30–180	46
Naphtha	50–200	46
Kerosene	160–300	42
Light oils	200–350	44
Heavy oils	over 350	42

Naphtha is also called crude gasoline and has a boiling point of about 30°C–200°C. Naphtha is used as a raw material for the chemical industry, a gasoline blended material, and a jet fuel.

Kerosene has a boiling point of around 160°C–300°C and is mainly used as household fuel.

Light oils have a boiling point of 200°C–350°C and are mainly used for high-speed diesel engines. The cetane number, which is an index of the ease of autoignition, has been adjusted to be suitable for diesel engines. Light oils for the fuel of the steam generation correspond to the No.1 or No.2 fuel oil category in ASTM (American Society of Testing and Materials) D396 standard.

Gasoline, naphtha, kerosene, and light oils are referred to as distillate fuels due to the stage of crude oil distillation. On the other hand, heavy oils are also called residual fuels. This fuels are adjusted by mixing the low cost residual oils with light oils from the distillation process. Heavy oils are used as fuels for the power generation, the steam generation and low-speed diesel engines. Heavy oils are classified into three categories as the order of its viscosity; bunker A (No.4 fuel oil), bunker B (No.5 fuel oil), and bunker C (No.6 fuel oil). Table 3.6 shows the classification and characteristics of ASTM D396 fuel oil.

The proven reserves of oil are 1729.7 thousand million barrels and the reserves-to-production ratio is 50.0 years [28]. Fig. 3.64 shows the change of proven reserves of oil from 1998 to 2018 [28].

3.4.1.3 Solid fuel

The fuel used in the solid state is called solid fuel, which is to consist mainly of coals, biomass, and their processed fuels. Table 3.7 shows typical solid fuels. Since the characteristics of coal vary greatly depending on the place of origin, different classification methods are used in each country. Coal is a carbon-based fuel formed by coalification of ancient plants under geothermal and pressure. Coals are classified into anthracite, bituminous, subbituminous coal, and lignite. Peat is excluded from the coal classification due to the low heating value.

Coal consists of carbon, hydrogen, oxygen, sulfur, and nitrogen, as well as moisture and ash. Ash is an impurity of inorganic substances. Ultimate analysis estimates the mass proportions of these chemical elements. The heating value of coal and the amount of the combustion air can be estimated by using the ultimate

Table 3.6 Characteristic of fuel oils by ASTM D396.

		No. 1	No. 2	No. 4	No. 5	No. 6
Elements wt.%	Sulfur	0.01–0.5	0.05–1.0	0.2–2.0	0.5–3.0	0.7–3.5
	Hydrogen	13.3–14.1	11.8–13.9	(10.6–13.0)[a]	(10.5–12.0)[a]	(9.5–12.0)[a]
	Carbon	85.9–86.7	86.1–88.2	(86.5–89.2)[a]	(86.5–89.2)[a]	(86.5–90.2)[a]
	Nitrogen	0–0.1	0–0.1	–	–	–
Specific gravity		0.825–0.806	0.887–0.825	0.966–0.876	0.972–0.922	1.022–0.922
Viscosity in Cs[b] (at 38°C)		1.4–2.2	1.9–3.0	10.5–65	65–200	260–750
Higher heating value MJ/kg		45.64–46.08	44.48–45.82	42.41–45.01	42.00–44.13	40.39–44.06

[a]Estimated.
[b]Cs is the centi-Stoke.

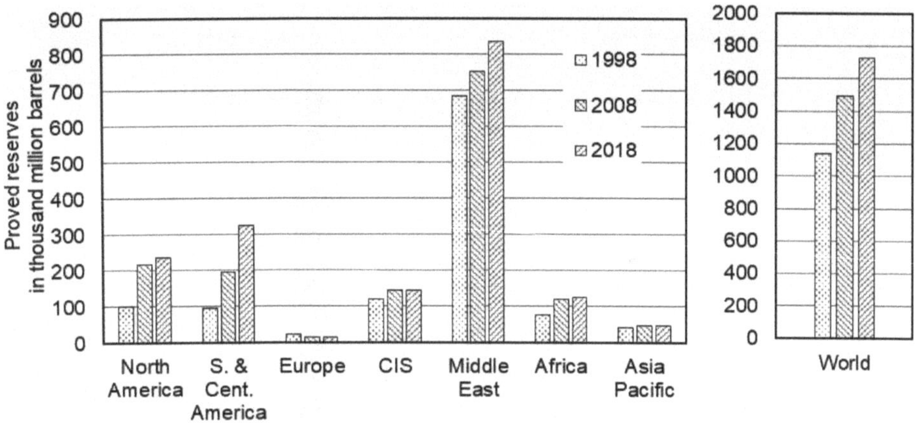

Figure 3.64 Proved reserves of oil from 1998 to 2018 [28].

Table 3.7 Characteristics of solid fuels.

Solid fuels		**Higher heating value (MJ/kg)**
Coal	Anthracite	26–35
	Bituminous Subbituminous Lignite	24–36 19–26 10–20
Peat Coke Wood		6–10 26–30 11–18

Higher heating value is evaluated by dry ash-free basis, except wood.

analysis. On the other hand, proximate analysis evaluates the characteristics of coal by dividing coal into four components: volatile substances, fixed carbon, ash and moisture. The ratio of fixed carbon to volatile matter is an index of the rank of coal. The amount of volatile matter in a coal indicates ignitability and stability on the combustion in burner. The ash content shows the load for ash collecting system.

Because of the varieties of the ash and moisture contents in coals, the heating values of coals are evaluated in three bases: as received, dry, and dry and ash-free. Fig. 3.65 shows the coal composition on the two analysis, and the difference of the bases of the heating value estimation. Table 3.8 shows an example of the ultimate and proximate analysis of a bituminous coal.

Anthracite is classified as the highest grade coal. The fixed carbon content is high, and the moisture content and volatile content are low. Bituminous coal is the most common industrial solid fuel. Fixed carbon content is 69% to 86%. Subbituminous coal is difficult to handle due to the high moisture content of 15%

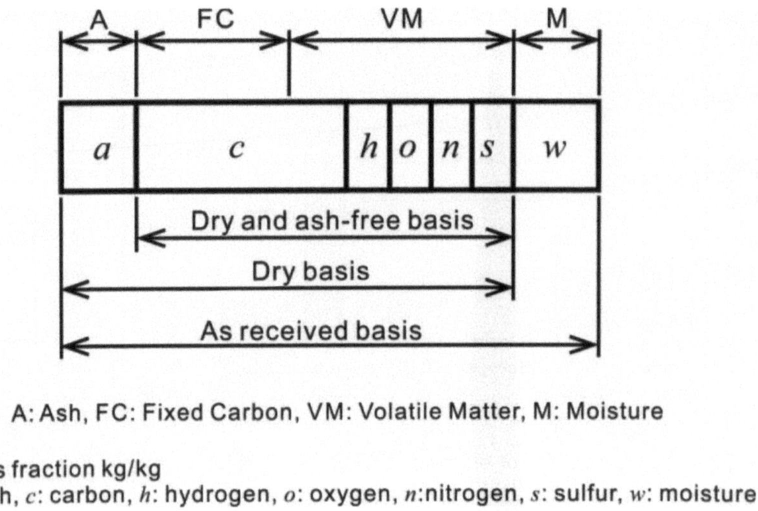

Figure 3.65 Composition of coal and analysis bases.

Table 3.8 Example of coal analyses on as-received basis of bituminous.

Proximate analysis		**Ultimate analysis**	
Component	**wt.%**	**Component**	**wt.%**
Moisture	2.9	Moisture	2.9
Volatile matter	28.2	Carbon	75.6
Fixed carbon	60.0	Hydrogen	4.4
Ash	8.9	Oxygen	6.8
		Nitrogen	0.8
		Sulfur	0.6
		Ash	8.9
Total	100.0	Total	100.0

to 30%. However, it has a good ignition property because of high in volatile matter content. Lignite has low fixed carbon content and high moisture and volatile matter contents.

The proven reserves of coal are 1055 billion tonnes and the reserves-to-production ratio is 132 years [28]. Fig. 3.66 shows proven reserves of anthracite and bituminous, and subbituminous coals in 2018 [28].

Peat has moisture content of 40%–70%. It cannot used as fuels without reducing the moisture content by drying or squeezing. Coke is produced by dry-distilling the coking coal in the absence of air at a temperature of about 1000°C. The main purposes of using cokes are the production of iron in blast furnace. Wood is a

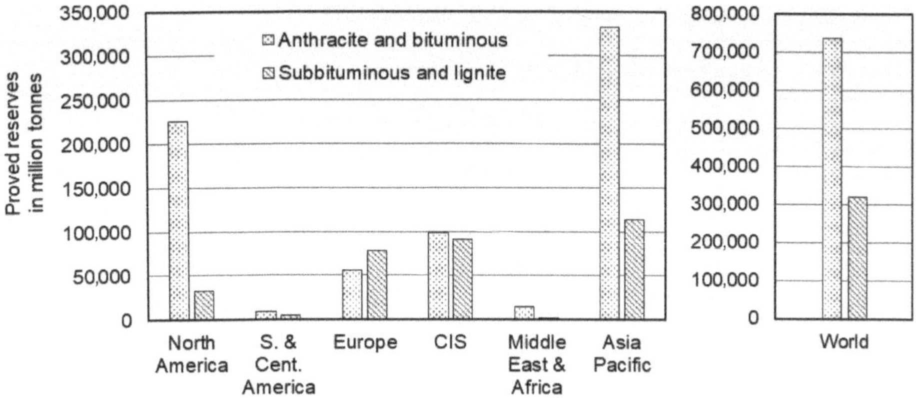

Figure 3.66 Proved reserves of coals in 2018 [28].

renewable biomass resource and is generated from forest land scraps and sawdust. They have many forms such as firewood, chips, and pellets. Ash content is 0.5%–12%. The heating value changes from 11 to 20 MJ/kg as received basis depending on its moisture content.

3.4.2 Stoichiometric calculation

3.4.2.1 Combustion air requirements

The combustion gas contains carbon and hydrogen as main components, and also contains a small amount of sulfur, oxygen, nitrogen and ash. Ash is noncombustible components of nongaseous residual. Water vapor is contained in combustion gas. The amount of air required or the product gas generated in combustion can be calculated by considering that the fuel is decomposed into these constituent elements and separately oxidized to produce the product gas. However, since the bonding energy of each element in the fuel is not taken into consideration in this method, an accurate evaluation of the amount of heat generation during combustion cannot be estimated.

Combustion is an exothermic oxidation reaction of fuel by oxygen. In industrial combustion equipment, oxygen comes from air. The amount of required air and the product gas in combustion per unit mass of fuel can be obtained from the complete combustion reaction-equation for each fuel constituent element. These amounts are the theoretical amount, which indicated by the subscript $(_0)$. The complete combustion table is shown in Table 3.9 [31]. m_N^3 is to read as "normal cubic meter." It indicates the amount of gas occupied 1.0 m^3 in the standard state at temperature of 0°C and the atmospheric pressure of 1.0 atm. The volume of 1 kmol of gas in the standard state occupies 22.41 m^3. Taking carbon C as an example, 1 kmol of carbon requires 1 kmol of oxygen molecule and produces 1 kmol of carbon dioxide in the complete combustion, expressed in Eq. (3.405). Thus, 12.011 kg carbon requires oxygen of 32.00 kg or 22.41 m_N^3.

$$C + O_2 = CO_2 \tag{3.405}$$

Table 3.9 Combustion table for combustible elements [31].

Elements	Atomic weight	Complete combustion reaction	Required air[a]	
			kg/kg	m_N^3/kg
C	12.011	$C + O_2 = CO_2$	11.51	8.907
H	1.008	$H + 1/4O_2 = H_2O$	34.30	26.53
S	32.065	$S + O_2 = SO_2$	4.313	3.336
O	15.999	$O - 1/2O_2 = 0$	− 4.322	− 3.343
N	14.007	$N = 1/2N_2$	0	0

[a]Dividing the required oxygen volume by mass fraction of oxygen in dry air of 0.2095 m_N^3/m_N^3.

Table 3.10 Air composition.

	Composition of dry air	
	m_N^3/m_N^3	kg/kg
Oxgen, O_2	20.95	23.14
Atmospheric nitrogen, N_{2a}	79.05	76.86

The atmospheric air consists of nitrogen, oxygen, and other minor gases. The dry air is composed of 0.78102 N_2, 0.20946 O_2, 0.00916 Ar and 0.00033 CO_2 in the mole fraction. In order to simplify the combustion calculation, nitrogen includes argon and other minute components. It is evaluated as atmospheric nitrogen N_{2a}, as shown in Table 3.10.

The theoretical amount of air required for complete combustion of a unit mass of fuel is

$$A_0 = \frac{32.00}{0.2314}\left(\frac{c}{12.011} + \frac{1}{4}\times\frac{h}{1.008} + \frac{s}{32.065} - \frac{1}{2}\times\frac{o}{15.999}\right) \tag{3.406}$$
$$= 11.51c + 34.30h + 4.313s - 4.322o \quad \text{kg/kg}$$

$$A_0' = \frac{22.41}{0.2095}\left(\frac{c}{12.011} + \frac{1}{4}\times\frac{h}{1.008} + \frac{s}{32.065} - \frac{1}{2}\times\frac{o}{15.999}\right) \tag{3.407}$$
$$= 8.906c + 26.53h + 3.336s - 3.343o \quad m_N^3/\text{kg}$$

where c, h, s, o and n are the mass fraction carbon, hydrogen, sulfur, oxygen and nitrogen in a unit mass of fuel. The mass fraction of moisture is represented by w. Oxygen contained in fuel reduces the amount of oxygen of $1/2O_2$ to be supplied from the outside.

On the gaseous fuel, the volume fractions m_N^3/m_N^3 indicated by surrounding with braces $\{CH_4\}$ of each component gas are used for the required air calculation. Table 3.11

Table 3.11 Combustion table for combustible elements.

Substance	Complete combustion reaction	Required oxygen O''_0	Required air[a] A''_0	Higher heating value H''_h	Lower heating value H''_h
		m_N^3/m_N^3	m_N^3/m_N^3	MJ/m_N^3	MJ/m_N^3
Hydrogen	$H_2 + 0.5\ O_2 = H_2O$	0.5	2.387	12.780	10.830
Carbon monoxide	$CO + 0.5\ O_2 = CO_2$	0.5	2.387	12.610	12.611
Methane	$CH_4 + 2\ O_2 = CO_2 + 2\ H_2O$	2.0	9.547	39.940	36.060
Ethane	$C_2H_6 + 3.5\ O_2 = 2\ CO_2 + 3\ H_2O$	3.5	16.71	70.470	64.550
Propane	$C_3H_8 + 5\ O_2 = 3\ CO_2 + 4\ H_2O$	5.0	23.87	101.40	93.390
Isobutane	$C_4H_{10} + 6.5\ O_2 = 4\ CO_2 + 5\ H_2O$	6.5	31.03	133.10	122.90
Alkane hydrocarbon	$C_mH_n + (m + n/4)\ O_2 = m\ CO_2 + (n/2)H_2O$	$m + \frac{n}{4}$	$\frac{(m + \frac{n}{4})}{0.2095}$	–	–
Oxygen	$O_2 - O_2 = 0$	− 1.0	− 4.773	0	0

[a]Dividing the required oxygen volume by mass fraction of oxygen in dry air of 0.2095 m_N^3/m_N^3.

shows the complete combustion table for each component gas. Taking methane CH_4 as an example.

$$CH_4 + 2O_2 = CO_2 + 2H_2O \tag{3.408}$$

According to Eq. (3.408), the theoretical volume of oxygen for a unit volume of methane is $O_0'' = 2.0\mathrm{m_N^3/m_N^3}$. The theoretical volume of air is $A_0'' = 2.0/0.2095 = 9.547\mathrm{m_N^3/m_N^3}$ by applying the volume fraction 0.2095 of oxygen in dry air. Hydrocarbon fuel of C_mH_n requires $(m + n/4)$ $\mathrm{m_N^3/m_N^3}$ of theoretical volume of oxygen, and $(m + n/4)/0.2095$ of theoretical volume of air. Oxygen contained in fuel reduces the amount of oxygen to be supplied from the outside. The required air for the mixture fuel gas is as follows:

$$\begin{aligned} A_0'' &= \frac{1}{0.2095}\left(\frac{1}{2}\{H_2\} + \frac{1}{2}\{CO\} + \sum\left(m + \frac{n}{4}\right)\{C_mH_n\} - \{O_2\}\right) \\ &= 2.387\{H_2\} + 2.387\{CO\} + 4.773\sum\left(m + \frac{n}{4}\right)\{C_mH_n\} - 4.773\{O_2\} \quad \mathrm{m_N^3/m_N^3} \end{aligned} \tag{3.409}$$

The theoretical air volumes calculated in Eqs. (3.406), (3.407) and (3.409) are the minimum air required for complete combustion. An excess air is added to compensate for the incomplete mixing of fuel and air in the actual burning equipment. The ratio of the actually supplied air volume A to the theoretical air volume A_0 is defined as the excess air ratio (sometimes called air ratio) α.

$$\alpha = \frac{A}{A_0} \tag{3.410}$$

Supplying a significantly excessive volume of air causes a decrease in temperature in the combustion chamber and a heat loss accompanied with the exhaust gas due to a increase of the combustion gas volume. It is not preferable desirable for the viewpoint of energy efficiency. As will be described in next section, the excess air ratio can be calculated from the flue gas analysis, so it is important to maintain the combustion state in the furnace appropriately to be minimum heat loss.

3.4.2.2 Combustion products

Combustion gas generally consists of CO_2, SO_2, O_2, N_2, and H_2O on the based on complete combustion. The gas volume generated from a unit mass of fuel is called combustion gas volume on a wet basis. The measurement of component analysis of combustion gas is carried out after removing water vapor. The volume excluding water vapor from the combustion gas is referred as combustion gas volume on a dry basis.

When liquid or solid fuel burns at an air ratio α of 1 or more, Table 3.12 shows the gas components in the combustion gas. CO_2 of 22.41 $\mathrm{m_N^3}$ is generated from 1 kmol of carbon, thus, carbon dioxide from c kg of carbon is expressed as (22.41/12.011)

Table 3.12 Combustion gas constituents from solid or liquid fuels.

Components	Complete combustion (m_N^3)	Incomplete combustion (m_N^3)
CO_2	$22.41 \times \frac{c}{12.011} = 1.866c$	$22.41 \times \frac{c}{12.011} \times (1-\xi) = 1.866c(1-\xi)$
CO	0.0	$22.41 \times \frac{c}{12.011} \times \xi = 1.866c\xi$
H_2O	$22.41 \times \frac{h}{2.016} + 22.41 \times \frac{w}{18.015} = 11.12h + 1.244w$	same as on the left
N_2	$0.7905\alpha A_0' + 22.41 \times \frac{n}{28.014} = 0.7905\alpha A_0' + 0.8000n$	same as on the left
O_2	$0.2095(\alpha - 1)A_0'$	$0.2095(\alpha-1)A_0' + \frac{1}{2} \times 22.41 \times \frac{c}{12.011} \times \xi = 0.2095(\alpha-1)A_0' + 0.9329c\xi$
SO_2	$22.41 \times \frac{s}{32.065} = 0.6989s$	same as on the left

$c = 1.866c$. H_2O from h kg of hydrogen in the fuel is expressed as $(22.41/2.016)h$. Furthermore, $(22.41/18.015)w$ of water vapor is added by evaporation of moisture w kg in the fuel. N_2 in the combustion gas is from the supply air of $0.7905\alpha A_0'$, and added nitrogen gas of $(22.41/28.014)n$ by gasification of nitrogen fraction n in fuel. The excess oxygen is $0.2095(\alpha - 1)A_0'$.

The combustion is sometimes incomplete, which may be at times occurred for the purpose of NOx reduction. At the incomplete combustion, unburned carbon, carbon monoxide, hydrogen, and unburned hydrocarbons are contained in the combustion gas. On the combustion equipment for power generation, hydrogen and unburned hydrocarbons are rarely generated and can be ignored. Therefore, CO is considered as the product of incomplete combustion. On the incomplete combustion, mass fraction of carbon c in the fuel either produces CO_2 according to Eq. (3.405) or CO according to Eq. (3.411).

$$C + \frac{1}{2}O_2 = CO \tag{3.411}$$

The conversion ratio ξ to carbon monoxide is defined as follows.

$$\xi = \frac{(CO)}{(CO) + (CO_2)} \tag{3.412}$$

where (CO) and (CO_2) are the volume fractions of CO and CO_2 in the dry combustion gas volume measured in the combustion gas in dry basis. The gas volumes of CO_2, CO and O_2 are modified by using the conversion ratio ξ.

Combustion gas volume in wet basis V_w' generated from 1 kg of fuel is expressed by the following equation.

$$\begin{aligned} V_w' &= (\alpha - 0.2095)A_0' + 22.41\left(\frac{(1+0.5\xi)c}{12.011} + \frac{h}{2.016} + \frac{n}{28.014} + \frac{s}{32.065} + \frac{w}{18.015}\right) \\ &= (\alpha - 0.2095)A_0' + (1.866 + 0.9329\xi)c + 11.12h + 0.800n + 0.6989s + 1.244w \quad m_N^3/kg \end{aligned} \tag{3.413}$$

Combustion gas volume in dry basis V_d' is as follows.

$$\begin{aligned} V_d' &= (\alpha - 0.2095)A_0' + 22.41\left(\frac{(1+0.5\xi)c}{12.011} + \frac{n}{28.014} + \frac{s}{32.065}\right) \\ &= (\alpha - 0.2095)A_0' + (1.866 + 0.9329\xi)c + 0.800n + 0.6989s \quad m_N^3/kg \end{aligned} \tag{3.414}$$

The sum of production volume of CO_2 and CO is expressed as $V_d'[(CO_2) + (CO)]$.

$$V_d'[(CO_2) + (CO)] = 1.866c(1 - \xi) + 1.866c\xi = 1.866c \tag{3.415}$$

By measuring the CO and CO_2 volume fractions in the combustion gas, the combustion gas volume in dry basis V_d' can be calculated from Eq. (3.415), if the carbon fraction c in the fuel is known.

$$V_d' = \frac{1.866c}{(CO_2) + (CO)} \tag{3.416}$$

The wet combustion gas mass G_w is calculated from the mass conservation law before and after the reaction.

$$G_w = 1 + \alpha A_0 = 1 + \alpha(11.51c + 34.30h + 4.313s - 4.322o)\ \text{kg/kg} \tag{3.417}$$

H_2O from h kg of hydrogen in the fuel is expressed as $h(2 \times 1.008 + 15.999)/2 = 9.007h$ from the complete combustion equation of hydrogen $H + 1/4O_2 = 1/2H_2O$. The dry combustion gas mass G_d is conducted by subtracting the water vapor mass $(9.007h + w)$ from Eq. (3.417).

$$G_d = 1 + \alpha(11.51c + 34.30h + 4.313s - 4.322o) - (9.007h + w)\ \text{kg/kg} \tag{3.418}$$

For gaseous fuel, the combustion gas components are estimated by using the volume fractions $\{H_2\}$, $\{CO\}$, $\{C_mH_n\}$ of the component in the fuel. Table 3.13 shows the gas components produced by the gaseous fuel combustion. CO_2 and CO are generated according to the number of C in the fuel component, and are represented as $\{CO\} + \Sigma m\{C_mH_n\}$. H_2O is generated from components containing H in the fuel and is represented as $\{H_2\} + \{H_2O\} + \Sigma n\{C_mH_n\}/2$. N_2 in the combustion gas is from the supply air of $0.7905\alpha A_0''$, and added $\{N_2\}$ in fuel.

Volume of oxygen in the combustion gas is the sum of oxygen supplied in the excess air and half of the volume of CO generated by incomplete combustion.

Combustion gas volume in wet basis V_w'' generated from 1 m_N^3 of gaseous fuel is as follows.

$$V_w'' = (\alpha - 0.2095)A_0'' + \left(1 + \frac{\xi}{2}\right)\left(\{CO\} + \sum m\{C_mH_n\}\right) + \left(\{H_2\} + \{H_2O\} + \frac{1}{2}\sum n\{C_mH_n\}\right) + \{N_2\}\ m_N^3/m_N^3 \tag{3.419}$$

Table 3.13 Combustion gas constituents from gaseous fuels.

Components	Complete combustion (m_N^3)
CO_2 and CO	$\{CO\} + \sum m\{C_mH_n\}$
H_2O	$\{H_2\} + \{H_2O\} + \frac{1}{2}\sum n\{C_mH_n\}$
N_2	$0.7905\alpha A_0'' + \{N_2\}$
O_2	$0.2095(\alpha - 1)A_0'' + \frac{\xi}{2}\left(\{CO\} + \sum m\{C_mH_n\}\right)$

Combustion gas volume in dry basis V_d'' is expressed as following.

$$V_d'' = (\alpha - 0.2095)A_0'' + \left(1 + \frac{\xi}{2}\right)\left(\{CO\} + \sum m\{C_mH_n\}\right) + \{N_2\}\ m_N^3/m_N^3 \tag{3.420}$$

Similar to Eq. (3.415), the sum of the production volumes of CO and CO_2 is expressed as follows.

$$V_d''[(CO_2) + (CO)] = \{CO\} + \sum m\{C_mH_n\} \tag{3.421}$$

Therefore, the combustion gas volume in dry basis can be obtained by analyzing the CO and CO_2 volume fractions in the combustion gas.

$$V_d'' = \frac{\{CO\} + \sum m\{C_mH_n\}}{(CO_2) + (CO)} \tag{3.422}$$

For effective usage of energy, thermal management technology to operate the furnace at an appropriate air ratio is important. The air ratio can be evaluated using the component analysis data of the combustion gas. The measurement of the component analysis is performed after removing the water vapor, and the volume fractions of oxygen, carbon dioxide, and carbon monoxide are obtained for the combustion gas volume in dry basis.

The volumes of CO, N_2 and O_2 generated from 1 kg of solid and liquid fuels are $V_d'(CO)$, $V_d'(N_2)$, and $V_d'(O_2)$, which are by the following formula from Table 3.12.

$$V_d'(CO_2) = 1.866c\xi \tag{3.423}$$

$$V_d'(N_2) = 0.7905\alpha A_0' + 0.8000n \tag{3.424}$$

$$V_d'(O_2) = 0.2095(\alpha - 1)A_0' + 0.9329c\xi \tag{3.425}$$

By solving these three equations for the air ratio α with assuming the small amount of nitrogen component in the fuel, the following equation can be obtained. Eq. (3.426) can be applicable for the combustion gas from the gaseous fuel.

$$\frac{1}{\alpha} = 1 - \frac{0.3773[(O_2) - 0.5(CO)]}{(N_2)} \tag{3.426}$$

Furthermore, by applying the $(CO) = 0$ and $(N_2) = 0.79$ approximation, it is further simplified as follows.

$$\frac{1}{\alpha} = 1 - \frac{(O_2)}{0.2095} \tag{3.427}$$

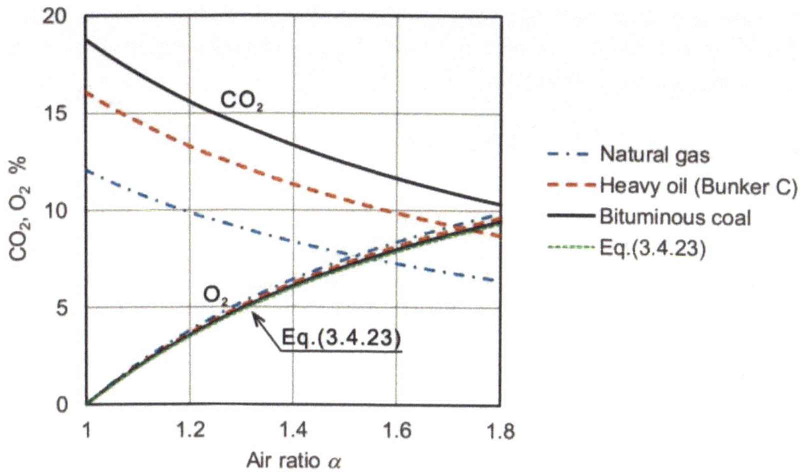

Figure 3.67 Variations of volume fractions of CO_2 and O_2 in combustion gas in dry basis.

Fig. 3.67 shows the variations of CO_2 and O_2 with air ratio for different fuels of natural gas, heavy oil of bunker C and bituminous coal. Volume fraction of CO_2 in combustion gas in dry basis differ between fuels. However, variation of fuel has little effect on the dependence of air ratio on the oxygen content of the combustion gas. The relationship between O_2 and air ratio of natural gas, heavy oil and coal exhibits a good agreement with Eq. (3.427). Therefore, the air ratio can be constantly monitored only by measuring the oxygen concentration in the flue gas.

3.4.2.3 *Heating value of fuel*

The heating value is the amount of heat extracted from 1 kg or 1 m_N^3 of fuels with completely combustion while the combustion products cooling down to the initial temperature. Water vapor is a product of combustion by reaction or evaporation in the combustion process. The heating values of a fuel depend on the water vapor state whether this vapor remains in the vapor state or is condensed to liquid. The amount of heat, including the latent heat, is called high heating value or gross heating value, assuming that all the water vapor contained in the combustion gas condenses. When the vapor contained in the combustion gas remains as the vapor state, the heating value does not include latent heat and is called low heating value or net heating value. A difference of about 10% occurs in the thermal efficiency depending on which heating value is used. In the industrial thermal equipment, the latent heat of condensation of water vapor is not used and low calorific value is often used. Heat exchanger in the thermal equipment will be corroded by condensed water, if the combustion gas is cooled below the saturation temperature of water vapor. However, high heating value is used to evaluate the thermal efficiency in some countries and fields.

The high and low heating values of component gases are also shown in Table 3.10. Gaseous fuel is often a mixture of many component gases, and its heating value is calculated as follows.

$$H_h'' = 12.780\{H_2\} + 12.610\{CO\} + 39.940\{CH_4\} + \cdots \ \mathrm{MJ}/m_\mathrm{N}^3 \tag{3.428}$$

$$H_l'' = 10.830\{H_2\} + 12.611\{CO\} + 36.060\{CH_4\} + \cdots \ \mathrm{MJ}/m_\mathrm{N}^3 \tag{3.429}$$

For example, the high and low heating values of city gas [32] with a composition of 89.60% methane, 5.62% ethane, 3.43% propane, and 1.35% butane are shown below.

$$\begin{aligned} H_h'' &= 39.940 \times 0.896 + 70.470 \times 0.0562 + 101.40 \times 0.0343 \\ &\quad + 133.10 \times 0.0135 = 45.02 \ \mathrm{MJ}/m_\mathrm{N}^3 \end{aligned} \tag{3.430}$$

$$\begin{aligned} H_l'' &= 36.060 \times 0.896 + 64.550 \times 0.0562 + 93.390 \times 0.0343 \\ &\quad + 122.90 \times 0.0135 = 40.79 \ \mathrm{MJ}/m_\mathrm{N}^3 \end{aligned} \tag{3.431}$$

For liquid and solid fuels, the binding energy of each element in the fuel should be considered, thus, the heating values cannot be evaluated precisely using the mass fraction of c, h, s and o of ultimate analysis data. For an accurate heating value of liquid and solid fuels, the actual measurement in a laboratory is needed. One of the most frequently used correlations is Dulong's formula although there is some margin of error of 0.4 MJ/kg. w is the mass fraction of total moisture in a fuel.

$$H_h = 33.83c + 144.3\left(h - \frac{o}{8}\right) + 9.420s \ \mathrm{MJ/kg} \tag{3.432}$$

$$H_l = H_h - 2.442(8.937h + w) \ \mathrm{MJ/kg} \tag{3.433}$$

3.4.3 Calculation of gas temperature

3.4.3.1 Theoretical adiabatic flame temperature

The adiabatic flame temperature is the maximum temperature of the combustion gas that can be reached by the combustion reaction under the no heat loss to the surroundings in the adiabatic condition. Especially, the theoretical maximum temperature assuming no thermal dissociation and complete combustion is called the adiabatic theoretical flame temperature T_{bt}.

By the complete combustion of 1 kg of fuel with no condensation of water vapor, low heating value H_l kJ/kg is released. All the heat generation is converted to sensitive heat of the combustion gas to increase the gas temperature. The theoretical adiabatic flame temperature T_{bt} can be expressed as Eq. (3.435) based on the heat balance.

$$G_w c_{pm}(T_{bt} - T_0) = H_l \tag{3.434}$$

$$T_{bt} = \frac{H_l}{G_w\, c_{pm}} + T_0 \tag{3.435}$$

T_0 is the initial temperature of 298.15K before combustion reaction, and c_{pm} is the averaged specific heat in kJ/kgK at constant pressure between T_0 and T_{bt}.

In the gaseous fuel, the lower heating value H_l'' is usually given as the value per unit normal cubic meter m_N^3. Therefore, T_{bt} can be evaluated from Eq. (3.435) by converting the lower heating value H_l'' into the value per unit fuel mass H_l. As the alternative approach, by using the specific heat at per unit normal cubic meter m_N^3, T_{bt} can be obtained as followings.

$$T_{bt} = \frac{H_l''}{V_w'' c_{pm}''} + T_0 \tag{3.436}$$

c_{pm}'' is the averaged specific heat at constant pressure in kJ/m_N^3K between T_0 and T_{bt}.

The specific heat of each component in the combustion gas changes drastically as the temperature, thus, the averaged specific heat of each component from T_0 to T_{bt} should be evaluated to satisfy the heat balance of Eq. (3.434). Furthermore, the averaged specific heat c_{pm} or c_{pm}'' must be calculated corresponding to the component fraction in the combustion gas in wet basis. The averaged specific heat at constant pressure c_{pi} and c_{pi}'' of each combustion gas components between T_0 and T are shown in Table 3.14 and Fig. 3.68. Table 3.14 are conducted by dividing the enthalpy, listed in the JANAF-NIST thermochemical tables [33], by temperature difference between T_0 and T.

The average specific heat at constant pressure $c_{\rm pm}$ and $c_{\rm pm}''$ are calculated by the following equations.

$$c_{\rm pm} = \sum_i (c_{\rm pi} \cdot m_i) \tag{3.437}$$

$$c_{\rm pm}'' = \sum_i \left(c_{\rm pi}'' \cdot v_i\right) \tag{3.438}$$

m_i is the mass fraction of each component in the combustion gas volume in wet basis, and v_i is the volume fraction. T_{bt} is calculated by Eqs. (3.425) or (3.426) using the averaged specific heat $c_{\rm pm}$ or $c_{\rm pm}''$ at a assuming T_{bt}. This calculation is iterated until the resulted T_{bt} matches the assumed one [31].

Table 3.15 shows the calculation of the theoretical adiabatic flame temperature for city gas with an air ratio of 1.3, as a example. The composition of city gas is referred to ref. (7). At the calculation step 18 in Table 3.15, T_{bt} is assumed to be 2000K, the average specific heat $c_{\rm pm}''$ is evaluated at steps 19 and 20, and the adiabatic flame temperature is

Table 3.14 Averaged specific heat at constant pressure c_{pi} and c''_{pi} between T_0 and T.

T K	CO_2		H_2O		N_2		O_2		SO_2	
	kJ/kgK	kJ/m_N^3K	**kJ/kgK**	kJ/m_N^3K	**kJ/kgK**	kJ/m_N^3K	**kJ/kgK**	kJ/m_N^3K	**kJ/kgK**	kJ/m_N^3K
1000	1.081	2.123	2.056	1.653	1.092	1.364	1.011	1.443	0.766	2.189
1100	1.102	2.164	2.090	1.680	1.102	1.378	1.022	1.458	0.777	2.221
1200	1.121	2.200	2.124	1.707	1.113	1.391	1.031	1.472	0.787	2.249
1300	1.137	2.233	2.158	1.734	1.122	1.403	1.040	1.485	0.796	2.274
1400	1.153	2.263	2.191	1.761	1.132	1.415	1.048	1.496	0.804	2.297
1500	1.167	2.291	2.224	1.787	1.141	1.426	1.056	1.507	0.811	2.317
1600	1.179	2.316	2.256	1.813	1.149	1.436	1.063	1.517	0.817	2.335
1700	1.191	2.339	2.287	1.838	1.157	1.446	1.069	1.526	0.823	2.351
1800	1.202	2.360	2.317	1.862	1.164	1.455	1.075	1.535	0.828	2.366
1900	1.212	2.379	2.346	1.886	1.171	1.464	1.081	1.543	0.833	2.380
2000	1.221	2.397	2.374	1.908	1.177	1.472	1.087	1.551	0.837	2.392
2100	1.229	2.414	2.401	1.930	1.184	1.479	1.092	1.559	0.841	2.404
2200	1.237	2.429	2.427	1.951	1.189	1.486	1.097	1.566	0.845	2.415
2300	1.245	2.444	2.452	1.971	1.195	1.493	1.102	1.573	0.848	2.425
2400	1.252	2.458	2.476	1.990	1.200	1.499	1.107	1.580	0.852	2.434

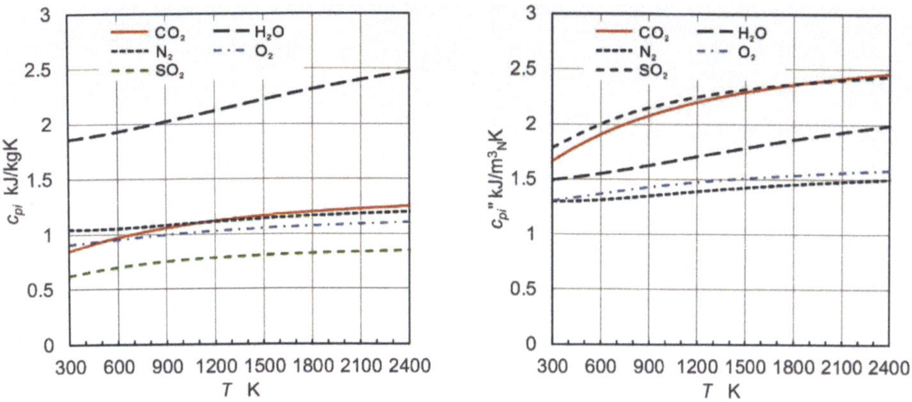

Figure 3.68 Averaged specific heat at constant pressure c_{pi} and c_{pi}'' between T_0 and T.

found to be 1984K. Then, the averaged specific heat of each combustion gas component at 1984K is calculated by linear interpolation from Table 3.14 at calculation steps of 22 and 23. Finally, T_{bt} of 1986K is obtained by iterative calculation.

3.4.3.2 Gas temperature with heat loss and incomplete combustion

In the combustion furnace, the gas temperature drops below the adiabatic flame temperature due to heat absorption by heating objects, heat loss to surroundings, and radiation heat loss from the flame. The heat loss Q_r from 1 kg of fuel can be considered as a decrease of the low heating value of the fuel.

By preheating the air-fuel mixture before combustion up to T_p, the sensible heat Q_p is expressed as follows.

$$Q_p = G_w c_{pp}(T_p - T_0) \tag{3.439}$$

where c_{pp} is the c_{pp} is the average specific heat from T_0 to T_p of the air-fuel mixture.

The gas temperature including these factors is given below.

$$T_{bt} = \frac{H_l - Q_r + G_w c_{pp}(T_p - T_0)}{G_w c_{pm}} + T_0 \tag{3.440}$$

When the flue gas contains unburned components such as CO, H_2, and soot, these are exhausted without generating heat from the combustion chamber, resulting in incomplete combustion loss. The ratio of the low heat value H_l to the heat value Q_c actually generated in the combustion process is called the combustion efficiency η_c.

$$\eta_c = \frac{Q_c}{H_l} \tag{3.441}$$

Table 3.15 Calculation of theoretical adiabatic flame temperature T_{bt} of city gas under a complete combustion condition with air ratio $\alpha = 1.3$.

	Combustion calculation of city gas			CH_4	C_2H_6	C_3H_8	i-C_4H_{10}
1	Volume fraction in fuel $\{C_mH_n\}$	m_N^3/m_N^3	Ref. [32]	0.8960	0.0562	0.0343	0.0135
2	Lower heating value in each component, H_l''	MJ/m_N^3	Table 3.11	36.06	64.55	93.39	122.90
3	Lower heating value of fuel, H_l''	MJ/m_N^3	Eq. (3.429)	40.80			
4	m of $\{C_mH_n\}$		Table 3.11	1	2	3	4
5	n of $\{C_mH_n\}$		Table 3.11	4	6	8	10
6	$m + n/4$		Table 3.11	2.0	3.5	5.0	6.5
7	Theoretical required air, $A_0'' = \Sigma(m + n/4)\{C_mH_n\}/0.2095$	m_N^3/m_N^3	Eq. (3.409)	10.73			
8	Air ratio, α			1.3			
9	Supplied air volume, $A'' = \alpha A_0''$	m_N^3/m_N^3	Eq. (3.410)	13.95			
10	CO_2, $= \Sigma m\{C_mH_n\}$	m_N^3/m_N^3	Table 3.13	1.165			
11	H_2O, $= \Sigma n\{C_mH_n\}/2$	m_N^3/m_N^3	Table 3.13	2.165			
12	N_2, $= 0.7905\ \alpha\ A_0''$	m_N^3/m_N^3	Table 3.13	11.027			
13	O_2, $= 0.2095(\alpha - 1)\ A_0''$	m_N^3/m_N^3	Table 3.13	0.6744			
	Combustion gas			CO_2	H_2O	N_2	O_2
14	Volume of each combustion gas component	m_N^3/m_N^3	Table 3.13	1.165	2.165	11.027	0.674
15	Combustion gas volume in wet basis, V_w''	m_N^3/m_N^3	Eq. (3.419)	15.03			
16	Volume fraction in each gas component, v_i	m_N^3/m_N^3		0.0775	0.1440	0.7336	0.0449
17	T_0	K		298.15			
18	Assumed T_{bt}	K		2000			
19	Averaged specific heat from T_0 to assumed T_{bt}, c_{pi}''	kJ/m_N^3K	Table 3.14	2.397	1.908	1.472	1.551
20	Averaged specific heat, c_{pm}''	kJ/m_N^3K	Eq. (3.438)	1.610			
21	$T_{bt,1}$ (first iteration)	K	Eq. (3.436)	1984			
22	Averaged specific heat from T_0 to $T_{bt,1}$, c_{pi}''	kJ/m_N^3K	Table 3.14	2.395	1.904	1.470	1.550
23	Averaged specific heat, c_{pm}''	kJ/m_N^3K	Eq. (3.438)	1.608			
24	$T_{bt,2}$ (second iteration)	K	Eq. (3.436)	1986			
25	Averaged specific heat from T_0 to $T_{bt,2}$, c_{pi}''	kJ/m_N^3K	Table 3.14	2.396	1.904	1.470	1.551
26	Averaged specific heat, c_{pm}''	kJ/m_N^3K	Eq. (3.438)	1.608			
27	T_{bt} (third iteration)	K	Eq. (3.436)	1986			

Combustion efficiency is usually nearly 100% for gaseous fuels and 90%–98% for pulverized coal and heavy oil. Including incomplete combustion, gas temperature can be expressed as follows.

$$T_{bt} = \frac{\eta_c H_l - Q_r + G_w c_{pp}(T_p - T_0)}{G_w c_{pm}} + T_0 \tag{3.442}$$

3.4.3.3 Equilibrium flame temperature

In the previous section, the gas components of N_2, O_2, CO_2, H_2O, and SO_2 in complete combustion are considered for the calculation of the adiabatic theoretical flame temperature. In this case, the composition of the combustion gas is determined by the conservation of C, H, O, N, and S. The theoretical flame temperature is evaluated by the heat balance of the low heating value and the sensible heat in the combustion gas. However, thermal dissociation occurs in high temperature combustion gas, and the concentration of intermediate products such as CO and OH during the combustion reaction process increases. Therefore, some of the chemical energy of the fuel cannot be released and the gas temperature shows lower than the adiabatic theoretical flame temperature. The complete combustion cannot be carried out at a high-temperature flame, and the combustion reaction shows the chemical equilibrium. The temperature in the chemical equilibrium state is called the adiabatic equilibrium flame temperature.

Thermal dissociations are expressed by the following equations.

$$CO_2 \rightleftarrows CO + \frac{1}{2}O_2 \tag{3.443}$$

$$H_2O \rightleftarrows \frac{1}{2}H_2 + OH \tag{3.444}$$

$$H_2O \rightleftarrows H_2 + \frac{1}{2}O_2 \tag{3.445}$$

$$\frac{1}{2}O_2 \rightleftarrows O \tag{3.446}$$

$$\frac{1}{2}O_2 + \frac{1}{2}N_2 \rightleftarrows NO \tag{3.447}$$

$$\frac{1}{2}H_2 \rightleftarrows H \tag{3.448}$$

$$\frac{1}{2}N_2 \rightleftarrows N \tag{3.449}$$

In these reactions, the forward and the reverse reactions are balanced in equilibrium state. The fraction of each gas component gas in the combustion gas can be obtained from the equilibrium constant, which depends only on the temperature.

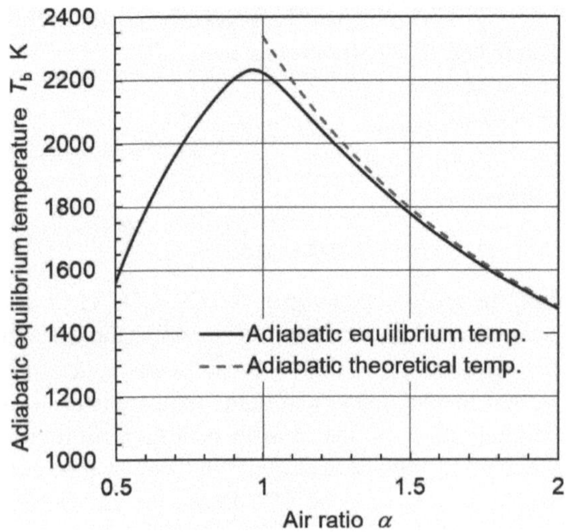

Figure 3.69 Adiabatic equilibrium temperature and adiabatic theoretical temperature of methane. Chemical equilibrium is calculated by using Gibbs free energy minimization method [34].

In the calculation of adiabatic equilibrium flame temperature, the composition of the combustion gas is obtained from the equilibrium constant by assuming the flame temperature. Enthalpy after the combustion reaction can be determined from the composition and temperature of the combustion gas. The flame temperature is calculated by iterative calculation to be coincide between the total enthalpies before and after the combustion reaction.

Fig. 3.69 shows a comparison between the adiabatic equilibrium temperature and the adiabatic theoretical temperature of methane. When the temperature of combustion gas is less than 2000K, the difference of the theoretical temperature and the equilibrium temperature is small. Estimation of the theoretical temperature is useful in practical thermal management of combustion equipment. However, above 2000K, the difference between these temperatures increases sharply due to the thermal dissociation.

3.5 Fundamentals of nuclear physics

3.5.1 Fission chain reactions and neutron multiplication

3.5.1.1 Fission chain reactions

It is of interest in the meaning of the "life" of a neutron in a multiplying system, assuming that the system is a thermal reactor, where most of the fission process takes place within the thermal energy region. The neutrons born due to fission have

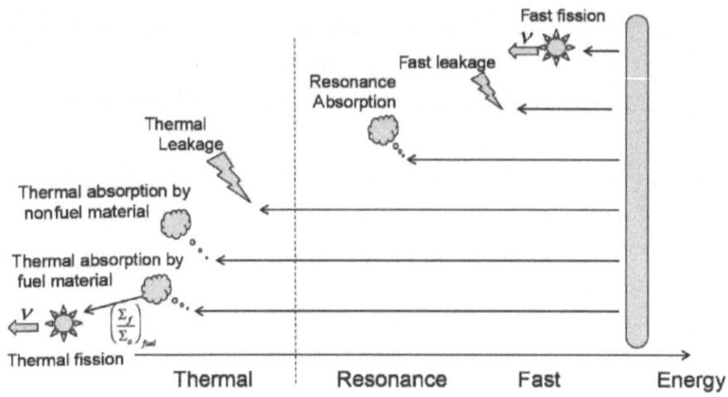

Figure 3.70 Slowing down, loss and production of neutrons [35].

an average energy of 2 MeV and appear in the fast energy range. These fast neutrons collide with the nuclides in the media and lose their energy. After losing considerable energy, the neutrons reach the thermal energy range, where they are absorbed in the fuel and trigger the next fission process with a certain probability. Also, the probability that a certain fraction of the fast neutrons should be considered by a trigger of a fission process within the fast energy range. Some of the neutrons are lost due to leakage from the system and absorption by nonfuel materials. This process, together with other possible phenomena, is depicted in Fig. 3.70.

Assuming that there are N_i initial fission neutrons in generation i. All of these neutrons will eventually be lost by the end of the generation, either by leakage or absorption (=capture + fission). The number of neutrons leaking out of the system is set as L and those absorbed as A, so that $N_i = A + L$. Furthermore, the total number of neutrons produced by fission as P, which form the initial group of neutrons for the next generation $i+1$, that is, $N_{i+1} = P$ is set. The ratio of N_i to N_{i+1} is a measure of the multiplication of neutrons through a fission chain reaction. As shown in Eq. (3.450), this ratio is defined as k_{eff} and designate it as the "effective multiplication factor":

$$k_{eff} = \frac{N_{i+1}}{N_i} = \frac{P}{A+L}. \tag{3.450}$$

As shown in Fig. 3.70, absorption A comprises resonance absorption, absorption by nonfuel material, and nonfission absorption (capture) by the fuel material. Leakage L comprises fast leakage and thermal leakage. Production P comprises of fast fission and thermal fission.

Assuming that N_0 neutrons are injected into the system at $t = 0$, time evolution of neutrons in the system is shown schematically in Fig. 3.71.

- If k_{eff} is less than unity, $N_{i+1} < N_i$, the neutron population will die out: this is designated as "subcritical."
- If $k_{eff} > 1$, $N_{i+1} > N_i$, the neutron population increases exponentially: this is designated as "supercritical."
- If $k_{eff} = 1$, $N_{i+1} = N_i$, the neutron population does not change with succeeding generations: this is designated as "critical."

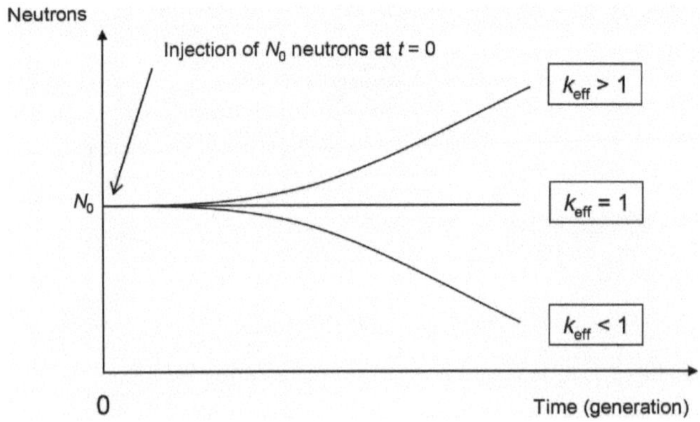

Figure 3.71 Time evolution of neutrons [35].

From another viewpoint of "life," "generation" of a neutron is of interest in an easily understanding of a fission chain reaction. In this process, neutrons emitted by fission nuclei induce fissions in other fissile or fissionable nuclei; the neutrons from these fissions induce fissions in still other fissile or fissionable nuclei; and so on. Such a chain reaction can be described quantitatively in terms of the multiplication factor, which is denoted by the symbol k_{eff}. This is defined as the ratio of the number of fissions (or fission neutrons) in one generation divided by the number of fissions (or fission neutrons) in the preceding generation. In equation form, this is

$$k_{eff} = \frac{\text{Number of fissions in one generation}}{\text{Number of fissions in preceeding generation}}. \tag{3.451}$$

If k_{eff} is greater than 1, in Eq. (3.451), the number of fissions increases from generation to generation. In this case, the energy released by the chain reaction increases with time evolution, and the system is said to be supercritical. However, if k_{eff} is less than 1, the number of fissions decreases with time and the chain reaction is called subcritical. Finally, in the special situation when k_{eff} is equal to 1, the chain reaction proceeds at a constant rate, and the system is said to be critical.

Devices that are designed so that the fission chain reaction can proceed in a controlled manner are called nuclear reactors. In a reactor, this control is accomplished by varying the value of k_{eff}, which can be done by the persons operating the system. To increase the power being produced by a reactor, the operator increases k_{eff} to a value greater than unity so that the reactor becomes supercritical. When the desired power level has been reached, he returns the reactor to critical by adjusting the value of k_{eff} to be unity, and the reactor then maintains the specified power level. To reduce power or shut the reactor down, the operator merely reduces k_{eff}, making the reactor subcritical. As a result, the power output of the system decreases

3.5.1.2 Neutron multiplication

When the system is subcritical (i.e., $k_{eff} < 1$), the fission chain cannot be sustained without supplying neutrons from outside the system, this is because the neutron population within the media eventually becomes zero. However, when there is a sufficient supply of neutrons, a constant neutron population is maintained. In the following discussion, the time axis is treated in a discrete manner and is separated by each neutron generation. The number of externally supplied neutrons per generation as S is designated.

The initial external neutrons S in the first generation undergo multiplication and become Sk_{eff} at the end of the generation. This Sk_{eff} group of neutrons multiply in the same manner in the next generation, and the resultant number of neutrons becomes Sk_{eff}^2 at the end of the second generation, Sk_{eff}^3 at the end of the third generation, and so on. Thus, the initial number of external neutrons S becomes Sk_{eff}^n after n generations.

Total number of neutrons should be considered in the system after considerable generations. Please keep in mind that S neutrons are supplied in each generation. The neutrons that exist in the present generation originate from each and every preceding generation combined, each undergoing neutron multiplication. Therefore, the total number of neutrons F existing in the present generation is a summation of neutrons originating from every preceding generation, and this can be expressed as follows:

$$F = S + Sk_{eff} + Sk_{eff}^2 + k_{eff}^3 + \cdots \frac{S}{1 + k_{eff}}. \tag{3.452}$$

In other words, the number of neutrons S from the neutron source becomes $F = S/(1 - k_{eff})$ with the multiplication rate of k_{eff} (<1), that is, it becomes $1/(1 - k_{eff})$ times S.

This means that the neutron multiplication M in this media is given by:

$$M = \frac{F}{S} = \frac{S}{S(1 - k_{eff})} = \frac{1}{1 - k_{eff}}. \tag{3.453}$$

As the media approaches criticality, that is, as $k_{eff} \rightarrow 1$, neutron multiplication M approaches infinity. To avoid the practical difficulties entailed with handling infinite neutron multiplication, the inverse of M is usually taken, denoted as $1/M$, and the inverse multiplication of M is referred, approaching zero as $k_{eff} \rightarrow 1$:

$$\frac{1}{M} = 1 - k_{eff}. \tag{3.454}$$

If neutron multiplication M is available to measure, then the critical mass can be predicted by plotting the $1/M$ value versus fuel mass (number of fuel plates, core size, etc.) and extrapolating the $1/M$ curve to $1/M = 0$ as shown in Fig. 3.72.

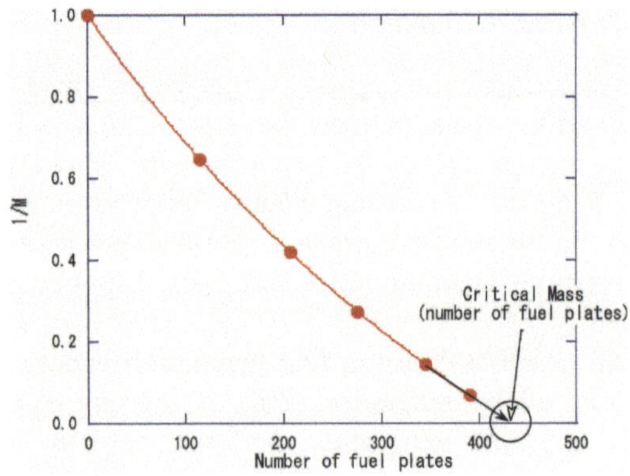

Figure 3.72 Plot of $1/M$ versus fuel mass [35].

3.5.2 Nuclear reactor fuel

3.5.2.1 Conversion

The η fission neutrons are emitted per neutron absorbed by a fissile or fissionable nucleus. To become critical, a reactor obviously must be fueled with a nuclide having a value of η greater than 1. If η were less than 1, a fission in one generation would necessarily lead to less than one fission in succeeding generations, and the reactor could never achieve a critical state. However, η cannot be exactly 1 since some neutrons inevitably lost either in nonfission absorption reactions or by escaping from the system.

A check of fission data shows that η is, in fact, substantially greater than 1 for the fissile nuclides at all incident neutron energies, and for the fissionable but nonfissile nuclei such as U-238, well above their fission thresholds. However, with the nonfissile nuclides, only those neutrons in a reactor with energies above the fission threshold are able to induce fission. For a reactor fueled only with fissionable but nonfissile nuclides, the value of η is equal to the actual value of η, multiplied by the fraction of the neutrons absorbed in the reactor with energies above the fission threshold. Regrettably this number is always less than 1, and it follows that the reactors cannot be made critical with nonfissile material alone. Fissile nuclides are therefore essential ingredients in all reactor fuels.

Only one fissile nuclide U-235, is found in nature, where it occurs with an isotopic abundance of 0.72%. The remainder of natural uranium, except for a trace of U-234, is U-238. Thus, only 1 uranium atom out of 139 is U-235. Despite this low concentration of fissile isotope it is possible to fuel certain types of critical reactors with natural uranium, and all of the early reactors were of this type.

As is the case for any natural resource, the supply of U-235 is finite. Indeed, if nuclear power was based on the fission of U-235 alone, the era of nuclear energy would be a comparatively short one. Fortunately, it is possible to manufacture certain fissile isotopes from abundant nonfissile material, a process known as conversion. The two most important fissile isotopes that can be produced by conversion are Pu-239.

The Pu-239 is obtained from reactions as follows:

$$^{238}\mathrm{U}(n,\gamma)^{239}\mathrm{U} \xrightarrow{\beta^{-1}} {}^{239}\mathrm{Np} \xrightarrow{\beta^{-}} {}^{239}\mathrm{Pu}. \tag{3.455}$$

The fertile isotope in this case is the U-238; Np-239 is the intermediate nucleus. To realize these reactions in practice, U-238 must be irradiated in a reactor. This, however, occurs automatically in most power reactors of current design. Thus, most present-day reactors are fueled with uranium that is only slightly enriched in U-235. Practically all of the fuel in these reactors is U-238, and the conversion of U-238 to Pu-239 takes place as a matter of course during the normal operation of these reactors. The plutonium is later extracted chemically from the fuel, uranium and plutonium being in different elements.

After Pu-239 has been formed in a reactor, it may absorb a neutron and undergo fission or be transformed into Pu-240. The Pu-240, which is not fissile, may in turn capture another neutron to produce the fissile isotope Pu-241. Finally, the Pu-241 may undergo fission or be transformed into Pu-242. Thus, the plutonium extracted from reactor fuel by reprocessing contains, in decreasing amounts, the isotopes Pu-239, Pu-240, and Pu-242. The fractional content of each isotope depends on the burnup of the fuel

The conversion process is described quantitatively in terms of the parameter C, which is called the conversion ratio or sometimes the breeding ratio. This is defined as the average number of fissile atoms produced in a reactor per fissile fuel atom consumed. Thus, when N atoms of fuel are consumed, NC atoms of fertile material are converted into new fissile atoms. However, if the newly produced fissile isotope is the same as the isotope that fuels the reactor, the new atoms may later be consumed to convert another $NC \times C = NC^2$ atoms of fertile material; these may be consumed to convert NC^3 fertile atoms; and so on, as similar with the neutron multiplication in Section 3.5.1.2. In this way, it is easy to see that the consumption of N fuel atoms results in the conversion of a total of as follows:

$$NC + NC^2 + NC^3 + \cdots \frac{NC}{1-C}. \tag{3.456}$$

Fertile atoms provide C is less than 1. When $C = 1$, an infinite amount of fertile material can be converted starting with a given amount of fuel.

3.5.2.2 Breeding

A most important situation occurs when C is greater than 1. In this case, more than one fissile atom is produced for every fissile atom consumed, which is a process described as breeding. Reactors that designed so that breeding will take place are called breeder reactors or simply breeders. Reactors that convert but do not breed are called converters; reactor that neither convert nor breed but simply consume fuel are called burners. Breeders are remarkable devices for, in addition to providing power through the energy released in fission, they actually produce more fissile fuel than they consume.

Needless to say, it is more difficult to design reactor that will breed than one that merely converts. For one thing, while η must be greater than 1 for conversion, it must be greater than 2 for breeding. This is because, in any reactor, one fission neutron must eventually be absorbed in fuel just to keep the reactor critical and maintain the chain reaction. If the reactor is to breed, more than one neutron must be absorbed in fertile material to produce then new fissile isotope. In actual fact, η must be substantially greater than 2, in any reactor, some neutrons are inevitably absorbed by nonfuel atoms or lost by leakage. In this connection, it is important to recognize that η is not a constant, but depends on the energy of the neutron that induces the fission. By contrast, the thermal neutron values of η for U-235 and Pu-239, 2.07, and 2.14, respectively, are not sufficiently greater than 2 to permit breeding.

3.5.3 Nuclear power plant

3.5.3.1 Steam power plant

Since the world's first reactor, which was composed of natural uranium and graphite, was brought to critical in Chicago on December 2, 1942, a large number of reactors have been designed and built for a variety of purposes; for the propulsion of ships, aircraft, rockets, and satellites; for medical irradiation; and for the generation of electrical power. Among several reactors, attention is made on reactors that are used in power plants for the generation of electricity. In all nuclear power plants, the fission energy obtained from the reactor is emphasized to produce steam either directly in the reactor or in auxiliary heat exchanger called steam generators. The combination of reactor and steam generator combination is called the nuclear steam supply system (NSSS). NSSS serves the same function as the steam boiler in a conventional fossil-fuel plant.

Fig. 3.73 schematically shows the steam system in a nuclear power plant from the point where the steam leaves the reactor or steam generators to the point where it returns as feed water to be reconverted into steam. As indicated in the figure, the steam is used to drive steam turbines, which are coupled to a generator to produce electricity. Turbines consist of a series of bladed wheels affixed to an axle, which rotates at high speed as steam at high temperature and pressure strikes the turbine blades. Steam is just the gaseous form of water, and it does not contain any liquid water. To emphasize this, steam is often referred to as dry steam; steam mixed with liquid water is called wet steam. Steam is always delivered to a turbine as dry as possible, but as it passes through successive turbine stages, its temperature and pressure are reduced, and some of the steam is condensed to droplets of liquid water. It has been found that these droplets impinging on the turbine blades lead to excessive erosion of the blades and hence to reduced turbine lifetime. In addition, the inertia of the liquid in the turbine tends to reduce the turbine efficiency.

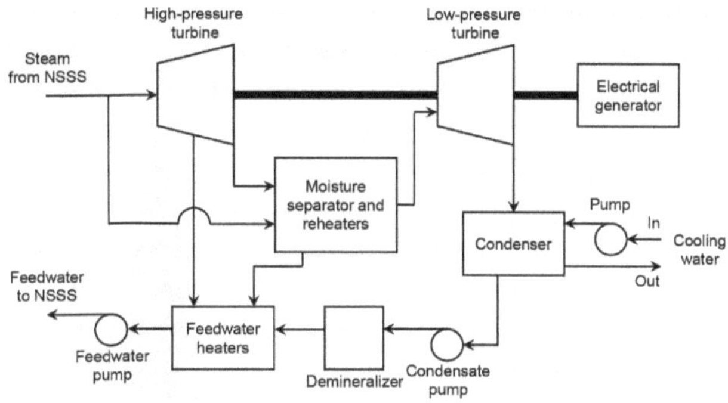

Figure 3.73 Simplified diagram of a nuclear steam power plant.

It is recalled from thermodynamics that water is converted to steam by adding heat (the latent heat of vaporization) to water at constant pressure and temperature. The steam so produced is said to be saturated. The extraction of any energy from such saturated steam, as occurs when it passed through a turbine, results in the condensation of some of the steam to liquid water. However, if the steam is heated above the temperature at which it was evolved from boiling water, some heat can be extracted without condensation taking place. Steam that has been heated in this way is said to be superheated. By superheating the steam before it enters a turbine, the total amount of water produced in the turbine can be significantly reduced. Superheating may be done in a separate unit designed for the purpose, in a portion of the steam generator, or in the reactor.

Another approach to this problem is to remove the wet steam from the turbine when the water content has reached a specified level (separate the steam from the water, which is done in a moisture separator), reheat this spent steam using the higher pressure and hotter steam from the reactor steam generators, and pass the reheated steam (which may be superheated at this point) to one or more lower pressure turbines as shown in Fig. 3.73. Moisture separation and reheating is usually performed in the same device: one for each low-pressure turbine. The wet steam passes over metal blades that draw off the water; then as dry steam, it passes around a series of tubes caning the hotter NSSS steam. Reheating is utilized in fossil as well as nuclear power plants.

From the low-pressure turbines, the spent steam passes into a condenser, where it is cooled by water from as suitable outside source, and the steam is condensed to water. In this part of the steam cycle, heat is rejected by the system. This rejection of heat is an essential part of the operation of the plant from a thermos dynamic point of view. Thus, it is not possible to skip the condenser portion of the cycle and return the spent steam directly to the reactor side of the system.

3.5.3.2 Core components

Before discussing specific reactors for the generation of steam, it is helpful to first review the principal component parts or regions of typical reactors. It must be emphasized that this drawing is merely schematic and does not depict an actual reactor.

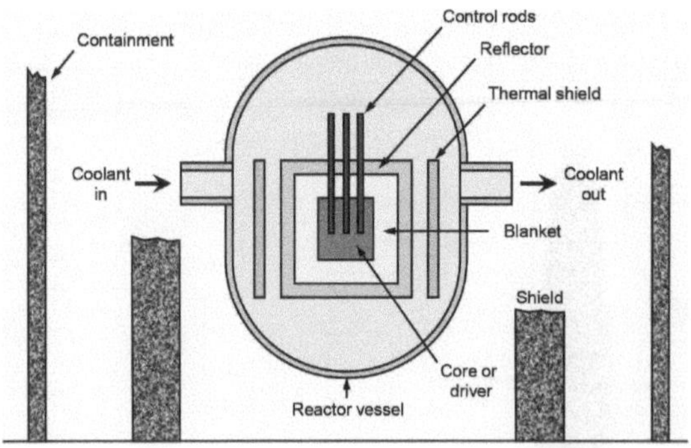

Figure 3.74 Main components of a nuclear reactor with radiation shield and containment.

The central region of a reactor shown in Fig. 3.74 is called the core. In a thermal reactor, the core contains the fuel, the moderator, and a coolant. The fuel includes the fissile isotope that is responsible for both the criticality of the reactor and the release of fission energy. However, the fuel may also contain large amounts of fertile material. Indeed, most modem power reactors (which are thermal) are fueled with uranium enriched to only a few percent U-235, so that most of the fuel is actually U-238.

The moderator, which is only present in thermal reactors, is used to moderate—that is, to slow down—the neutrons from fission to thermal energies. It was shown that nuclei with low-mass numbers are most effective for this purpose, so the moderator is always a low-mass-number material. Water (two-thirds of atoms of which are hydrogen), heavy water, and graphite (the common form of carbon) are often used as moderators. Beryllium and beryllium oxide (BeO), a white ceramic material, have been used occasionally, but they are very costly.

The coolant, as the name implies, is used to remove the heat from the core and other parts of the reactor where heat may be produced water, heavy water, and various gases are the most commonly used coolants for thermal reactors. In the case of water and heavy water, these coolants also frequently serve as the moderator.

The region adjacent to the core, or to the blanket if one is present, is called the reflector. This is a thick of moderating material whose function can be understood in the following way. Suppose the core (or blanket) were bare—that is, open and exposed to the air. In this case, all of the neutrons that escaped from the surface of the core would be lost from the reactor—none would return. By placing a region of moderator around the reactor, however, some of these neutrons are returned to the core or blanket after one or more collisions in the reactor. Not all of the neutrons return, of course, but some do, thus saving neutrons for the chain reaction.

The control rods shown in Fig. 3.74 are movable pieces of neutron-absorbing material; as their name suggests, they are used to control the reactor. Since they absorb neutrons, any movement of the rods affects the multiplication factor of system. Withdrawal of the rods increases k_{eff}, whereas insertion decreases k_{eff}. Thus, the reactor can be started up or shut down, or its power output can be changed by the appropriate motion of the rods. Control rods must also be adjusted to keep a reactor critical and operating at a specified power level, when, in the course of time, fuel is consumed and various neutron-absorbing fission products accumulate in the reactor. Several materials have been used for control rods: boron steel: since boron has a high neutron absorption cross-section; hafnium or cadmium, metallic elements that are strong thermal neutron absorbers; silver; and various alloys of these metals. The rods may be cylindrical in shape (rods, in the true sense of the word) or may be sheets, blades, or crossed blades, which are called cruciform rods.

3.5.3.3 Reactor components

The various reactor components just described are all located within the reactor vessel, which, if the components are under pressure, is also called the pressure vessel. To reduce the thermal stresses in the reactor vessel caused by the absorption of γ-ray emanating from the core, it is necessary in some reactors to place a thermal shield, a thick layer of γ-ray absorbing material (usually iron or steel), between the reflector and the inner wall of the vessel. Since the thermal shield absorbs considerable energy, it ordinarily must be cooled along with the core and blanket.

The reactor vessel and all other components of the NSSS that contain sources of radiation are surrounded by radiation shielding in varying amounts for the protection of plant personnel during normal operation of the reactor; the amount of shielding varies according to the design of the reactor. To protect the general public from the consequences of a reactor accident—in particular, one involving the release of fission products from the reactor—entire reactor installation is enclosed in a containment structure. In some plants, this takes the form of a heavily constructed building housing the entire NSSS; in other installations, the containment is split, with a portion surrounding the reactor (the primary containment), and the remainder coinciding with the reactor building (the secondary containment).

In addition to the several components illustrated in Figs. 3.73 and 3.74, elaborate safety systems must be included in a nuclear power plant for use in emergencies. Provisions must also be made for the loading of fuel and storage of radioactive spent fuel prior to its shipment from the plant. Various sensing devices must be located at several points inside and outside the reactor vessel to monitor the operation of the system. Finally, a considerable amount of structural material is necessary to give support and integrity to the plant.

3.5.4 Light-water reactors

The most widely used reactor in the world today for producing electric power is the thermal reactor, which is moderated, reflected and cooled by ordinary (light) water. As noted earlier, water has excellent moderating properties. In addition, its thermodynamic properties are well understood, and it is readily available at little cost. However, water has a high vapor pressure, which means that light-water reactors

(LWRs) must be operated at high pressures. Water also absorbs thermal neutrons to such an extent that it is not possible to fuel a LWR with natural uranium; it simply would never become critical. Uranium in water reactors must always be enriched, at least to some extent.

There are basically two types of LWR now in use: the pressurized-water reactor (PWR) and the boiling-water reactor (BWR). Both of these types of reactors are well established in the United States and abroad.

3.5.4.1 Pressurized-water reactor

The PWR was one of the first types of power reactors developed commercially in the United States. This reactor has also become standard on nuclear-powered ships and naval vessels throughout the world.

Fig. 3.75 shows a cross-sectional view of a typical PWR. As indicated in the figure, water enters the pressure vessel at a temperature of about 290°C, flows down around the outside of the core where it serves as a reflector, passes upward through the core where it is heated, and then exits from the vessel with a temperature of about 325°C. The water in a PWR is maintained at a high pressure: approximately 15 MPa. At this pressure, the water will not boil, at least not to any great extent.

Since the water does not boil in the reactor, the steam for the turbines must be produced external to the reactor. This is done in steam generators, which are heat exchangers with pressurized water on the hot side. A typical steam generator is shown in Fig. 3.76. High-pressure, heated coolant water from the reactor enters at the bottom and passes upward and then downward through several thousand tubes each in the shape of an inverted U. The outer surfaces of these tubes are in contact with lower pressure and cooler feed water returning from the turbine condenser.

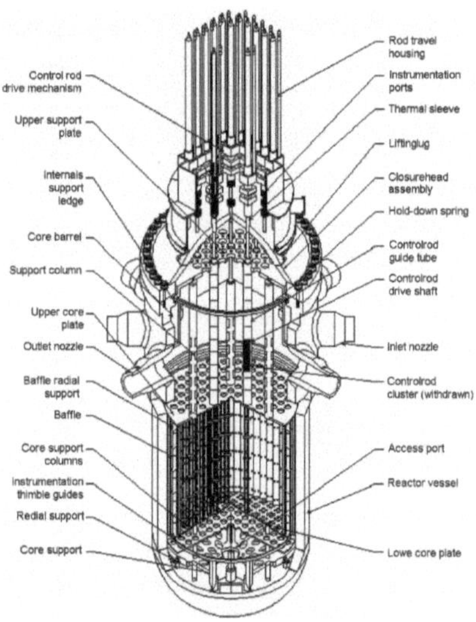

Figure 3.75 Reactor vessel in PWR (Wikipedia commons: PWR reactor pressure vessel).

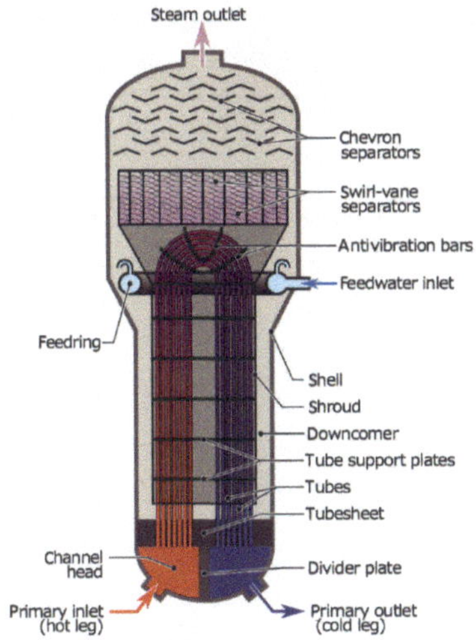

Figure 3.76 Steam generator in PWR (Wikipedia commons: vertical recirculating-type steam generator (typical of Westinghouse and combustion engineering-designed reactors) and components).

Heat transferred from the hot water inside the tubes causes the feed water to boil and produce steam. The lower section of a steam generator where this boiling occurs is called the evaporator section. The wet steam produced in the evaporator passes upward into a portion of the steam generator known as the steam drum section. Here the steam is dried in various moisture separators before exiting to the turbines. Steam generators are also manufactured with straight tubes rather than U tubes. Large PWR systems utilizes as many as four steam generator, which produce steam at about 293°C or 560°F and 5 MPa or 750 psi. This gives an overall efficiency of between 32% and 33% for a PWR plant.

As indicated in Fig. 3.77, a pressurizer consists of a tank, containing steam and water in its upper and lower sections, respectively, with a pressure-actuated spray nozzle at the top and pressure-actuated immersion-type heaters at the bottom. The device operates in the following way. Suppose that the power output of the turbine is reduced in response to a drop in the electrical load on the plant. This leads to a temporary increase in the average temperature of the reactor coolant and a corresponding increase in coolant volume. The expansion of the coolant raises the water level in the pressurizer, which raises the pressure of the steam and actuates the valves to the spray nozzle. Water from one of the cold legs of the reactor coolant system sprays into the top portion of the pressurizer and condenses some of the steam. This quenching action reduces the pressure and limits the pressure rise. When the electrical load is increased instead, the attendant decrease in coolant volume drops the water level and reduces the pressure in the pressurizer. This, in turn, causes some of the water to flash to steam, again limiting the pressure change. At the same time, the pressure drop actuates the heater to further limit the pressure reduction

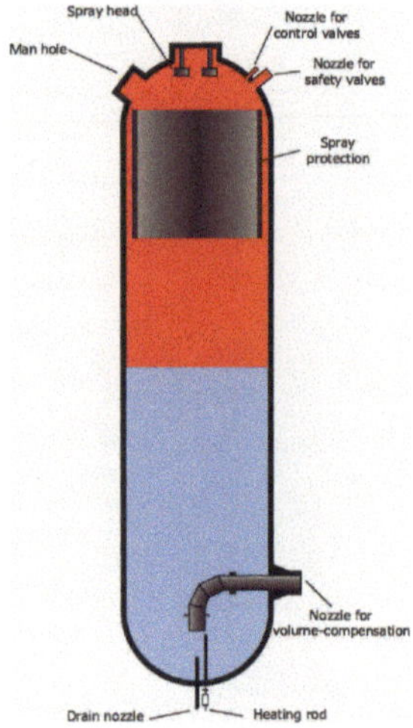

Figure 3.77 Pressurizer of PWR (Wikipedia commons: sectional view of a pressurizer).

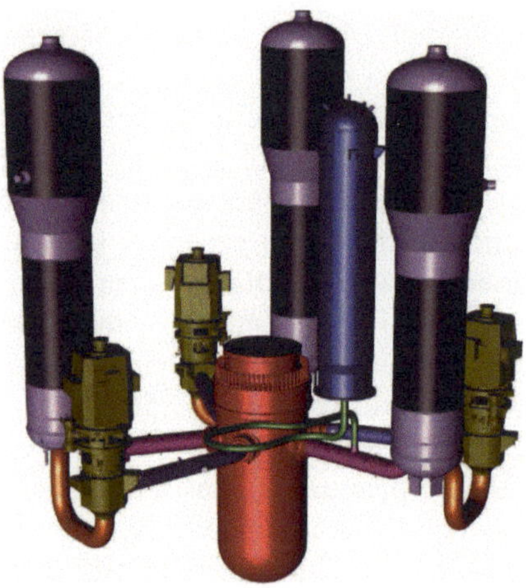

Figure 3.78 Reactor coolant system of four loops in PWR (Wikipedia commons: schematic of reactor coolant system for PWR).

The major components of a PWR steam supply system are shown in Fig. 3.78. It should be noted that there are four coolant pumps, one for each coolant loop, but only one pressurizer for the entire system.

The fuel in PWR reactors is slightly enriched (from 2 to 5 w/o uranium dioxide, UO_2, which is a black ceramic material with a high melting point of approximately 2800°C or 5,070°F). The UO_2 is in the form of small cylindrical pellets, about 1 cm in diameter and 2 cm long, which are usually concave on the ends. The pe llets are loaded into sealed stainless steel or zircaloy (Zr) tube about 4 m long. The result is a solidly packed fuel rod or fuel pin. In addition to providing support for the fuel pellets, the Zr fuel tubes also prevent the escape to the passing coolant of fission products, especially fission product gases that are released from the pellets during the operation of the reactor. In this context, the fuel tubes are also known as fuel cladding. PWR fuel of the type described here is nominally capable of delivering about 30 MWd per kg of contained uranium before it must be replaced.

On a few occasions, it has been observed that some fuel pellets, although expanding initially as the fuel reaches its operating temperature, later contract due to a gradual increase in the density of the uranium dioxide and corresponding reduction in its specific volume (cm^3/g). This densification of the fuel, as the phenomenon is called, is evidently caused by the migration and combination of small voids within the ceramic. The decrease in the pellet volume leaves void spaces within the fuel tubes. Because of the high pressure of the mode1ator coolant, this produces large stresses across the fuel tubes, increasing the likelihood of fuel-tube rupture. To circumvent this problem, the fuel tubes are pressurized with helium at about 3.4 MPa. As fission product gases accumulate in the tubes over the life of the core, the pressure in the tubes gradually builds up to about 14 MPa near the end of the core life.

The completed, pressurized fuel rods are next arranged in a square lattice structure called a fuel assembly, the fuel assemblies are then arranged in a near-cylindrical array to form the core. The fuel rods in the fuel assemblies are kept a part by various spacers. This is important since rods that come into contact may overheat and release fission products.

Control of the PWR is accomplished by the use of control rods, which normally enter the core from the top, and a chemical shim system. In this latter system, the value of the multiplication factor is changed by varying the concentration of a neutron absorber (usually boric acid) dissolved in the coolant water. Fuels capable of higher burnup have gadolinium included to offset the higher enrichment of the fuel during the early stages of core life.

3.5.4.2 Boiling water reactor

For a long time, it was thought that if water were permitted to boil within a reactor, dangerous instabilities would result because of uneven formation and movement of the steam bubbles. Experiments carried out in the early1950s (the famous BORAX experiments) showed that this was indeed the case if the boiling occurred at low pressure. However, when the pressure was raised, the boiling became stable and the reactor was controllable. Since these first demonstrations of the feasibility of boiling water reactors, the BWR has reached a high state of development. The BWR and PWR are now competing on a neck-and-neck basis for the commercial nuclear power market in LWR.

There are obvious advantages to a BWR. For one thing, the steam is formed in the reactor and goes directly to the turbines; steam generators in separate heat transfer loops are not necessary as they are with the PWR. For this reason, the BWR is said to operate in a direct cycle. Furthermore, it is recalled that, for a given amount of water, more heat can be absorbed as latent heat—that is, the heat necessary to vaporize a liquid—than as sensible heat, which, in the PWR, only changes the temperature of the fluid. Therefore, it follows that less water must be pumped through a BWR per unit time than through a PWR for the same power output. However, the water becomes radioactive in passing through the reactor core. Since this water is utilized in the electricity-producing side of the plant, all of the components of the steam utilization system: the turbines, condenser, reheaters, pumps, piping, and so on must be shielded in a BWR plant.

The pressure in a BWR is approximately 7 MPa: about one-half the pressure in a PWR. As a result, the wall of the pressure vessel for a BWR need not be as thick as it is for a PWR. However, it turns out that the power density (watts/cm^3) is smaller in a BWR than in a PWR, and so overall dimensions of a pressure vessel for a BWR must be larger than for a PWR of the same power. As far as the cost of the pressure vessel is concerned, these two effects more or less tend to offset one another.

Starting with the low chamber or plenum, the water moves upward through the core; as it does, it receives both sensible and latent heat. By the time, it reaches the top of the core and enters the upper plenum, a portion of the coolant has been vaporized. This mixture of steam and liquid water next passes through steam separators, which remove most of the water. The steam then goes through a driver assembly, which removes the remaining water; it then exits from the reactor via a steam line to the turbine. The residual water from the separators and the dryer mixes with feed water returning from the condenser and passes downward through an annular region external to the core, between the core shroud and the reactor vessel, known as the downcomer, and returns to the lower plenum.

The driving force behind the flow of the coolant through the core is provided by a recirculation system. This system consists of two loops external to the reactor vessel, each containing are circulation pump. These pumps withdraw water near the bottom of the downcomer and pump it at a higher pressure through a pipe manifold to a number of jet pumps (18–24 depending on the reactor power) located within the downcomer. A BWR of the type described produces saturated steam at about 290°C and 7 MPa. The overall of efficiency of a BWR plant is on the order of 33%–34%. The fuel for a BWR is essentially the same as for a PWR, that is, slightly enriched UO_2 pellets in sealed tubes, and the core configurations of the two types of reactors are more or less identical. However, BWR control rods are always placed at the bottom of the reactor rather than at the top, as in the case of the PWR. The reason for this is that much of the upper portion of the core of a BWR is normally occupied by steam voids, and movement of the rods in this region does not have as large an effect on the value of k_{eff} as rod motions in the lower, water-filled part of the core. The rods are thus placed near the part of the reactor where they will do the most good.

References

[1] J.A. Beattie, O.C. Bridgeman, A new equation of state for fluids: I. Application to gaseous ethyl ether and carbon dioxide, J. Am. Chem. Soc. 49 (1927) 1665–1667.

[2] J.A. Beattie, O.C. Bridgeman, A new equation of state for fluids: II. Application to helium, neon, argon, hydrogen, nitrogen, oxygen air and methane, J. Am. Chem. Soc. 50 (1928) 3133–3138.

[3] M. Benedict, G.B. Webb, L.C. Rubin, An empirical equation for thermodynamic properties of light hydrocarbons and their mixtures: I. Methane, ethane, propane, and n-butane, J. Chem. Phys. 8 (4) (1940) 334–345.
[4] F.P. Incropera, A.S. Lavine, T.L. Bergman, D.P. DeWitt, Fundamentals of Heat and Mass Transfer, Wiley, 2007.
[5] J.P. Holman, J.P. Holman, Heat Transfer, McGraw-Hill Companies, 1976.
[6] W.M. Rohsenow, [1951] A Method of Correlating Heat Transfer Data for Surface Boiling of Liquids, MIT Division of Industrial Cooporation, Cambridge, MA, 1951.
[7] S.S. Kutateladze, On the transition to film boiling under natural convection, Kotloturbostroenie 3 (1948) 10–12.
[8] N. Zuber, On the stability of boiling heat transfer, Trans. Am. Soc. Mech. Engrs. (1958) 711–720.
[9] J.G. Collier, J.R. Thome, Convective Boiling and Condensation, third ed., Clarendon Press, Oxford, 1994.
[10] J.H. Lienhard, A Heat Transfer Textbook, Courier Dover Publications, 2019.
[11] Chapter 3-CHF—transition boiling, in: Y. Koizumi, M. Shoji, M. Monde, Y. Takata, N. Nagai (Eds.), Boiling, Elsevier, Boston, MA, 2017.
[12] M.T. Cichelli, C.F. Bonilla, Heat transfer to falling liquids under pressure, Trans. AIChE 41 (1945) 755.
[13] L.A. Bromley, Heat transfer in stable film boiling, US Atomic Energy Commission, Technical Information Division, 1949.
[14] J.S. Tong, Y.S. Tong, Boiling Heat Transfer and Two-Phase Flow, second ed., Taylor & Francis, London, 1997.
[15] P.G. Kroeger, N. Zuber, An analysis of the effects of various parameters on the average void fractions in subcooled boiling, Int. J. Heat Mass Transfer 11 (1966) 211–233.
[16] K. Roko, M. Shiraha, Heat transfer in post-dryout region in vertical steam generator (2nd report simplified calculation), Trans. JSME 47 (422) (1981) 2014–2020 (in Japanese).
[17] V.E. Schrock, L.M. Grossman, Florced convection boiing in tubes, Nucl. Sci. Eng. 12 (4) (1962) 474–481.
[18] H. Umekawa, M. Ozawa, N. Ishida, Dryout and post-dryout heat transfer in a natural circulation loop of liquid nitrogen, Heat Transfer – Japan. Res. 26 (7) (1998) 449–458.
[19] Y. Katto, A generalized correlation of critical heat flux for the forced convection boiling in vertical unifomly heated round tubes, Int. J. Heat Mass Transfer 21 (1978) 1528–1542.
[20] Y. Katto, An improved version of the generalized correlatoin of critical heat flux for the forced convective boiling in uniformly heated vertical tubes, Int. J. Heat Mass Transfer 27 (1984) 1641–1648.
[21] G.P. Celata, M. Cumo, A. Mariani, M. Simoncini, G. Zummo, Rationalization of existing mechanistic models for the prediction of water subcooled flow boiling critical heat flux, Int. J. Heat Mass Transfer 37 (1994) 347–360.
[22] J. Weisman, B.S. Pei, Prediction of critical heat flux in flow boiling at low qualities, Int. J. Heat Mass Transfer 26 (10) (1983) 1463–1477.
[23] E.M. Sparrow, J.L. Gregg, A boundary-layer treatment of laminar-film condensation, J. Heat Transfer 81 (1) (1959) 13–18.
[24] J.C.Y. Koh, E.M. Sparrow, J.P. Hartnett, The two phase boundary layer in laminar film condensation, Int. J. Heat Mass Transfer 2 (1) (1961) 69–82.
[25] W. Nusselt, The surface condensation of water vapour, Zeitschr. Ver. Deut. Ing. 60 (1916) 541–546.

[26] D. Tanner, D. Pope, C. Potter, D. West, Heat transfer in dropwise condensation at low steam pressures in the absence and presence of non-condensable gas, Int. J. Heat Mass Transfer 11 (2) (1968) 181–190.
[27] G.F. Hewitt, J. Barbosa, Heat Exchanger Design Handbook, Begell House, New York, 2008.
[28] BP Statistical Review of World Energy 2019, 68th ed., BP p.l.c., 2019.
[29] Technically Recoverable Shale Oil and Shale Gas Resources, U.S. EIA, 2013.
[30] LNG Small Scale Plant (Technical Guide Line), The Japan Gas Association, 2000.
[31] Y. Mizutani, Combustion Engineering, Morikita, 1977, pp. 35–58.
[32] Tokyo Gas Co., Ltd., <https://home.tokyo-gas.co.jp/gas/userguide/shurui.html>, 2020.
[33] W.M. Chase Jr (Ed.), NIST-JANAF Thermochemical Tables, fourth ed., American Chemical Society, 1998.
[34] JSME Combustion Handbook, The Japan Society of Mechanical Engineers, 1995.
[35] T. Misawa, H. Unesaki, C.H. Pyeon, Nuclear Reactor Physics Experiments, Kyoto University Press, Kyoto, Japan, 2010. Chapter 2.

Further reading

C.F. Colebrook, Turbulent flow in pipes, with particular reference to the transition region between smooth and rough pipe laws, J. Inst. Civ. Eng. 11 (4) (1939) 133–156.
A.N. Kolmogorov, The local structure of turbulence in incompressible viscous fluid for very large reynolds numbersTranslated by Levin, V., 1991 Proc. R. Soc. A 434 (1991) 9–13.
P. Kundu, I. Cohen, Fluid Mechanics, fourth revised ed., Academic Press, 2008.
O. Ladyzhenskaya, The Mathematical Theory of Viscous Incompressible Flows, second ed., Gordon and Breach, New York, 1969.
J.R. Lamarsh, A.J. Baratta, Introduction to Nuclear Engineering, third ed., Prentica-Hall Inc., Upper Saddle River, NJ, 2012. Chapter 4.
P.H. Oosthuizen, W.E. Carscallen, Introduction to Compressible Fluid Flow, second ed., CRC Press, 2013.
L. Prandtl, O. Tietjens, Applied Hydro-and Aero-mechanics, McGraw-Hill, New York, 1934.
H. Tennekes, J.L. Lumley, A First Course in Turbulence, The MIT Press, Cambridge, London, 1972.

Power generation and society

4

Motonari Haraguchi[1], *Hiroshi Miyano*[2] *and Takao Nakamura*[3]
[1]Mitsubishi Power, Ltd., Yokohama, Japan, [2]Hosei University, Chiyoda, Tokyo, Japan, [3]Osaka University, Suita, Osaka, Japan

Japan's electrification rate is about one-quarter. Power generation plays an important and highly influential role in; society. How the power generation system works has a great impact on society. Systems responsible for power generation in Japan are roughly divided into thermal power generation and nuclear power generation. Fossil fuels are the source of the thermal power generation. After describing various fossil fuels, the utilization of fuels and the power generation systems for the various fuels with each characteristic are described. A series of steps from the location determination to construction of a typical thermal power plant is briefly introduced. Important regulation and standards in that process will be outlined. As for the nuclear power generation, safety is an important issue. Based on the lessons learned from the experience of the Fukushima Daiichi Nuclear Power Plant accident, which was caused by the huge tsunami triggered by the Great East Japan Earthquake, improvements for nuclear safety assurance regulations and framework were performed. The important points and ideas are introduced.

4.1 Thermal power generation

4.1.1 Important fundamentals

4.1.1.1 Various types of fuel

There are various types of fuel. The combustible components in the fuel are carbon (C), hydrogen (H), and sulfur (S), where the main components are carbon and hydrogen. Fig. 4.1 shows the molar ratio of hydrogen to carbon atoms (number ratio) H/C for various fuels. Combustion reaction generates carbon dioxide (CO_2), hydrogen (H), and water (H_2O). Since oxygen dioxide is a global warming substance whose emissions are restricted, it is desirable that the H/C in fuel should be as large as possible. Hydrogen is the largest fuel with a molar ratio of H/C, and coal is the smallest. In this sense, hydrogen is the cleanest fuel and attracts attention from the viewpoint of protecting the global environment. Methane (CH_4), which is attracting attention as new shale gas and methane hydrate, is liquefied at low temperature ($-162°C$) under atmospheric pressure and transported as liquefied natural gas (LNG). And liquefied

Fundamentals of Thermal and Nuclear Power Generation. DOI: https://doi.org/10.1016/B978-0-12-820733-8.00004-2

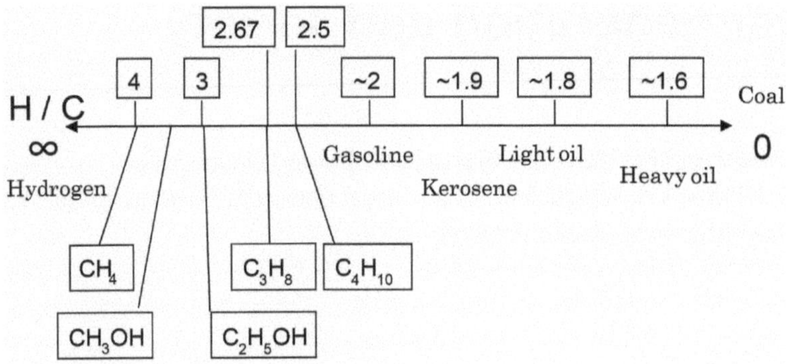

Figure 4.1 Hydrogen and carbon of fuel.

petroleum gas, which can be also transported, is a liquefied form of propane C_3H_8 or butane C_4H_{10} fuel. Petroleum gas is a gas at normal temperature and pressure, but it can easily be liquefied at a relatively low pressure of less than 10 bar with normal temperature.

There are two types of calorific value of fuel. The maximum value obtained when steam in exhaust gas is cooled down to water is called high heating value (HHV), and the value obtained with steam condition is called low heating value assuming that none of the water vapor resulting from the process is condensed out so that its latent heat is not available. The former generates more heat than the latter by the latent heat of vaporization. The calorific values of the two types differ by about 10% for petroleum fuels and 6% for natural gas. Boilers and combustion engine efficiency, which are often defined per calorific value, are higher when lower calorific values are adopted.

The reduction of carbon dioxide emissions through fuel change is being carried by converting the fuel into a large H/C fuel. LNG-fueled ferries and tugboats have also been used on ships. By switching from liquid fuel with a molar ratio H/C of about 2 into gas fuel with a molar ratio H/C of about 4, we can reduce CO_2 emissions by about 30% on an HHV basis. Human beings used solid fuels such as coal and charcoal over the course of recorded human history, and have succeeded in developing high-performance power sources with liquid fuels since the 20th century. And this 21st century is to be an era of clean gas fuels with high H/C ratio in consideration of harmony with the environment. However, because methane has 21 times as much the global warming effect as carbon dioxide, we must be very careful about its release into the atmosphere.

Heavy oil, which is currently widely used for industrial purposes, is relatively inexpensive, but unfortunately, its H/C ratio is considerably low among liquid fuels. The heavy oil contains small amount of metal components such as vanadium, which brings on corrosion on high-temperature heat transfer surfaces in boilers. In addition, the sulfur content of heavy oil is transformed into toxic sulfur dioxide gas SO_2 through combustion, and portion of which becomes anhydrous sulfuric acid SO_3.

SO_3 is not only a harmful substance but also a cause of corrosion on the low-temperature heat transfer surfaces.

Alternative energy sources to achieve fossil fuel free goal are solar power, wind power, geothermal power, and tidal energy, which do not emit carbon dioxide during consumption process, and carbon-neutral biomass fuels such as wood and grass, whose emitted CO_2 can be recovered by cultivating again. In particular, bioethanol derived from corn, etc., is an ethyl alcohol with an H/C of 3, which has attracted attention in recent years because of its ease handling as a substitute for petroleum. However, it is very important not to damage the environment. Further, it is doubtful that food resources such as maize are used as energy resources, while food shortages are expected in the near future due to global population growth. For this reason, research and development has been carried out to produce biofuels from untapped marine algae, as the ocean accounts for 70% of Earth's surface area.

4.1.1.2 *Electric power*

The definition of electric power Watt (W) is instantaneous power to perform the work to move upward 102 g object like a cell phone in Fig. 4.2 by 1 m for 1 second, where 102 g object makes 1 Newton (N) gravity force onto a hand and this 1 W is the work to lift the cell phone against 1 N gravity force. And electric appliances, such as microwave ovens and washing machines, are marked with power consumption of kW, which means required instantaneous electric energy.

By the way, a large passenger plane takes off at an output of about 100 MW, which means that 100 million people lift their mobile phones simultaneously at a speed of 1 m/s, which is an incredible amount of power. 100 million mobile phones weigh as much as 10,200 tons.

The expression "Amount of power generated by 3000 households for 1 year" is sometimes used in newspaper and magazine articles introducing renewable energy such as solar and wind power. Some people think that 3000 of them can independently live on renewable energy, but that is not true. Solar and wind power are fluctuating, and it is impossible to be independent without electricity storage facilities. It may be correct to say that 30,000 houses can generate electricity for about 2.4 hours a day for a year. In short, we have to think about the time period when electricity does not come.

It is not a dream that 3000 homes will become independent once electricity storage facilities are in place and a smart grid effectively using renewable energy is created.

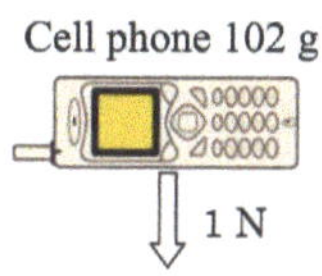

Figure 4.2 Power to lift 102 g cell phone.

4.1.1.3 Power (kW) versus work/energy (kWh)

The unit of Watt (W) comes from James Watt famous for an improved steam engine. Thomas Newcomen, also from England, was the inventor of the first steam engine, and about 50 years later, James Watt improved Newcomen's engine and greatly increase its efficiency. Horsepower is used as a unit of energy in ships and cars, and this horsepower is expressed in units of Pferde Stärke (PS) expressed in German language. It is said that the horsepower unit was introduced because many factories had introduced steam engines in James Watt era and it was necessary to quantitatively determined power charge based on horsepower values, while James Watt tried to measure how much work the horse did. The horse walked 188 ft/min at 175 pounds power, which has been decide to be 1 horsepower (PS). And 188 ft/min is just about 1 m/s, and the horse probably walked with care to make the calculations easier. Currently, 1 horsepower (PS) = 0.736 kW, which is a slightly modified version from the previous British definition.

In the case of electric appliances, 1.2 kW microwave ovens, rice cookers and dryers consume 1.6 horsepower. With this 1.6 horsepower, a luxurious carriage can move smoothly. Please note that these electric appliances generally consume more than the energy of a horse. On top of kW, there is another unit of kWh, which is often confusing. This 1 kWh refers to the total energy when 1 kW of electricity is used for 1 hour. The unit kW is just instantaneous energy, and kWh is cumulative energy, mathematically speaking, kWh is 1 hour integral value of kW.

For example, to boil 2 L water, it takes about 4 minutes with a 3 kW heater, but about 12 minutes with 1 kW heater. However, the cumulative electric power of both systems is same 0.2 kWh. The energy required to boil 2 L water is 0.2 kWh anyway, which means it does not matter whether it is quickly heated or slowly heated.

Electric power companies generally purchase electricity of photovoltaic power generation for relatively higher price, which will bring higher electricity price for general consumers. In that sense, the more solar power generation, the more consumers must pay for electricity.

4.1.1.4 Electric power generation

Solar energy creates wind, rain, and ocean currents on Earth. Hydroelectric power generation works by storing rainfall on mountains in a dam lake, turning the falling water energy into a rotating force of hydro turbine blades, and this rotating power generates electricity. Both wind current and ocean current power generates electricity by rotating turbine blades. All of them come from the solar energy.

Photovoltaic power generation directly converts sunlight into electricity using photovoltaic cells, a type of semiconductor device. On the other hand, fuel cells directly generate electricity in the process of converting hydrogen into water through electrochemical reaction. This fuel cell is different from ordinary batteries which can only storage and discharge electricity energy.

Recently, the performance of lithium batteries as secondary batteries has been remarkably improved, and they have been installed in electric vehicles (EV) and hybrid vehicles.

The most common method of power generation is to convert heat energy into kinetic energy by a thermal engine and then turn a generator to generate electricity. Thermal engines are roughly divided into two types; one is reciprocal type piston engines and rotating turbine with blades. Piston engines are also used in automobiles and are usually called gasoline engines or diesel engines except for automobile rotary engine. Normally, gasoline engines ignite by spark, on the other hand, diesel engines ignite by compression. Turbine engines are used in thermal power plants and are called steam or gas turbines. High-temperature steam is generated by boiler combustion or by heat recovery steam generator (HRSG), and then introduced into steam turbines for rotation generators. In the case of gas turbine, combustion gas directly rotates gas turbines.

A gas turbine is a combination of a compressor and a turbine. Intake air is compressed to high temperature as well as heated to high temperature, and fuel combustion makes even higher temperature for rotating a turbine and generator to generate electricity. Theoretically speaking, higher temperature brings greater power and higher efficiency. Surprisingly, increasing only the performance of the compressor does not substantially increase the power, whereas increasing the performance of both the compressor and turbine greatly increases the power.

Solar thermal power generation, in which oil and other materials are heated by solar light and steam is generated from this heat to drive steam turbines, is also active in desert areas with little rain. Geothermal power generation utilizes high-temperature steam from underground or steam generated from hot spring water. Biomass power generation used biomass fuel such as wood and processed pellets for steam turbine operation. Perhaps unexpectedly, steam turbines generate about 80% of the world's electricity at present.

4.1.1.5 Peak power

As an example of peak power situation, Japan's record peak power consumption in 2005 was 180 GW (equivalent to180 large nuclear power plants) on August 5. Since electricity cannot yet be stored in large quantities, power companies must construct and operate lots of power plants to meet the peak demand of the year, which means that many power plants must be operated with partial load operation except peak load time. On the other hand, power plants are the most efficient at rated power operations, or generally 100% load operation. Therefore, plant efficiency is greatly reduced during low load operation, thus this peak load condition brings total plant efficiency decrease as a whole. For this reason, it is important to reduce peak power consumption and construct reasonable number of power plants to achieve appropriate efficiency.

The peak power season is contingent upon regions, industrial conditions and demographics.

In cold areas, the peak season is winter when heating demand increases, whereas in hot areas, the peak season is summer when cooling demand increases. The peak power consumption can be reduced, for instance in Japan, the peak power consumption of 2017 was decreased to the level of 155 large nuclear power plants. This decrease is largely attributable to the adoption of LED lighting in companies and households, as well as the awareness of saving electricity. In the future, it will become more important to slightly raise the cooling temperature during peak hours and to charge EV during off-peak hours. An increase in household batteries is also expected to reduce peak power consumption.

4.1.1.6 Combined cycle power generation

From the viewpoint of thermodynamics, it is waste of energy to boil water at 100°C by natural gas flame whose temperature is over 2000°C. It is very important to sequentially utilize thermal energy from high temperature down to low temperature.

The theory that higher working gas temperature brings higher the thermal efficiency can be explained using an ideal engine called the Carnot cycle, and it is similar to the fact that the high altitude water can generate more power up until reaching sea level. Fig. 4.3 shows a conceptual diagram of operating gas temperatures for various types of engines. In fact, the working gas temperature of diesel engines is the highest, followed by gas turbines and steam turbines. Waste heat turbines use low-boiling-temperature fluids such as cooling media and ammonia used in air conditioners. Some of these systems operate even at temperatures as low as 70°C. Basically, if water is kept at low pressure, it boils at low temperature, and so it can be used as a waste heat turbine. The cascade use of energy means the process of using higher temperature gas heat, medium temperature gas and low temperature gas one after another. The image is that water coming from a hydroelectric power plant at higher altitude can operate a power plant at lower altitude.

The combustion process of diesel engines is explosive, and the maximum temperature of the combustion gas in the cylinder reaches about 2000°C–2500°C. Though the maximum combustion gas temperature is 2000°C or higher, the temperature of cylinder surface is about 500°C even for a large diesel engine, and the cylinders and pistons have sufficient strength for this surface temperature. As the average combustion gas temperature reaches about 1700°C, the thermal efficiency of diesel engines becomes very high.

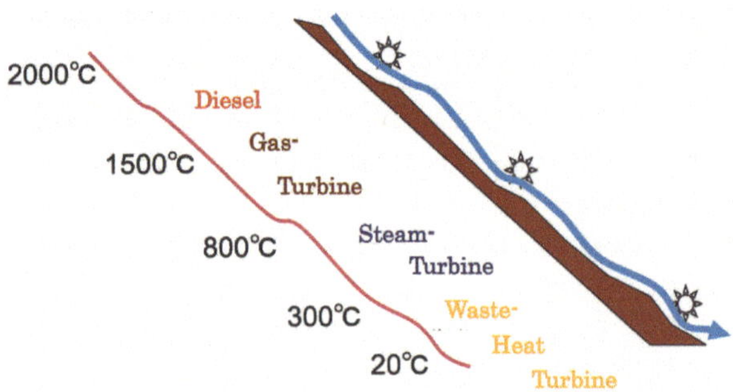

Figure 4.3 Cascade use of energy.

Steam turbines use steam produced by boilers to drive turbine blades then generators, and exhausted steam is returned to water by cooling water of seawater or other means, and sent to the boiler again by a boiler feedwater pump. A waste heat turbine is one kind of steam turbines, and is used for hot spring power generation because it can operate at temperatures as low as 70°C. Ocean Thermal Energy Conversion (OTEC) generation uses ammonia and other gases to operate at an even lower temperature of about 40°C.

In a gas turbine, the intake atmospheric air is compressed to high pressure at a compressor. And then compressed air is introduced to a combustor for mixing with fuel. By means of combustion, high temperature and high pressure gas is produced.

High-temperature combustion gas directly impinges the turbine blades and rotate turbine blade then generator as well as compressor. As the turbine blade surface temperature approaches the combustion gas temperature, the maximum combustion gas temperature is determined by the allowable temperature of the blade material. Increasing the gas temperature is essential to increase thermal efficiency. However, the allowable temperature of metallic materials is around 1000°C, as can be seen from the fact that the temperature of iron melted to bright red in steel works is around 1500°C. Therefore, it is necessary to protect the blades from this high-temperature environment, and various countermeasures have been made. As a result of these efforts, turbine blades currently can withstand high-temperature gas of about 1650°C for high efficiency as of 2020.

Gas turbine is a kind of turbojet engine, which is widely adopted for aircraft engines. In the case of jet engines, propulsion power is obtained not from the turbine shaft but also from reaction kinetic energy of the turbine exhaust jet.

A gas turbine is a clean power source with small size and high power output and relatively low emission of pollutants. However, the thermal efficiency is not high because the pressure ratio is not high enough to effectively convert thermal energy to electrical energy. Therefore, as shown in Fig. 4.4, a combined cycle has emerged in which steam is generated from gas turbine exhaust gas, and the steam turbine is operated to obtain additional power. By recovering exhaust gas energy with HRSG for generating steam, the fuel consumption rate can be improved by more than 20%. Diesel engines are sometimes used in lieu of gas turbines particularly for ships and distributed power sources.

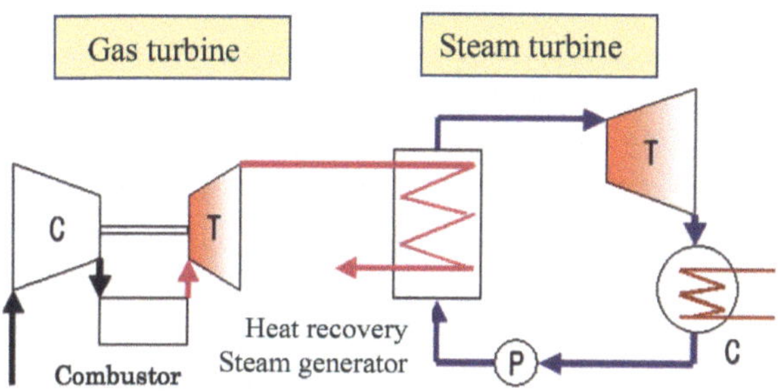

Figure 4.4 Combined cycle.

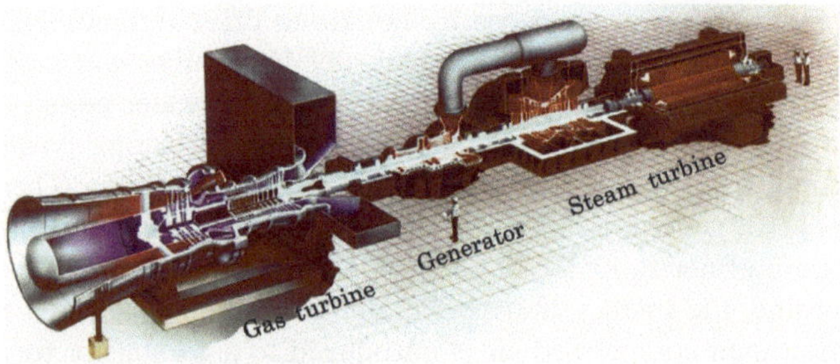

Figure 4.5 Single shaft type gas turbine and steam turbine combined cycle plant.

When gas turbine efficiency is 40% and steam turbine efficiency is 30% in the combined cycle generation, the overall efficiency is about 58%. In this case, gas turbine can convert 40% of fuel energy to electricity, and remaining 60% is converted into electricity with 30% steam turbine efficiency, which means that the total efficiency becomes $40\% + 60\% \times 0.3 = 58\%$.

Fig. 4.5 shows single shaft type gas turbine and steam turbine combined cycle. This combined cycle is considered to be one of the most advanced large-scale thermal power generation systems. Gas turbines, steam turbines, and generators are arranged in a tandem shaft.

In general, the NO_x concentration tends to increase as the combustion gas temperature increases. As one of the countermeasures for higher combustion temperature, a dry low NO_x combustor for preventing high-temperature combustion spots is usually adopted to achieve environmental requirement. And the simple countermeasure is to adopt a wet combustor where water or steam is injected, which lower the combustion temperature for lower NO_x level.

In addition, the first and second stages of the gas turbine employ air or steam cooling to lower the blade and nozzle surface temperature, furthermore the first stage blades and nozzles are usually made of single crystal material with enhanced high-temperature strength.

At present, steam turbines and gas turbines are the main onshore power generators. From the viewpoints of environmental conservation, combined cycle power generation system has become more popular due to higher efficiency and less CO_2 emission level.

4.1.1.7 Heat pump

According to the second law of thermodynamics, heat falls from a hot place to a cold place like going down a slope, as shown in Fig. 4.6. Thermal engines such as diesel engines and turbines are driven by this heat falling force. Therefore, the heat in the high-temperature place can do much more work as the falling force becomes larger.

On the other hand, a heat pump such as an air conditioner is similar to the automobile that transports heat from a low place to a high place, which is reverse

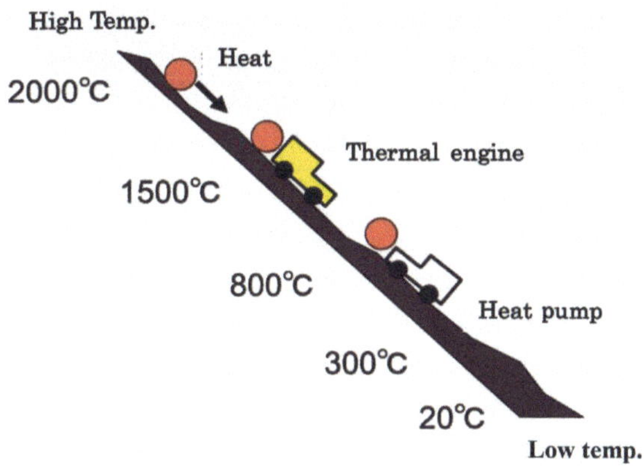

Figure 4.6 Heat pump movement against downslope.

movement of thermal engines. In this case, the heat cannot be transported without energy. This process also uses lots of energy when the temperature difference is large. Therefore, if the outside air temperature is much different from inside temperature, the energy required for air conditioning becomes much larger.

When heating a room in general, a heat pump is more efficient than an electric heater. Heat pumps with COP = 3 (coefficient of performance) can heat the room with one-third energy in comparison with electric heater, or can transport three times heat of the electric heater case. However, the COP decreases as the temperature difference between the hot and cold regions increases. As for popularity of heat pump in Japan, the heat pump become much more popular since Great East Japan Earthquake in 2012 and it is expected that about 20% of households in Japan will have heat pump facility by 2025.

At present, a heat pump commonly utilizes the heat of air to make hot water while it is also possible to utilize water for making cold water, which is useful for aquariums and hotels where there is a demand for cold water as well as hot water. Aquariums supply hot water to tropical fish tanks and cold water to cold water fish tanks, while hotels supply both cold water for cooling and hot water for bathing at the same time. Fig. 4.7 shows an example of heat pump water heater.

Fig. 4.8 shows heat transport capacity over power consumption of electric air conditioners for household use in 2004. COP is obtained by dividing the heat transport amount by the corresponding power consumption. The straight line shows COP = 3 and the broken line shows COP = 6. In Fig. 4.8, the temperature difference between the outside air and the room inside is 13°C for heating and 8°C for cooling. Due to the characteristics of heat pumps, the greater the temperature difference, the lower the efficiency.

Refrigerants, the working media of heat pumps, have passed through a difficult history. After the adoption of the Montreal Protocol in 1987, Freon refrigerant had been replaced by hydrofluorocarbon (HFC) as an alternative refrigerant of no ozone depletion potential since 2000. However, it has turned out that typical HFC

Figure 4.7 Heat pump water heater.

refrigerant R410A has a global warming potential 2000 times that of carbon dioxide. For this reason, new HFC R32 with a global warming potential of around 700 has been developed and has been used since around 2012. Because these alternative refrigerants have an adverse effect on climate change as greenhouse gases, the 28th Montreal Protocol in 2016, decided to regulate those alternative refrigerants in a step-by-step manner. The ultimate goal of development is to use natural refrigerants such as water and carbon dioxide, and even binary power generation systems that use alternative refrigerants have begun to use water.

Fig. 4.9 shows COP in 2015, which is 11 years after the previous survey. In recent years, the COP values of household air conditioners of less than 4 kW is

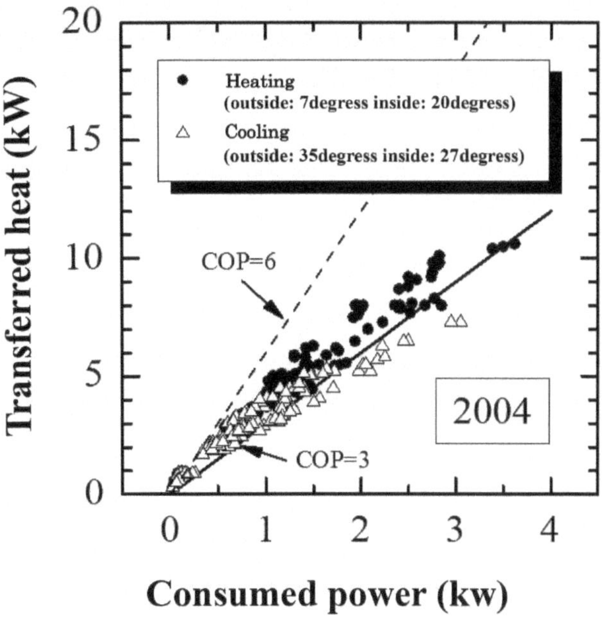

Figure 4.8 Coefficient of performance (COP) for air conditioner in 2004.

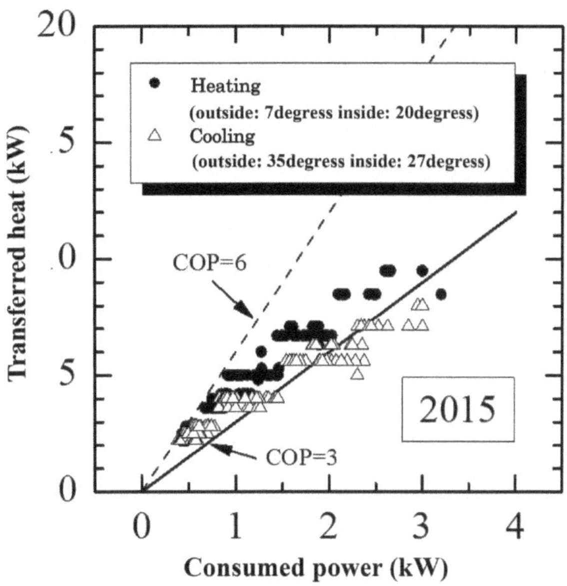

Figure 4.9 Coefficient of performance (COP) for air conditioner in 2015.

approaching COP = 6 line. Incidentally, additional functions such as air cleaning function and temperature distribution adjustment are enhanced and matured, and the power consumption tends to be different even for the same transportation heat quantity.

Although COP values have been substantially improved, the ultimate target is much more than current COP. As an example, if higher heat source temperature is 308K and the lower is 300, the cooling COP of an ideal heat pump is 37.5. The current COP is less than half the ideal level, and there is further expansion room resting on future technological innovation.

In comparison with heat-driven absorption refrigerator (COP = 0.7–1.2), air conditioners are extremely efficient owing to using high-equality electric energy. As the COP values of these heat pumps depend on temperature difference between the outside air and the inside, actual COP values may decrease in practical operation. For example, if the outside air temperature drops below 0°C, the heating COP will significantly decrease.

Electric heat pump water heaters using carbon dioxide (CO_2) as a refrigerant have recently been available and can supply hot water at 90°C with COP of 3 in winter and over 5 in summer. The annual average COP of this water heater is now 3–4 in consideration of heat loss in hot water storage tanks, etc.

4.1.1.8 Cogeneration

Cogeneration (combined heat and power supply) is a system in which power generation equipment is distributed in regions where energy is consumed and waste heat is also used together with electricity. In this case, the power generation scale becomes smaller, thus the power generation efficiency becomes lower. The common gas engine for cogeneration is a spark ignition reciprocating engine with natural gas. And those gas engines are of 300–1000 kW rating, and have 30%–35% thermal efficiency which is much lower than that of combined cycle generation as a typical centralized power sources by electric power utilities. Actually, this lower thermal efficiency is covered by extracted or exhausted heat utilization.

The power transmission loss of standard large-scale centralized power source is about 5%. On the other hand, in the case of distributed power source, electric power is consumed in the vicinity of power stations resulting in close to zero transmission loss. In addition, exhaust heat of about 85°C from engine jacket cooling water can be used and steam of about 0.1–0.8 MPa pressure can be obtained from exhaust gas. Power generation using this low-temperature heat is being actualized.

Some gas engines adopt so-called Miller cycle, where the valve open/close timing is adjusted so that the expansion stroke is longer than compression stroke, and thermal efficiency is improved. In the case of 300 kW power generation, thermal efficiency of about 38% on HHV basis has been achieved. Though the compression stroke and the expansion stroke are usually same, this valve adjustment increase expansion stroke working on the outside, then larger work can be obtained.

Small gas turbines are sometimes used for cogeneration. The efficiency based on HHV basis is about 17% for 1–2 MW class gas turbines. The efficiency of the regenerative cycle gas turbine has been improved to about 36% by utilizing exhaust gas heat.

Recently, gas engines capable of up to 8 MW have appeared with thermal efficiency of 40% to record 45%. There are two types of gas engines; one is spark ignition engines using only gas fuel, and the other is diesel engine using gas and liquid fuels. The latter system can be operated using only liquid fuel stored in tanks in the event of an earthquake or other emergency situations without gas supply.

In 2016, Japan installed about 260 MW of cogeneration systems, bringing the cumulative total to 10,500 MW. This generation capacity is equivalent to 10 large nuclear power plants. If compact gas engines (Miller cycle) with maximum efficiency of 38% based on HHV standard are used, the carbon dioxide emission coefficient is 0.426 kg-CO_2/kWh, which means that 0.426 kg of carbon dioxide is emitted to produce 1 kWh. This emission coefficient level is smaller than large power utilities emission coefficient of 0.49 kg-CO_2/kWh level.

4.1.1.9 Centralized power versus distributed power

As shown in Fig. 4.10, the efficiency of the combined cycle as a large-scale centralized power source is close to 50% on HHV basis. Therefore, if 30% of the generated electricity is used for lighting and power and the remaining 20% is used for an electric heat pump with COP = 3, 60% of the thermal energy can be utilized as a consequence. On the other hand, although the power generation efficiency of distributed power is as low as 30% at present, 60% of the waste heat energy can be converted into almost the same amount heat or cold energy by absorption refrigerators. In summary, total efficiency of centralized power system with latest combined cycle is almost the same as the total efficiency of the cogeneration of distributed power system.

However, it is not practical to install an absorption refrigerator for waste heat at each household, in consideration of equipment, operation and maintenance costs. For this reason, it would be practical to use waste heat from distributed power sources for cooling in summer at large facilities. In that sense, centralized power supply is more advantageous when air conditioning is used frequently in extremely

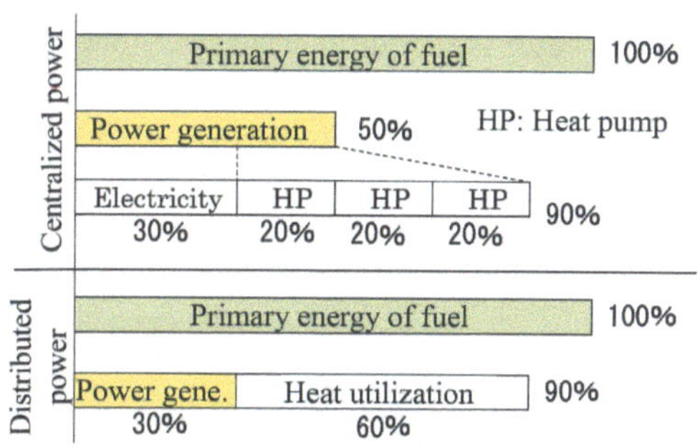

Figure 4.10 Centralized power versus distributed power.

hot summers. In addition, when the outside air temperature in winter is very low, the efficiency of electric heat pumps tends to substantially decrease, thus cogeneration with waste heat usage may have more advantage.

The comparison in Fig. 4.10 is to be limited to waste heat being used for heating and hot bath in the ambient temperature without substantial heat pump efficiency decrease. If COP of heat pumps would increase and sewage/seawater heat would be used in the future, current cogeneration advantage in winter would disappear.

4.1.1.10 SMART grids

SMART grids visualize and control electricity consumption by means of Information and Communication Technology (ICT) and Internet of Things (IOT) technology to bidirectionally connect the power supply side and the consumer side and to exchange information and power with each user. While conventional electric power systems are technical systems that supply electric power unilaterally with simultaneous same capacity to consumers, SMART grids are social systems that exchange electric power and information for flexible power supply.

The concept of SMART grids was proposed in 2004 at Tokyo University of Marine Science and Technology in Japan (SMART Study Group). The name SMART stands for Small Advanced Regional Energy Technology. A few years later, the US-based SMART grid became more widely known, appearing in President Obama's inaugural address in 2009. The SMART Concept proposed by Tokyo University of Marine Science and Technology has proposed the following three important functions.

1. Produced local renewable energy should be easily consumed locally. And the appropriate circumstances for this target should be established. It is required to take countermeasures against large fluctuations of renewable energy.
2. The tools should be provided to make use of civilian power. Demand control at the user side becomes more advanced and further promotes energy conservation. SMART Grids are focusing on the power adjustment of the entire community of about 400 households. And it is important to take voluntary action to save electricity.
3. In case of emergency or disaster, electric power is to be supplied from distributed power sources including ships without centralized power system.

The conventional power grid only unilaterally supplies power to the user as shown in Fig. 4.11. In this case, the electric power company must supply the same amount of electricity to each user at any time. This is called simultaneous same capacity, and if this is not achieved, the frequency will fluctuate, and in the worst case blackout will occur. The difference between supply and demand will cause electric frequency fluctuation. In case that grid frequency shifts more than 5%, turbine generators stop to protect its own equipment, which may sequentially stop other turbine generators, and eventually results in widespread power failure. In particular, if we introduce a large amount of renewable energy, such as photovoltaic power generation, which generates power when the sky is clear but the power output decreases drastically when the sky becomes cloudy. Thus major concern is whether simultaneous same capacity is secured or not.

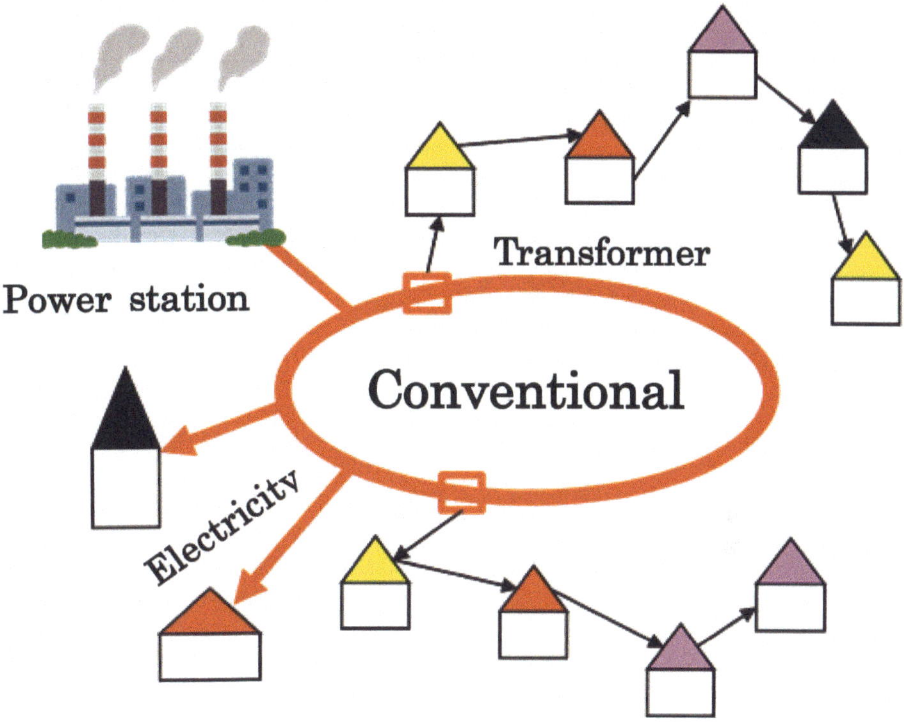

Figure 4.11 Conventional power grid.

By the way, I often see cylindrical machines on utility poles, which are pole transformers. The transformer reduces the voltage from the electric wire to 100–200 V, which can be used at home. Wires from the pole transformer are connected in series to about five houses. It has been pointed out that if these houses introduce photovoltaic power generation and sell electricity, if the amount of electricity generated is large, electricity will flow through the power grid, and the voltage of these five houses will rise above the standard. The conventional system using a pole transformer is not considered to be compatible with renewable energy.

In contrast to conventional systems, SMART Grids are linked to each other via ICT technology, as shown in Fig. 4.12. They exchange information and power with each user, and users' power consumption is visualized and controlled for flexible power supply and consumption. In addition to solar and wind power generation, users can also use EVs and fuel cell vehicles (FCV). Statistics show that 95% of cars are parked in parking lots, suggesting that power interchange is feasible. In particular, the vehicle to house (V2H) system has recently been available at low price with the catchphrase that cars can supply power in case of disaster. At present, households are connected by means of electric hard wires, while noncontact power interchange method is becoming more feasible. With this noncontact electricity transfer method, automobiles in parking lots can easily function as batteries for society as a whole.

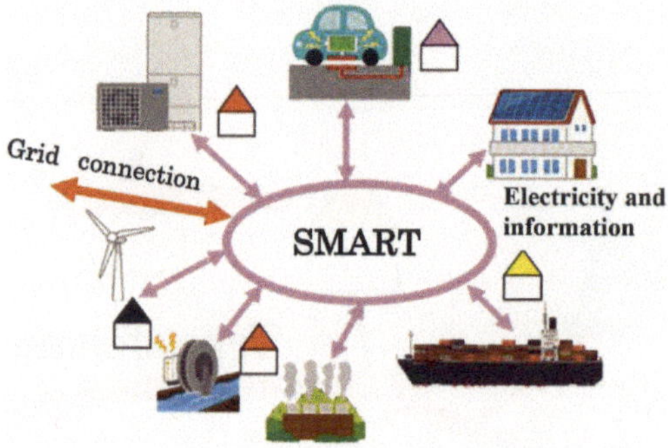

Figure 4.12 SMART grids.

It is also important to control heat pumps over wide area, which can compensate large fluctuation of renewable energy. These thermal storage devices can also serve as a water supply source in the event of a water shortage during disasters.

In New York in 2017, there was an attempt to lend or borrow electricity between buildings. If one building does not use peak power, this surplus electricity can be transferred to other buildings for sale. This attempt attracts lots of attention for the purpose of eliminating battery needs.

Human society is made up of a variety of industries and people. Some people need electricity for artificial breathing devices, and some industries cannot uniformly save electricity. SMART grids can provide an appropriate way to save power in consideration of each condition.

4.1.1.11 Storage batteries

Batteries are expected to be the savior of renewable energy; however, it is necessary to take into account about 10% charge loss and about 10% discharge loss even for new batteries. The loss of the inverter that converts AC and DC is relatively large; furthermore the battery loss itself increases as charge and discharge are repeated. Generally speaking, when the electricity charged by thermal power plants is used again, the net efficiency is reduced by more than 20%.

Storage batteries are suitable for storing variable renewable energy sources such as solar and wind power, but it should be noted that the CO_2 emission factor may be increased when batteries are used in the power network with many thermal power plants.

4.1.1.12 Fuel cells

Common combustion is the reaction between fuel and oxygen to produce heat. In contrast, fuel cells directly convert chemical energy into electrical energy through electrochemical reaction between fuel (hydrogen) and oxygen. And theoretical efficiency of fuel cells decreases as the operating temperature increases. This is opposite to thermal efficiency characteristics of heat engines,

Fig. 4.13 shows the general structure of a fuel cell. In this system, Hydrogen (H_2) converts to hydrogen ions (H^+) at the cathode and releases electrons (e^-). The electrons generate power at an external generator on the way to the anode. Hydrogen ions travel through the electrolyte to the anode, where hydrogen ions combine with oxygen taking up electrons from an external circuit to produce water (H_2O).

Various types of fuel cells with different electrolytes have been developed along the way, and their characteristics are outlined in Table 4.1. The generation efficiency is based on HHV, and energy not used for power generation is discharged as high-temperature heat. Therefore solid oxide fuel cells can be heat source for microgas turbines in combined cycle systems. In the case of polymer electrolyte fuel cell (PEM), as the working temperature is low, PEM is used for ships and automobiles.

Fig. 4.14 shows a combined system of fuel cell and microgas turbine. Solid oxide fuel cells operate at high operating temperature of about 1000°C, and their power generation efficiency exceeds 50%, where the residual energy is released as waste heat. Using this waste heat, methane and other fuels are reformed to produce

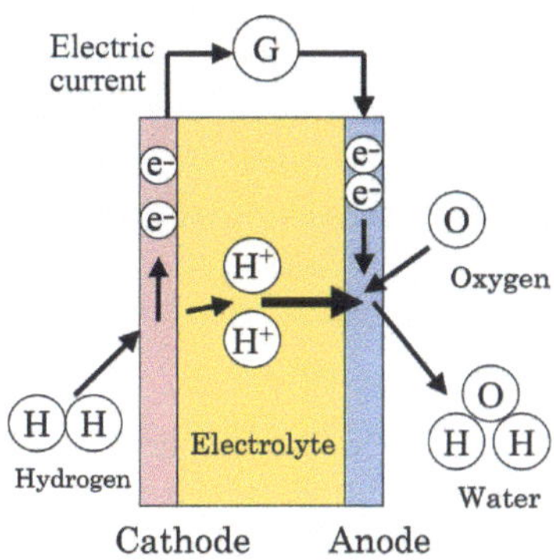

Figure 4.13 Fuel cell.

Table 4.1 Kinds of fuel cells.

Electrolyte	**Polymer electrolyte**	**Phosphoric acid**	**Molten carbonate**	**Solid oxide**
Name	PEM	PAFC	MCFC	SOFC
Working temp. (°C)	20–100	170–200	600–700	900–1000
Efficiency (% HHV)	36–45	36–45	45–60	50–60
Fuel	H_2	H_2	H_2, CO	H_2, CO
Mobile ions	H^+	H^+	CO_3^{2-}	O^{2-}

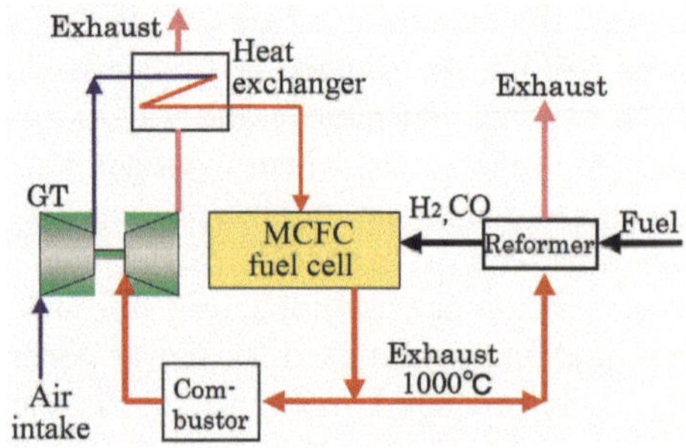

Figure 4.14 Fuel cell and microgas turbine combined system.

Table 4.2 Life cycle of CO_2 emission (kg/kWh) (2007).

Generation type	PSI (Switzerland)	IAEA	CRIEPI (Japan)
Coal	1.071	0.978	0.975
Oil	0.855	0.811	0.742
Natural gas	0.605	0.696	0.519–0.608
Solar power	0.144–0.189	0.097	0.053
Wind power	0.036	0.036	0.029
Nuclear	0.016	0.021	0.022
Hydro power	0.004	0.016–0.023	0.011

CRIEPI, Central Research Institute of Electric Power Industry; *IAEA*, International Atomic Energy Agency; *PSI*, The Paul Scherrer Institute.

hydrogen and carbon monoxide for microgas turbine driving. The generation efficiency of this system is expected to be over 60%.

In 2017, ceramic solid oxide fuel cells with microgas turbines began to be put on the market, where these fuel cells operate at high temperature of 900°C. The fuel cells do not require reformers but use natural gas.

4.1.1.13 Life cycle carbon dioxide emission

There are plenty of methods of power generation using various fuels. Table 4.2 shows how much carbon dioxide is emitted to generate 1kWh of electric power. The figures in Table 4.2 include all CO_2 emission from the construction, operation and up to dismantling power generation facilities. CO_2 emission decreases as hydrogen content increases from coal, oil to natural gas. It is also obvious that solar, wind, nuclear and hydroelectric power generation emit less carbon dioxide than fossil fuel power generation cases. In the case of nuclear power, CO_2 emission required for nuclear waste disposal and long-term storage is not included. It should be noted that these estimates are very complex and vary substantially depending on

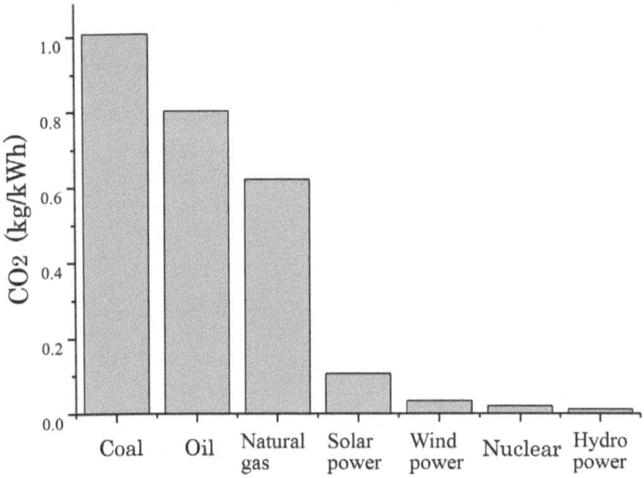

Figure 4.15 Life cycle CO_2 emission versus generation system.

institutions. And please note that H_2 as a fuel must be produced using other energy, because H_2 does not exist as a natural resource.

Fig. 4.15 shows the average CO_2 emissions of the three research institutes for each power generation system. Again, it is clear that nuclear and renewables emit less CO_2 than coal, oil, or natural gas fuels. In comparison with thermal power plants, nuclear power has less capability in changing power output to compensate renewable energy fluctuations. For this reason, SMART Grids using ICT and IOT have been proposed to develop efficient use of these energies.

4.1.2 Site selection to operation with relevant regulations and laws

4.1.2.1 Overview

Legal procedures on power station construction from site selection up to commercial operation start are wide-ranging depending on countries; they are basically promoted in accordance with Fig. 4.16 as an example in Japan.

During those processes, permissions and approvals based on many laws and ordinances such as Electric Business Act are necessary. Concretely, what kind of permission or approval is required differs depending on power output rating, fuel species, geographical, natural and social conditions of the site, etc.

There are a wide variety of laws and regulations and licenses related to power plant construction procedures. Persons in charge of power station construction should have thorough familiarity with these laws and regulations.

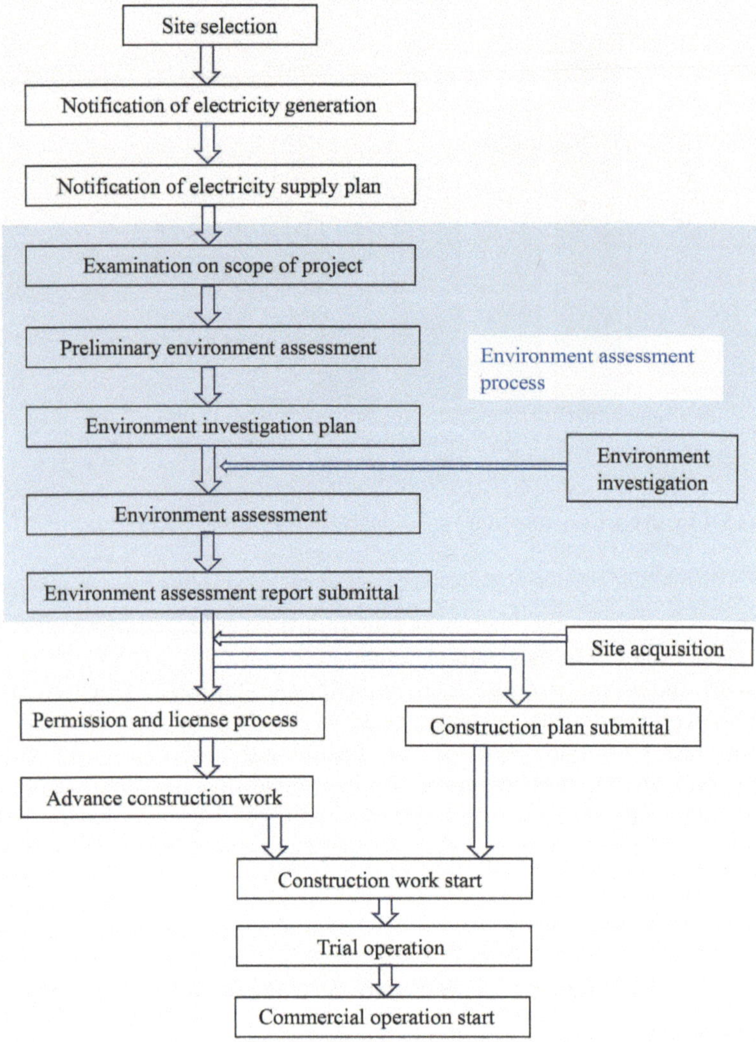

Figure 4.16 Procedure flow for power station construction.
Reproduced from [1].

4.1.2.2 Site selection and environment assessment

1. Site selection of power station

First off, it is necessary to select a site location, considering the following conditions:

a. Area space necessary for power station facilities.

b. Port facilities for receiving fuel, or alternative facilities such as railroad station.

c. Condenser cooling water availability, or alternative cooling measures such as air-cooling.

d. Roads and/or ports for the delivery of construction machines/materials and power station equipment.
e. Strength of ground foundation.
f. Construction labor availability and other construction factors such as labor union conditions.

In the selection of the site, these conditions are to be satisfied, and various regulations in Table 4.1 must be observed.

2. Notification of electricity generation business

The entity who plans to engage in power generation business must submit a notification of power generation business pursuant to the related laws to government or local municipality.

And power generation business is defined by related laws. For example in Japan, 1000 kW or over is power generation business category, in general.

In addition, notification to a power grid company or an organization is required for electricity connection.

3. Notification of electricity supply plan

Electricity generation entity must prepare electricity supply plan on a long-term basis such as for 10 years including overhaul and renovation schedule, as well as any play for increase or decrease electricity power generation capability.

4. Environment assessment

It is necessary to evaluate the environmental effect in advance and to make proper consideration on the environment conservation, because the large-scale power plant would affect the environment. The execution procedure of the environment assessment for the power plant is described below.

a. Scope of the project

The projects covered by the environment assessment are stipulated in related laws and ordinances. In general, more than 150 or 100 MW power stations are subject to environment assessment to properly evaluation environmental impact.

b. Environment consideration report

At the planning stage, in order to consider the environment influence by the project, measure for the environmental conservation must be examined.

The generation entity must submit this environment consideration report to the government and local municipality for review. And there is a case that the generation entity must gather public opinion as well.

The ministry in charge and/or local government reviews the environment consideration report and evaluate environment impact and decides whether or not the power plant can be constructed, taking public opinion into consideration as well.

c. Evaluation items

The following items are to be reviewed and evaluated for environmental impact.

- Atmosphere pollution by boiler exhaust gas.
- Treatment of combustion ashes.
- Cooling water usage and heated water discharge, especially for fishery impact.
- CO_2 emission amount.
- Noise and vibration to surrounding area.
- Scenery harmony with surrounding area.
- Wild animal conservation.
- Land conversion/excavation, etc.

- Forest conservation.
- Public road use and traffic.
- Port area activities.
- Structure construction permission.
- Strength of ground foundation.
- Measures against natural disasters.
- Waste treatment measures.

4.1.2.3 Construction

1. At the time of construction start

When the process of the environment assessment is completed, the construction work for the power station will start.

a. Notification of construction plan

It is necessary to notify the government of the construction plan which describes the detailed design of power station facilities prior to the construction start.

b. Environment related laws and regulations

When farmland or meadow/pastureland is acquired or developed as a power plant site, application for permission for conversion of farmland is necessary.

When using agricultural land for purposes other than agricultural land, it is necessary to submit an application for conversion of agricultural land to the local municipality.

In the case of reclamation of public waters such as ocean water, river water and lake water, it is necessary to submit application for a license for reclamation of public waters.

It is necessary to obtain the permission of the local governor or the Minister of Agriculture when reclamation, dredging work, etc. are carried out in the water area specified in the protected water surface area

2. Construction process

The following process work is required for construction work and completion.

a. Electric equipment regulation

In order to observe safety regulation for construction work of electric equipment, it is necessary to appoint a chief electricity engineer in charge of safety and inspection of electric equipment pursuant to the provision of safety regulations.

b. Fire-fighting regulation

The application and procedures for fire-fighting must be made and submitted to local government.

In the case of storing or handling hazardous materials in a designated quantity or more, it is necessary to submit application for permission for hazardous materials storage and handling facilities.

c. High pressure gas safety regulation

It is necessary to submit application for permission of establishment of high pressure gas storage facility and specific high pressure gas consumption notification. And chief engineer in charge of high pressure gas handling must be appointed.

d. Building act

Objects to be regulated in a power plant are buildings, building equipment, and structures.

And it is necessary to submit application for building certification to government and/or local municipality.

e. Waste management and public cleansing law

It is necessary to submit application for permission for disposal of general waste for the case where waste disposal facilities at the plant are larger than a certain scale.

f. Civil aeronautics regulation

For chimneys, steel towers, flare stacks, and other tall objects such as 60 m or more above the surface of the ground or the surface of the water, or for which there is a risk of serious damage to the safety of aircraft navigation, the installation of obstacle light and obstacle marking must be installed.

And it is necessary to submit notification of installation of obstacle lights and notification of installation of obstacle markings to the ministry in charge of aviation control.

g. Radio act

When establishing radio stations such as power supply systems between power stations, transceivers, radio paging, etc., and when installing high-frequency equipment such as power line transport equipment, and/or high-rise buildings such as turbine generator building, it is necessary to submit radio station license application and application for permission for high-frequency equipment, and notification of scheduled work for high-rise buildings to the ministry in charge and/or local municipality.

h. Noise and vibration regulation

Since noise and vibration are regulated by various regulations, the electric generation entity must take necessary procedures and pay attention to ordinances.

i. Industrial safety and health law

Since there are various rules and the procedures, the electric generation entity must take necessary procedures and pay attention to ordinances, while the procedure is mainly to report. The following are major regulation items:

- Industrial Safety and Health Regulations
- Safety Regulations for Pressure Vessels
- Safety Ordinance on Cranes, etc.
- Ordinance on the Prevention of Organic Solvent Poisoning
- Ordinance on Prevention of Hazards due to Specified Chemical Substances
- Ordinance on Prevention of Ionizing Radiation Hazards
- Ordinance on Office Sanitation Standards
- Ordinance on Prevention of Dust Hazards

4.1.2.4 Equipment installation process

Thermal power plant mainly consists of steam turbine/generator and boiler.

Both equipments have lots of auxiliary facilities. Hereafter some general installation procedures are described for turbine, generator and boiler.

1. Steam turbine installation

A steam turbine consists of rotor with moving blades, casings, bearings, diaphragms (or blade rings) and other parts. In factories, radial and axial clearances between rotating parts and stationary parts are adjusted either by shop assembling or computer simulated assembling process, so that proper clearances are secured. For that clearance adjustment, adjustment keys/pins or diaphragms are machined to the adequate dimensions.

After that, each part is shipped to the power station. At the power station building, the assembly condition should be reproduced by adjusting key/pins and/or

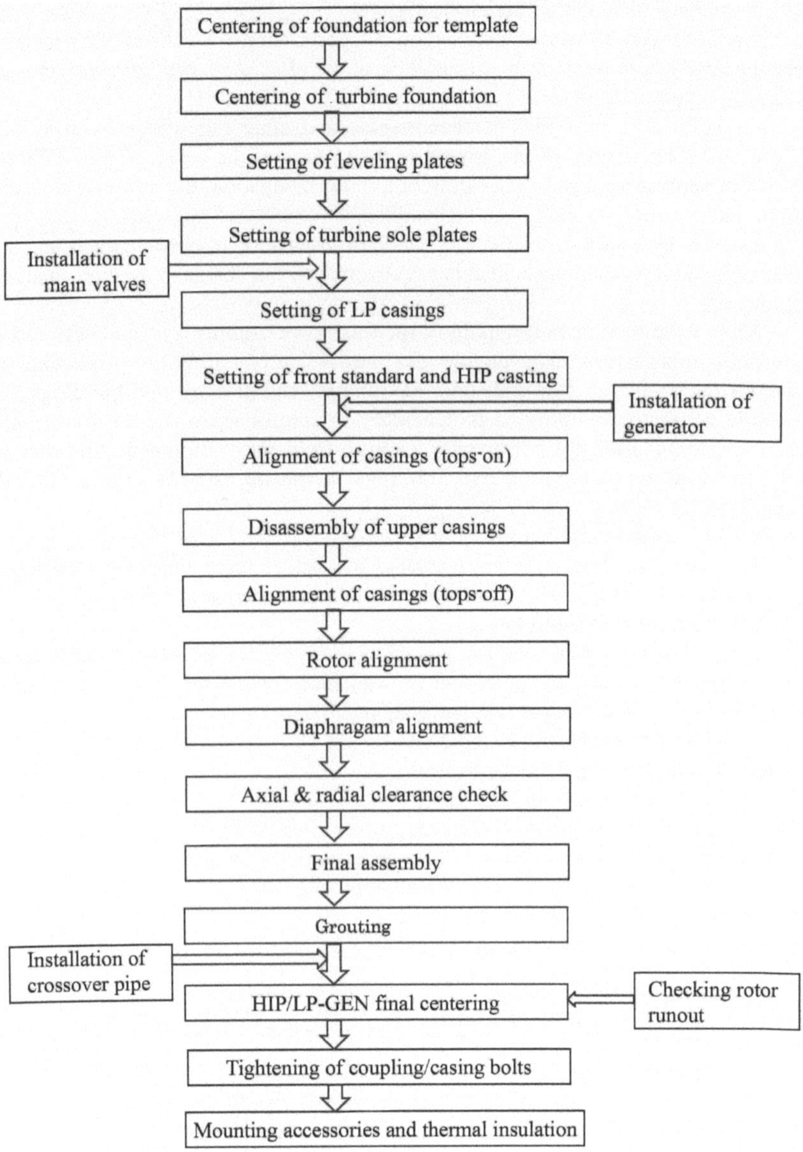

Figure 4.17 Typical steam turbine installation flow.

diaphragms to absorb the physical condition differences between the manufacturing factory and actual power station. An example of turbine installation process is shown in Fig. 4.17.

a. Foundation bolt setting

Before the machine is installed, the benchmarks of reference height and center location are set in the turbine building column or other baseline parts. The foundation bolt and the imbedded parts must be set by means of temporary templates prior to the final concrete placement for the turbine foundation.

The foundation bolts literally fix the equipment to the foundation. Since it is impossible or quite difficult to correct the equipment positions and heights after concrete placement for the foundation, close attention is necessary.

The setting tolerances are generally within 2 mm for the entire portions and foundation bolt pitches are within ± 1.5 mm, and the heights are set in consideration of bearing alignment sinking during concrete replacement.

b. Setting of base/sole plates

After the concrete placement and temporary template removal, concrete surface laitance should be removed by chipping, whose thickness is about 20–30 mm. Next, base/sole plates are set in accordance with marked proper locations and adjusted to predetermined heights using level adjusting bolts.

c. Lower half casings and bearing standards installation

The alignment benchmark is the low-pressure casing, which is positioned on the base/sole plates with axial keys. The axial keys are the fixed point in the axial direction.

After adjusting the casing horizontal surface and the dimensions between adjacent casings, attach anchor blocks for positioning in the radial direction and tie beams between casings.

The front and middle bearing standards are to slide on the base/sole plates in the axial direction. Those front and middle standards are installed on the bearing pedestals, and the bearing pedestal and casing are connected by center beams.

Though the temperature and pressure of the low-pressure turbine are low, the temperature drop is relatively large. Thus low-pressure casing consist of outer casing and inner casing to alleviate temperature gradient. Each casing's position is fixed by supports and liner keys.

d. Rotor and upper half casing assembly

After the rotor is inserted onto the lower casing, the diaphragms or blade rings are firstly assembled, then inner casings are assembled, thirdly outer casings are assembled in order. Make sure that the rotor position is same as the oil ring bore of each bearing. After assembling outer casings, rotor alignments are adjusted for each rotor.

e. Grout

The grout is filled between the foundation concrete and the underside of the base/sole plates, integrating foundation and base/sole plates which connect with turbine casings and front/middle standards. Great attention should be paid for grout shrinking by water evaporation. At present, excellent grout material with very little shrink are available for shorter installation time.

2. Generator installation

As for setting foundation bolts and base/sole plate and grouting method, the procedures are same as steam turbine installation procedure.

The generator stator is usually heaviest among turbine building equipment. When the overhead traveling crane in the turbine building has sufficient capability, the generator stator is lifted by the overhead traveling crane. If not, the generator stator is to be lifted by temporary crane to the turbine operation floor, and to be set on the steel rail or roller, then to be moved to the proper position.

Next, the generator rotor is carefully and slowly inserted by means of overhead traveling crane and chain blocks.

3. Boiler installation (water tubulous type)

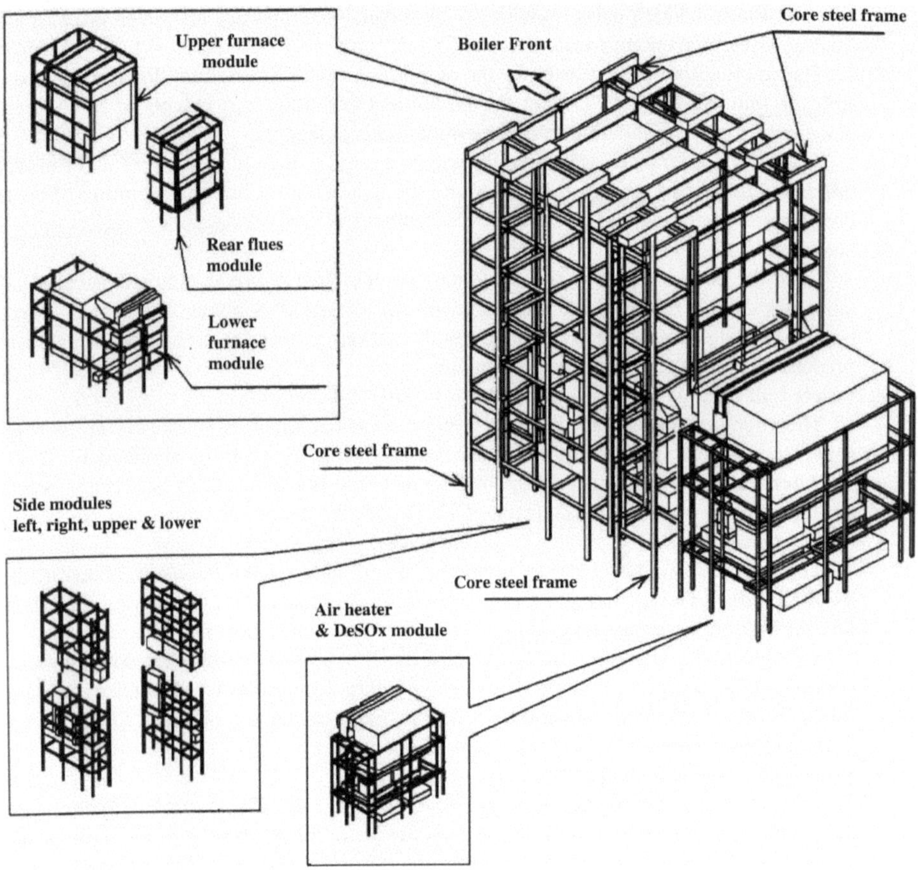

Figure 4.18 Example of boiler modules (eight modules) [2].

There are mainly three boiler installation methods, which are zone module method, synchronization, and floor block method. Those three methods are described hereafter. Recently, in order to secure safety and reliability and to shorten the installation period as well as due to manpower shortage, large blocks are pre-assembled and integrated in the factory and shipped to the site. Fig. 4.18 shows an example of boiler modules (eight modules).

a. Zone module method

Steel frame, duct, piping and auxiliaries are assembled for each floor at the location other than boiler building, and transferred to the boiler building in sequence then lifted to the proper place by means of hydraulic jacks. The advantages of this method are as follows:

i. Installation of boiler pressure parts and facilities can be implemented in parallel with steel frame construction work. (early installation work start).

ii. To shorten the installation period.

iii. Reduction and leveling of labor work force.

iv. Improve safety of work (reduced opportunities of high-place work).

b. Synchronization method

In order to shorten installation period and to reduce high-place work as well as to improve safety and quality. In parallel with boiler steel frame construction work, pressure parts, duct, coal bunker and other parts are transferred and assembled.

c. Floor block method

First off, steel columns, beams, braces, gratings, handrails and other ancillaries are fabricated to an integrated large floor block in the factory. At field, large floor blocks are lifted by cranes and carefully welded to be boiler steel frame etc. This method can substantially shorten the boiler steel framework at field.

4.1.2.5 Test and trial operation

After equipment installation, various tests are carried out, and then trial operation starts followed by acceptance tests. When acceptance tests are passed, commercial operation will start.

1. Tests and inspections before trial operation (commissioning)

The following tests should be implemented prior to plant trial operation.

a. Turbine oil flushing for cleaning lubrication oil piping
b. Electric hydraulic control (EHC) piping flushing
c. Turbine piping hydraulic pressure test or alternative nondestructive test (NDT)
d. Single device tests for auxiliary facilities such as valves, motors, and pumps
e. Boiler pressure parts' hydraulic test
f. Boiler piping hydraulic pressure test or alternative NDT
g. Boiler free blowing to check boiler scales and slugs

2. Trial operation

a. Trip and interlock test
b. Boiler safety valve test
c. Condenser vacuum test
d. Turbine ramping test to 2000–2400 rpm for vibration check
e. Turbine overspeed trip test
f. Load-up test and boiler silica purge operation
g. Turbine load dump tests at partial loads and full load
h. Boiler combustion test
i. Boiler dynamic characteristics test
j. Continuous operation test
k. Plant performance test (boiler and turbine)

4.1.2.6 Codes and standards

1. International codes and standards

As international codes and standards, there is International Electrotechnical Commission (IEC), and American Society of Mechanical Engineers (ASME) standard is also adopted on a worldwide basis. As an example, current status of IEC standard for steam turbines is shown in Table 4.3.

Table 4.3 Current status of IEC standard for steam turbines.

IEC standard number	Title	Published/ revised year	Current status
60045-1	Steam turbines Part 1: Specifications	1991	Revise ongoing To be published in 2019
60953-1	Rules for steam turbine thermal acceptance test Part 1: Method A—High accuracy for large condensing steam turbines	1990	Revise ongoing To be revised in 2019 or later
60953-2	Rules for steam turbine thermal acceptance test Part 2: Method B-wide range of accuracy for various types and sizes of turbines	1990	Revise ongoing To be revised in 2019 or later
60953-3	Rules for steam turbine thermal acceptance test Part 3: thermal performance verification tests of retrofitted steam turbines	2001	Revise ongoing To be revised in 2019 or later
10494	Turbines and turbine sets Measurement of emitted airborne noise —engineering/survey method	2018	Substituted for IEC61003
61064	Acceptance tests for steam turbine speed control systems	1991	No revise work
61370	Steam turbines—steam purity	2002	To be abolished IAPWS-TGD to be applied
Planning	Cogeneration combined heat and power	–	Committee to be organized

Reproduced from [3].

2. Codes and Standards by Academic Society

In the case that academic societies create the technical standards based upon actual field conditions, it is possible to shoulder a part of truly necessary safety and reliability regulation.

Academic societies can establish academic standards with fairness and openness by obtaining the consensus of academic/engineering experts, thereby the highest level of academic knowledge and technology in each field are gathered for improving safety and reliability, which can contribute to power industry development.

In addition, academic societies can play a central role in the latest findings reflection without delay, which can actualize more improved and comprehensive regulation on power industry.

4.2 Safety assurance system of nuclear power plants in Japan

This section summarizes the concept of ensuring safety in the use of nuclear power in Japan and the standardization mechanism for ensuring it.

In the standardization of the nuclear power field, strict regulations are important, but the strict stance on the establishment of standards by the private sector that governs specific operations is most important. Nuclear technology has been introduced as a national policy, and codes and standards have been developed and operated as a national role. However, more than 50 nuclear power plants have already been constructed and operated in Japan, and as with other conventional power plants, voluntary consensus codes system have been set by the private sector in consideration of the speed and flexibility of progress.

Here, we think that it is useful to summarize the entire system for formulating the voluntary consensus codes system in order to make future activities more effective.

4.2.1 Basic concept to ensure safety

4.2.1.1 Ensuring safety at the design stage

1. Ensuring safety at the design stage Process of nuclear power plant

   Nuclear power plants, as same as general industrial products, are designed based on the various requirements of the specifications prepared by the utilities, and are provided for construction and operation. The reliability and safety of nuclear power plants are ensured by the design based on the latest technical knowledge, laws, and regulatory standards. Fig. 4.19 shows the process of designing, constructing and operating a nuclear power plant. These processes are based on the philosophy of "safety first" and the concept of "defense-in-depth."

   As shown in this figure, the utility determines the specifications of requirement and places an order with the vendors (manufacturer, general constructors, etc.), and the vendors designs, constructs, and delivers the nuclear power plant to the utility. The utility operates the nuclear power plant. In the process, laws and regulations, standards, guidelines, etc. are applied to each based on the basic philosophy of "safety first," thereby ensuring "safety." Utility and vendor must recognize this process and play their role in their respective organizations.
2. Challenges at design stage
   a. Concept to ensure safety in design

      Safety design is performed together with core design as part of system design. Fig. 4.20 shows the mechanism of safety design activities based on the concept of "defense-in-depth." In the past design, the third layer was targeted up to plant design

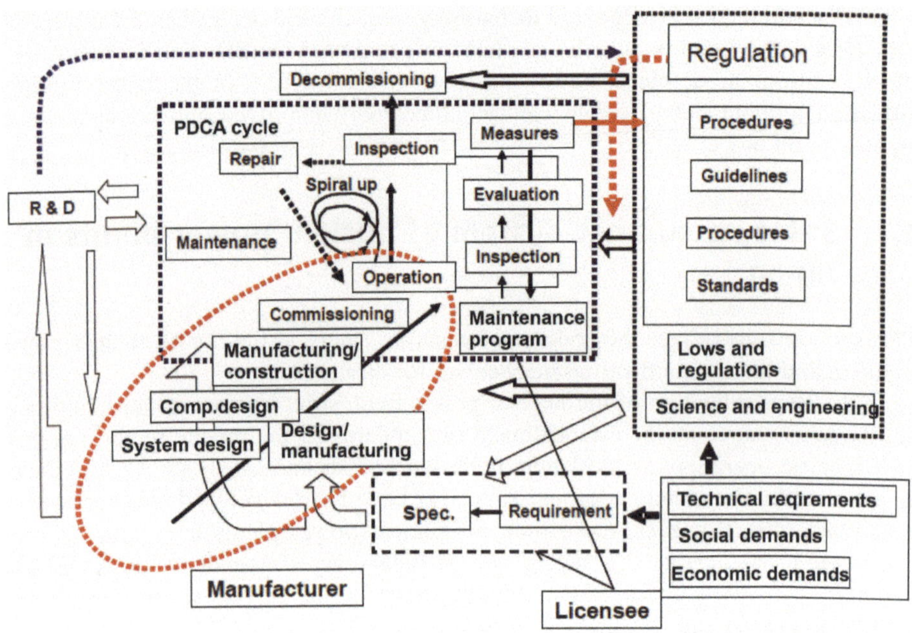

Figure 4.19 The process of designing, constructing and operating a nuclear power plant.

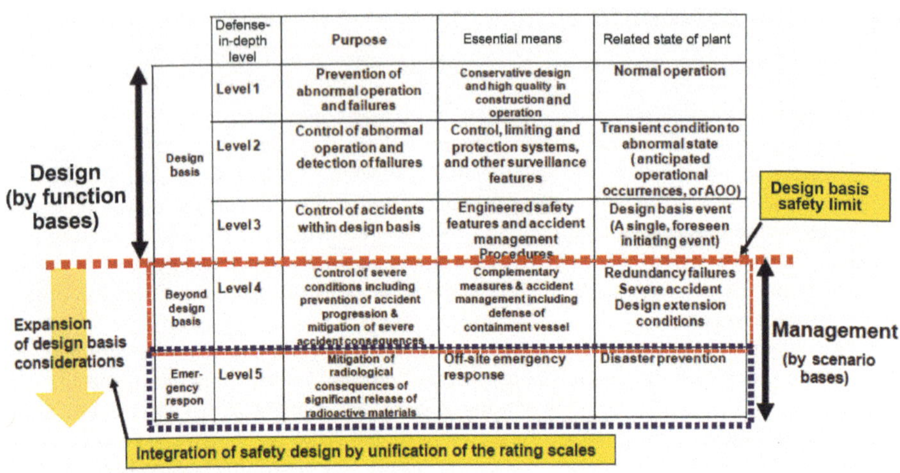

Figure 4.20 Safety assurance design concept based on IAEA's defense-in-depth idea.

(level 3) and the fourth layer was responding to severe accidents countermeasures. And the fifth layer, disaster prevention plan has been outside the scope of plant design. Therefore, it has been aimed to ensure safety by equipment-based design. The utility and manufacturers who design and construct nuclear power plants are responsible for up to level 3 and have constructed and commissioned the equipment according to the required specifications of level 3 design. Subsequent operations are the responsibility of the utility. The activities performed by the manufacturers to ensure the safety of nuclear power plants are to ensure that the required functions of individual equipment and systems are satisfied under the conditions specified by the design. Therefore, strict quality control activities and a large design margin are required.

b. Design issues clarified by lessons learned on the Fukushima Daiichi accident

i. Countermeasure to situations that exceed design standards

In some plants of Kashiwazaki-Kariwa Nuclear Power Station, the ground motion during the Chuetsu-oki Earthquake exceeded three times of the standard ground motion. Even in such a case, it was confirmed that the soundness of the equipment was sufficiently maintained, and it was recognized that there was a great margin in the structural design for earthquake motion. After that, the reference ground motions were reviewed at nuclear power stations nationwide, and back-checks confirmed that although the reference ground motions were reset to higher values, there was still enough margin for soundness. This is an evaluation of only the margin of structural soundness, but it is presumed that the margin to ensure function is even larger.

From this, the meaning of "beyond the design standard" must be reconsidered. In other words, during the Great East Japan Earthquake, seismic acceleration exceeding the standard ground motion was observed at many nuclear power plants on the Pacific coast of eastern Japan, but there were no observation data indicating changes in the operating conditions of nuclear power plants. It is considered that the soundness of equipment during the earthquake was secured from the results of analysis and evaluation and the performance of other plants. Therefore, instead of revising the reference value of acceleration response to a larger value simply as in the past, wide range of evaluation methods should be considered from a perspective of the development of a method that can evaluate realistic behavior and the introduction of seismic isolation devices for situations exceeding the reference ground motion.

ii. How design responds to natural disasters

It is said that such a large tsunami hit by the Great East Japan Earthquake, was "unexpected" occurrence. However, many say that "unexpected" is not allowed in nuclear safety. On the other hand, it is not easy to predict a large tsunami about once every 1000 years. Tsunami-resistant design is a comprehensive work on structural design of buildings and facilities and electrical instrumentation design after evaluating various phenomena such as wave generation, propagation, run-up, inundation, etc. Many simulations of the magnitude of the tsunami have now been performed, and various estimation results have been reported. Even if a standard tsunami is set and equipment is designed to respond, it is unlikely that a tsunami exceeding the standard value will come, but no one can deny it. Even if the degree to be assumed is different, it is necessary for equipment design to respond to the assumption that the tsunami will exceed the reference value as in the case of earthquake motion. Natural disasters, like the tsunami, cannot deny to exceed expectations. Imaginative responses to various natural disasters are required.

iii. Severe accident management and preparation

In the case of the Fukushima Daiichi Nuclear Power Station accident of Tokyo Electric Power Company (TEPCO; hereinafter referred to as the Fukushima

Daiichi accident), a situation occurred in which all power sources were lost and all safety cooling systems were lost. It is the situation that exceeds the design standard, and it is pointed out that the fourth layer of defense-in-depth, level 4 is inadequate. Level 4 response requires severe accident management in various scenarios. This is the response to an "unexpected" situation, and management is to take advantage of the remaining equipment and survive the situation with human wisdom. There are few findings nor knowledge that specifically examines how events progress in such unexpected situations. Here, preparations and training to prepare for various scenarios under various situations are important. It has been reported that the response to unexpected situations prepared in the United States after the 9/11 terrorist attacks was also effective as a response to the Fukushima accident. It is necessary to refer to how the preparation for responding to unexpected situations was derived.

iv. Fail-safe and robustness

The Fukushima Daiichi accident resulted in the release of large quantities of radioactive material from the containment to environment.

Here, I would like to address two issues of emergency response in the accidents. One is the isolated condenser (IC). Another is the containment vent.

ICs, also called emergency condensers, are effective equipment for cooling the reactor core when all AC power is lost. A fail-safe design has been designed in which the isolation valve closes when a primary cooling pipe rupture signal is detected and the path for cooling the core is cut off. This isolation valve is installed before and after the wall of the containment in the piping that penetrates the containment, and the valve in the containment cannot be manually opened when AC power is lost. In spite of the fact that the primary cooling pipe did not break at the time of the Fukushima accident, when the DC power of the instrumentation control system was cut off, an erroneous signal was sent out and AC power that was still active was supplied, The isolation valve was closed despite the design that the AS-IS (as it was) was turned off when the DC power was turned off.

On the other hand, the containment vent is designed so that the rupture plate automatically breaks and the pressure is released when the containment vessel pressure is high, and the vent valve must be operated manually when necessary. However, it was speculated that the manual operation of the valve could not be easily performed, which caused a subsequent hydrogen explosion and the spread of accidents.

Originally, there is always a possibility that a certain amount of leakage will occur in the containment vessel itself, and the sealing performance is not kept strictly. As a designer, designers must carefully consider the equipment, what is considered to be fail-safe and what emphasizes robustness. You must not lose sight to look the purpose of the design as what the safety design aims for.

c. Initiatives for new design

Conventional design was based on the philosophy of safety first, and the idea to secure the safety of the system by ensuring the specifications and design standards required for equipment and the structural design. Such designs have the limitations that cannot be avoided. Future design must also take into account the consequences of exceeding the design criteria, severe accident management, and the impact on disaster prevention activities beyond the availability of the power plant. Only the activity to design power plants is not enough to ensure safety. It is necessary to establish a new design approach based on a concept of safety assurance centered on plant life cycle, which spans different areas beyond designs.

4.2.1.2 Risk management

1. Risk implications and risk awareness

    Recognition of the word "risk" varies from person to person. It seems that risk is used only in a bad sense, such as "I hate risk" or "I want to be free of risk." In Japan, "risk" seems to be generally used as "danger."

    In the real world, nothing can be gained without a challenge to risk. Everyone in the world knows that Columbus has taken the danger and embarked on a voyage to accomplish the feat of discovering the Americas. The challenge is the essence of a human being, living with life. The essence of taking the risk and gaining the benefit is the Latin word for sailing "go voyage between the rocks" which means trying without fear. In other words, risk is a virtual threat, and if it succeeds, it will benefit, and if it fails, it will result in loss of life or loss of property. Much of the history has been the result of the hero's challenge to risk, while the consequences of the challenge have disappeared. But for any person, living can be risky. There is a risk in all actions and activities, and as a result, everyday life is established, and various benefits are obtained. It can sometimes face great losses and dangers.

2. Scientific concept of risk and quantification of risk

    In recent years, risks that have been vaguely recognized in the past have been formulated as investment risks and evaluation methods from an economic perspective, and risk management in risk-based business administration has been established.

    In Japan, the definition of risk is defined in the JIS standard "Guidelines for Establishing a Risk Management System" established in 2001 as "combination of the likelihood of a situation and its consequences, or the probability of occurrence of a situation and its combination." Here, the result of the risk is simply referred to as the "result of the situation," but generally, the risk of taking risk is referred to as taking "damage." In other words, the most serious loss is loss of life, and the loss of property and wasted time are other damages. This loss is expressed as a probability of the possibility of causing the damage, and the value quantified as the risk is multiplied by the probability.

    That is, [(risk = scale of damage × frequency of occurrence)].

    The easiest example to understand is the risks of transportation such as aircraft. The risk of death due to an aircraft crash, for example, is 0.2 deaths per million persons per year, which means that a passenger on an aircraft may be killed by accident with a probability of 2×10^{-7} year^{-1}. It can be seen that this is considerably smaller than 1.1×10^{-4} year^{-1} due to a car accident. On the other hand, the risk of death from lung cancer caused by tobacco smoking is 3.82×10^{-4} year^{-1}, and the risk of being forced into suicide is 2.5×10^{-4} year^{-1}, It turns out that both of which are quite higher than the risk of death from an aircraft accident. Thus, the risks can be compared quantitatively.

3. Management of various risks

    To understand the risks, you first need to see what the benefits are. In the case of illness, there is a risk of lung cancer corresponding to the benefits of smoking, such as pneumonia mentioned above. On the other hand, liver disease caused by excessive drinking, pneumonia living in an urban city due to air pollution, and side effects due to various medications, etc., are cases in which they put themselves at risk to obtain benefits.

    Regarding the use of transportation, few people will be aware of the risk even if they are aware of the benefits. Although the risks of airplanes and cars have been shown above, there are risks to driving and walking, as well as using railroads. Benefits are convenience, and they buy and use their time and comfort to find out how quickly they can reach their destination, but they are actually taking risks. For reference, the risk of death from a railway accident is the probability of 2×10^{-7} year^{-1}, 1.4×10^{-5} year^{-1} for motorcycles, 0.84×10^{-5} year^{-1} for bicycles, and 3.4×10^{-5} year^{-1} for walking. The risk of walking is unexpectedly high. When going out, it is safer to use aircraft or railway.

On the other hand, the companies' activities have more than 50 different risks when counted, and it is important to constantly evaluate those risks and to decide which risks and how to take them. A typical example of a company's risk aversion activities is to manage business risks by analyzing and responding to demand for summer beer and winter whiskey in the production of beer and whiskey. In order to reduce the risks and obtain the benefits, companies must always consider the management of this return on investment.

The Fukushima Daiichi accident provided an opportunity to think about the risk society receives from nuclear power. In society as well as individuals, risks must be taken to gain benefits. In society, choosing the combination of benefits and risks is an important issue, and one must look ahead. This choice is the leader's role and responsibility.

There have been conflicts and many disputes over whether to take environmental protection or take precedence over the economy. Air pollution is invisible. However, the Earth's atmosphere has definitely been polluted. Taking a private car use or public transportation as a society is another heavy choice. In Japan, energy issue is another serious issue, not just only environmental issue. Japan, which imports almost 100% of fossil fuels, is directly linked to the question of whether a socio-economic situation will be realized. What risk measures should we choose to get benefit from a prosperous society?

4. Benefits and risks of nuclear power

It is a universal recognition that nuclear power plants have the advantage of providing stable and affordable electricity. It is also known that greenhouse gas emissions are extremely small, including the cost of manufacturing construction materials. The risk of gaining such benefits has been considered internationally to be the risk of mortality of nearby residents due to a nuclear accident.

However, in the Fukushima Daiichi accident, there were no direct human casualties, but evacuation of residents is still ongoing, and the economic loss, including the issue of land pollution, has become extremely large.

Conventionally, the risk of death due to a nuclear accident per a nuclear power plant has been said to be 10^{-10} year^{-1}, but the Fukushima Daiichi accident raises the problem that it sufficient to use the risk of death as an indicator so far as a measure of the safety of a nuclear power plant.

5. Balancing costs and benefits

Securing energy, like drinking water and treating sewage, is not something that can be achieved without struggling. There is a need to assess the economic impact on society of the increasing costs of importing fossil fuels for energy. In the coming society, it will be necessary to ensure the costs and benefits assessment of energy supply and to decide which type of energy sources are selected.

We as individuals, companies, and society need to take risks and benefit. The important point is how not to make this risk apparent. In other words, how do we avoid risks and threats from becoming reality, and how do we actually pay the costs? On the other hand, there is a risk itself and a virtual cost to be paid for a virtual threat, and energy selection must be determined in consideration of a balance between the total cost for these risks and the obtained benefit.

4.2.1.3 Response to external events such as earthquake and tsunami

In the Great East Japan Earthquake that struck on March 11, 2011, a huge tsunami beyond imagination hit the Fukushima Daiichi Nuclear Power Station, causing a serious nuclear accident. The dangers of nuclear power are not virtual, but we have come to the realities that have surfaced. There are various debates as to why this has happened, but here I would like to reconsider how to ensure safety against earthquakes and tsunamis, the biggest threat to our nuclear power in Japan.

1. Origin of Fukushima Daiichi accident

The Tohoku-Pacific Ocean earthquake was the largest magnitude of M9.0 in the history of seismic observation in Japan and one of the largest earthquakes in the world. The largest earthquake in the world was M9.5, a 1960 Southern Chile earthquake. Many M8 class earthquakes have occurred in Japan in the past, but it was new to memory that the magnitude of the damage indicates that the Hyogoken-Nanbu Earthquake was the M7.3 earthquake occurred directly underneath the ground and the damage to nuclear power plant indicates that the Chuetsu earthquake was M6.8 short-range earthquake. The magnitude indicates the energy of the earthquake at the epicenter. If it differs by "1," the energy generated by the earthquake will differ by about 32 times. Therefore, it can be seen that the Tohoku-Pacific Ocean earthquake was an extremely large ground motion.

Neither the Fukushima Daiichi Nuclear Power Station nor the Onagawa Nuclear Power Station, which is closer to the epicenter, has been found to have damaged any important facilities of the power plant due to the earthquake. The reactor was ready to enter cold shutdown mode.

On the other hand, the tsunami caused by the earthquake was also stipulated from the beginning as an event that must be sufficiently considered at nuclear facilities, but the tsunami that hit the nuclear power plant in the Fukushima area was much larger than expected before. The progress of the severe accident could not be stopped.

Based on the above situation, I would like to consider how to prepare for earthquakes and tsunamis in the future.

2. Preparation for an earthquake

Since ancient times, Japan has been hit by a number of earthquake disasters, and the records of earthquakes have been in ancient document for about 1600 years, beginning in CE 416 recorded in Chronicles of Japan. It has been reported that about 20% of M6 or larger earthquakes occurring in the world occur in Japan, and research and countermeasures for earthquakes have long been a matter of concern. However, the real discussion of earthquake resistance began in Japan since the Great Kanto Earthquake in 1923, the standards were adopted in 1924, and its history is by no means old.

Seismic measures are a top priority for nuclear facilities. In the construction of nuclear power plants, the national seismic design review guidelines, which had been taken up as one of the design requirements, were established in 1981 and the standards for seismic performance of nuclear power plants were clarified. In addition, a review was made based on the Hyogoken-Nanbu Earthquake, and an evaluation was required based on the latest knowledge. In this review, (1) Investigation of faults since the late Pleistocene (dating back to about 130,000 years ago in Japan) (2) Evaluation of standard seismic ground motion by the latest fault model method, (3) Evaluation of vertical motion, and (4) Review and reinforcement of equipment significance classification, (5) Probabilistic safety assessment assuming ground motion exceeding the standard ground motion, etc. were incorporated, and revised in 2006.

In Japan, ensuring the safety of nuclear power plants in an earthquake is one of the most significant issues. The goal of ensuring the safety of nuclear power plants is to prevent the effects of radiation on the surrounding environment and residents. The goal is to confine the radioactivity of a nuclear power plant in the fuel of a nuclear reactor normally, so as not to significantly affect the surrounding area in the event of an accident. First, when an earthquake is detected, seismic design requires that the operating reactors should be shut down safely and promptly, and then the reactors should be cooled and quickly put into a cold shutdown state. In addition, if a situation that exceeds the design state occurs, the occurrence of troubles and nonconformities shall be prevented according to the concept of defense-in-depth, and if it does occur, prevent the development to accidents and minimize the effect of radiation on the environment as much as possible. It is necessary to consider the equipment and operation mechanism to make it smaller.

3. Preparation for tsunami

A tsunami is a natural external force that occurs even more rarely than an earthquake. There are still many unclear points about the type and scale of the damage. What has happened to a coastal nuclear power plant in the event of a huge tsunami has been unprecedented. Although tsunami countermeasures have been taken based on design assumptions without a precedent in the past, it has been pointed out that it is necessary to respond to cases that exceed expectations.

Even in Japan, which has suffered a great deal of damage from tsunamis since ancient times, research on tsunami only showed signs of development in the late 1970s. Various investigations have been conducted so far, and based on the results, the "Safety Design Review Guideline for Light Water Reactor Facilities for Power Generation" was decided by the Nuclear Safety Commission. According to the guidelines, the location of the site was to be based on the safety evaluation for each nuclear facility, and in each case, the design tsunami water level was set by calculating the largest historical tsunami and the tsunami with the greatest impact expected from active faults by numerical calculation, etc., and safety design was performed based on this.

However, the experience of tsunamis exceeding the expectations of the Central Sea of Japan Earthquake Tsunami and the Hokkaido Nansei-Oki Earthquake Tsunami has led to a flexible incorporation of developing technologies and various knowledge to further enhance the safety and reliability of power plants. As it was thought important to proceed, the Japan Society of Civil Engineers worked on systematization of tsunami evaluation technology, and in 2002, Tsunami Evaluation Technology was reported. This report has been compiled as a tsunami evaluation technology in which various technical fields cooperate with earthquakes which are the causes of tsunami occurrence, tsunami numerical calculation technology, hydraulic experiments, coastal disaster prevention, tsunami disaster prevention, and response at power plants.

4. Basic concept of ensuring safety against external events

TEPCO's Kashiwazaki-Kariwa Nuclear Power Station was damaged by the Niigata-ken Chuetsu-oki Earthquake in July 2007. However, the facility did not cause any serious damage, and the necessary safety functions were secured. In the design, there are various margins beyond the requirement to maintain the function for the reference seismic motion, and in reality, it is indicated that the equipment and structure can sufficiently withstand the large seismic motion exceeding the standard seismic motion and maintain the functions to be fulfilled.

It is important to ensure the soundness of the equipment against the standard ground motion in this earthquake, but it is also indicated to be more important to prepare for the case where the standard is exceeded.

Natural phenomena often exceed expectations, as indicated by the height of the tsunami in Fukushima. Although the standard value, the tsunami height in the present example, is set at the value that exceeds what experienced in the past, safety is not jeopardized immediately if the actual value exceeds the standard value. It is necessary to review the evaluation criteria based on the actual occurrence, but more importantly, how to take measures to prevent the occurrence of a serious accident in preparation for a situation exceeding this assumption.

One of the means to evaluate progress leading to a serious accident objectively and quantitatively is to apply risk assessment. The seismic guidelines revised in 2006 state that risks exceeding the standard should be evaluated as "residual risks" and measures to reduce those risks should be derived.

a. Always remember to prepare for a serious accident

The lessons learned from the Fukushima Daiichi Nuclear Power Station accident showed that ensuring the safety of the public, even in the occurrence of a serious potential accident, is an important issue as a safety measure in a nuclear power plant accident. In addition to responding thoroughly to serious accidents, we must never forget to establish a system that can ensure public safety even in the event of a serious accident.

4.2.2 Application of voluntary consensus code system in safety regulations

4.2.2.1 Background

Approximately 450 nuclear power plants are in operation around the world, and their total power generation is as high as about 2.5PWh, but accounts for at most 10% of the world's total power generation. Nevertheless, nuclear power is becoming indispensable to taking measures in global warming. Internationally, large-scale power generation facilities are expected to use nuclear power generation because they recognize that it has become more difficult to effectively respond to global environmental issues with thermal power generation other than nuclear power generation. New construction plans have been pursued in Europe, the United States and Asia, especially China.

On the other hand, in Japan, more than 50 light water nuclear power plants used to be operated, supplying about one-third of the total power generation before the Great East Japan Earthquake. In Japan, because of the small size of the land and the difficulty in establishing a new site location, safe and stable operation of the nuclear power plants (existing reactors) that are under operation, and efficient operations such as aging, advanced fuel, and increased power have been required. In addition, these existing reactors will soon be replaced (renewed), and technological development for ensuring safety during decommissioning of nuclear reactors is necessary. Furthermore, the development of new light water reactors is also expected to maintain competitiveness with other energy sources.

However, all nuclear power plants in Japan were shut down in the Fukushima Daiichi accident in 2011, occurred by a huge tsunami in the 2011 off the Pacific coast of Tohoku Earthquake. Since then, safety measures have been strengthened and efforts have been made to restart the operation, but at 2020 only 9 plants have been approved for restart and 24 plants have been decommissioned.

To ensure the safety of nuclear power plants that will be operated in the future, it is necessary to develop not only new technologies but also systematic organization to utilize them.

In order to promote systematic development, it is necessary to summarize these development needs in the form of a technology strategy map and roadmap. In other words, it is necessary to sort out concrete measures and clarify the goals and timing of their achievement, and the roles of the implementing bodies. In addition, in order to advance this activity, it is important that the safety regulatory organization and the technical development organization always collaborate and proceed with the development of codes and standards.

4.2.2.2 Performance-based regulation of regulatory standards and role of voluntary consensus code system

On nuclear safety regulations in Japan, performance-based regulation was implemented. This is a good opportunity to promote the use of voluntary consensus code system in regulations, and it is necessary to clarify the relationship with the regulations and voluntary consensus code system.

Fig. 4.21 shows one example of the concept to use of voluntary consensus code system in regulations. In developing a code system, rapid adoption of new technologies and effective use of resources are required. The necessary response is performance-based regulation, and the government endorses the voluntary consensus code system as a concrete example of the technical standards of the country where the performance is regulated (conduct technical evaluation and permit actual application). By utilizing voluntary consensus code system, it makes possible to

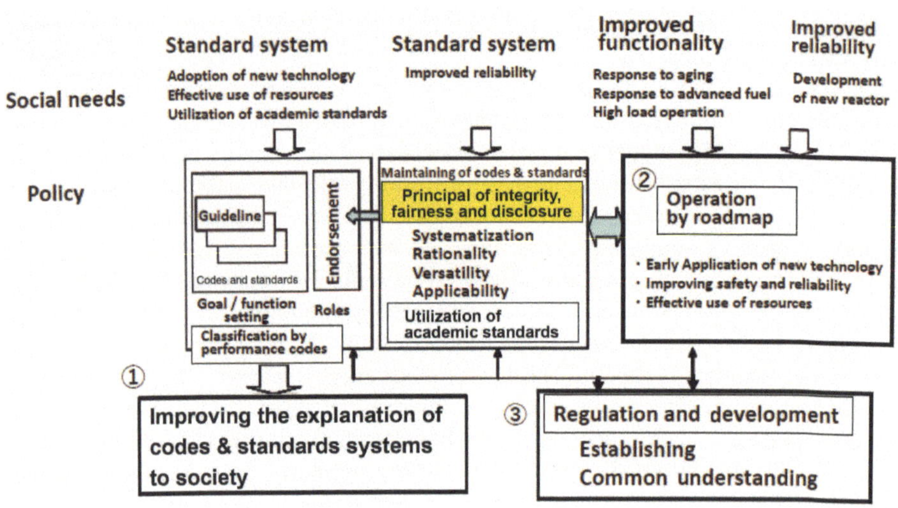

Figure 4.21 Role of academic standards.

improve accountability to the public. In order for the voluntary consensus code system to get the trust of the people, it is important that they are formulated on the principle of "integrity (justice)," "fairness," and "public disclosure."

In addition, the voluntary consensus code system has a role of promoting and revitalizing industry and improving international competitiveness in addition to the most important role of protecting the safety and health of the people. The main purpose of using voluntary consensus code system for nuclear regulation is to ensure safety. Here, we discuss how to develop voluntary consensus code system with a focus on ensuring safety.

4.2.2.3 Organization for formulating voluntary consensus code system

The government policy has stated that the performance will be regulated by the standards set by the regulatory organization, and that detailed specifications will be utilized code system developed by academic societies after technical evaluations and endorsement of regulatory organization. Endorsing code system are required that their development activities be conducted by an organization based on the principles of integrity, fairness and disclosure. The Atomic Energy Society of Japan (Standards Committee), the Japan Society of Mechanical Engineers (Power Equipment Standards Committee), and the Japan Electrical Association (Atomic Energy Standards Committee) are organizations that formulate code system that meet these requirements. Here, the code system, formulated by these three academic societies are called "voluntary consensus codes."

The Atomic Energy Society of Japan, based on nuclear safety, which is a characteristic area of nuclear power, sets standards for nuclear physics-related matters, rules for handling cores, nuclear fuels, and risks, guidelines, procedures, and ideas as a standard.

The Japan Society of Mechanical Engineers mainly develops standards and standards in the field of mechanical engineering, such as structural design, materials, and welding technology.

The Japan Electrical Association has been establishing standards in a wide range of fields, partly because it has begun the activities of formulating standards from the earliest days. The main areas are standards related to maintenance, such as monitoring, testing, and inspection, and areas related to operations such as management, classification, quality control, and operation, and specialized technical areas such as fire, lightning, and earthquake resistance.

In addition, many academic societies, such as the Atomic Energy Commission of the Japan Society of Civil Engineers, the Architectural Institute of Japan, the Japan Welding Association, and the Thermal and Nuclear Power Technology Association, share their roles in their respective fields and their department also establishes codes and standards.

The Nuclear Affairs Association Standards Council was established by the three above-mentioned academic societies in order to smoothly and efficiently formulate nuclear society-related standards. Related organizations such as the academic societies and the Nuclear Regulatory Commission, the Agency for Natural Resources and Energy, the Federation of Electric Power Companies, the Japan Electrical Manufacturers' Association, etc. also participate in the coordination and consultation on information exchange and establishment of voluntary consensus codes under the public.

4.2.2.4 Codes and standards system and challenges

1. Systematic development of codes and standards and sharing of roadmap

   Fig. 4.22 shows the concept of basic system (hierarchical structure) of nuclear regulatory standards and the position of voluntary consensus codes there. Voluntary consensus codes that define detailed specifications corresponding to the performance-specified regulatory standards are systematically formulated in the basic hierarchical structure shown in this figure as those that specifically define "acceptable implementation methods" and are maintained.

   The three academic societies have been conducting to publish voluntary consensus codes to ensure the safety of nuclear power in Japan. Based on the business flow from development, planning, construction to operation and decommissioning of light water nuclear power plants, the basic system of voluntary consensus codes, including its relation to regulatory standards, was organized. Although more than 100 required voluntary consensus codes have already been published, it is desirable to efficiently formulate the missing parts in the future.

   By formulating the performance-based regulatory standards and utilizing voluntary consensus codes, it is expected to improve the efficiency of regulatory administration and licensing, and enabling quick and flexible response to new technologies.

   On the other hand, by utilizing voluntary consensus codes, academic societies will play an important role in ensuring nuclear safety. Voluntary consensus codes are

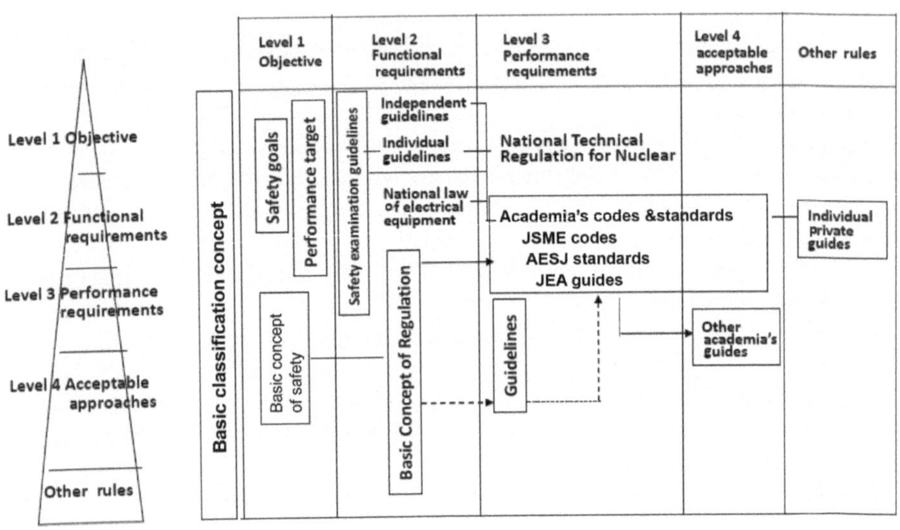

Figure 4.22 Basic concept of regulatory roles and codes and standards.

formulated based on technical grounds, so their formulation activities are supported by research and development to ensure safety and reliability. For such research and development, the support of its stakeholders, industry and government, is important.

In addition, from the standpoint of a regulatory organization responsible for ensuring the safety of the people, the regulatory organization needs to confirm and verify that voluntary consensus codes fully satisfy the regulatory requirements, endorsing voluntary consensus codes. Therefore, it is important to cooperate properly with academic societies.

In this way, in order to systematically and efficiently develop and utilize voluntary consensus codes, it is important in which fields and how industry–government–academia collaboration and research and development should be promoted.

In order to efficiently achieve the goal of systematic development of voluntary consensus codes while effectively using limited human resources, the needs of standards, the industry that provides new technologies, and the academics responsible for standard development associations and countries that undertake safety-based research to utilize the standards and, where appropriate, provide their technical basis, share a roadmap to clarify the regulatory and academic standards development plans. It is important to promote each activity.

In the development of new reactors such as next-generation light water reactors, it is required that the development body, the national government, and the academic society work together to develop cords and standards in parallel from the stage of reactor development.

From the standpoint of formulating voluntary consensus codes, the following shows an idea to share the roles of industry, government, and academia.

The industry has a primary responsibility for ensuring safety, and it is important not only to conduct voluntary safety activities, but also to operate voluntary consensus codes and to have a voluntary management system centered on peer reviews. However, in areas important for safety, it is necessary to cooperate with the government.

On the other hand, the government should take responsibility for confirming the conformity of the voluntary consensus codes to the regulatory standards set by the government from the viewpoint that it has a serious impact on safety and accountability to the public.

As the voluntary consensus codes must be formulated in accordance with the principles of integrity, fairness, and disclosure, they should be led by academic societies, which are institutions of the third party under the support of the government.

Therefore, industry–government–academia stakeholders must maintain a roadmap, understand each other's activities and procedures, and carry out activities such as research and development (R&D) and standard formulation to ensure the achievement of results according to the roadmap.

2. Issues in formulating codes and standards

In the previous section, we described the necessity of systematic development of voluntary consensus codes and the necessary sharing of the roadmap by related organizations. It is necessary to work on a solution to achieve the goal under the cooperation of industry–government–academia because there are a lot of issues as shown below.

[Issues directly related to systematic development of voluntary consensus codes]

a. Establishment of a standard system for safety regulations (Thorough performance regulation)

It is fundamental process that public and private sectors work together to ensure nuclear safety. It is necessary to clarify and review the entire safety regulatory standard system related to safety reviews, clarify the roles of the government, academic societies, and industry and to continue efforts to build a rational system based on consensus.

b. Participation of academic societies in planning, promotion and evaluation of R&D activities related to the formulation of codes and standards

It is desirable to establish a rational system in which academic societies consistently take the initiative in reviewing all functions of R&D activities related to the formulation of codes and standards, etc. from planning to promotion and evaluation.

c. Establish a system to reflect R&D results in codes and standards, etc.

From the viewpoint of the effective use of Japan's funding and human resources, the government, academic societies, and the industry, based on the common recognition of the mechanism for reflecting the results of R&D on safety in the field of nuclear energy in voluntary consensus codes, it is hoped that a system to work consistently will be established.

d. Building a knowledge base (fundamental database) for codes and standards

For nuclear technology, the accumulation and maintenance of technical capabilities is important. In the situation where the development and maintenance of codes and standards are entrusted to academic societies in accordance with the performance-based regulatory standards, it is necessary to ensure that detailed basis data is stored, as well as the basis for judgments and records of the development process. They must be stored, compiled into a database, and kept available. This database should be established in cooperation with the government, academic societies and industry. The mechanism for its construction must be established.

[Issues related to certification system and management foundation of academic society]

e. Develop a qualification certification system for code engineers, etc. and review the inspection–examination system

Along with the specification of performance-based regulatory standards, detailed specification rules have been left to voluntary consensus codes. Therefore, in order to confirm conformity with voluntary consensus codes, a system for accreditation and certification in industry must be established in cooperation with the government. At the same time, in order to secure and train engineers who formulate and operate codes and standards, discussions must be made with a view to establishing a qualification system as a code engineer.

f. Strengthen cooperation with overseas organizations in formulating codes and standards

The need for energy use of nuclear technology will intensify in the future and will be more widely used in the international community. Therefore, in formulating future codes and standards, it is necessary to strategically consider the relationship with the international community. In that sense, cooperation with ISO and ASME becomes increasingly important. It is necessary to promote international standardization by coordinating with ASME standards departments and ISO in the field of structural strength and related certification.

Ensuring the safety of nuclear power in Asia is a common issue not only in Japan but also throughout Asia and the world. From that perspective, internationalization is of great significance.

g. Strengthening the management base of the organization that formulates codes and standards

Regulatory standards for nuclear power have been performance-based, and the role of academic societies in formulating codes and standards has grown. At academic societies, especially the Atomic Energy Society of Japan and the Japan Society of Mechanical Engineers, the activities of formulating codes and standards are supported by the voluntary activities of individuals belonging to their respective societies. In order to proceed to the next stage, it is necessary to raise the public awareness of the standard-setting activities of academic societies and, at the same time, strengthen the business foundation of the academic societies.

4.2.2.5 Future issues in standard formulation activities

Focusing on the activities of the three academic societies, the Nuclear Affairs Association Standards Council is examining the basic approach to setting plans for codes and standards. The activities of academic societies related to safety regulations to formulate codes and standards should basically be carried out by sharing the roles under the equal cooperation of industry, government and academia.

The main issues to be sorted out in the future are:

1. Systematic development of voluntary consensus codes associated with the performance-based regulatory standards.
2. A mechanism to promote the safety basic research necessary for the development and maintenance of voluntary consensus codes in cooperation with the government, academic societies, and industry and to reflect them in the standards.
3. Sharing the roadmap as a means of coordination between the government, academic societies and industry.

4.2.3 Review of safety assurance activities in the countermeasures of the Fukushima Daiichi accident

4.2.3.1 Causes and lessons learned from the accident

The tsunami that occurred on March 11, 2011, following the magnitude 9.0 Great East Japan Earthquake (Pacific Ocean Tohoku Earthquake), made it impossible to stably cool the reactors at the Fukushima Daiichi Nuclear Power Station. A core damage accident (hereinafter, severe accident) occurred. In addition, hydrogen explosions were induced in the reactor buildings of Units 1, 3, and 4. As a result, large quantities of radioactive material were released into the surroundings, forcing many residents to evacuate for long periods. This greatly undermined public confidence in nuclear power.

In the various reports on the Fukushima Daiichi accident, one of the major factors behind the accident was that if serious measures were taken to implement a periodic safety review and take precautionary measures, the results would be significantly different from the current situation. The necessity and significance of continuous improvement for nuclear safety are given as lessons and recommendations.

In the light of the lessons learned and recommendations from these reports, the government revised the Reactor Regulation Law in September 2012 and (1) to introduce a system that reflects the latest technical knowledge on existing facilities and operations. (2) Rules were added to limit the operating period to 40 years in principle, (3) and to clarify the responsibility of utilities to improve their own safety, and (4) to strengthen measures against severe accidents. The Agency for Natural Resources and Energy's "Working Group on Voluntary Nuclear Safety Improvement" issued a report in May 2014, (1) implemented risk management under an appropriate risk governance framework, and (2) suggested actions (exhaustive risk assessment, enhancement of defense-in-depth), etc. that require practical use starting from the lessons learned from the Fukushima Daiichi accident. In response, the Agency for Natural Resources and Energy's "Voluntary Safety Improvement, Technology and Human Resources Working Group" presented issues from a national perspective, and the Atomic Energy Society of Japan's Special Expert Committee on Advanced Technology for Safety Measures shared the knowledge of the Society through a catch-ball that brings together and presented comprehensive solutions, the "2014 LWR Safety Technology and Human Resources Roadmap Final Report" that was compiled and prioritized based on the importance of elemental issues and expanded to the roadmap in May 2015.

In efforts to ensure the safety of nuclear power plants, it is necessary to adopt safety measures that can be implemented rationally and to continuously improve safety to reduce risks. Existing nuclear power plants may not be able to accommodate all of the new knowledge and design criteria that have emerged since their operation. On the other hand, equipment design is not immediately dangerous because safety margins are included. It is important to take into account the new knowledge and the difference between the new standards and the current status of the existing plant, and to implement additional safety measures in line with the purpose.

Decisions on the adoption of these safety measures, whether voluntary or mandatory, take into account deterministic and probabilistic evaluations, as well as good engineering practices, operational performance and control measures. It is important to make consistent and balanced choices. Specific steps include clarifying issues, clarifying judgment criteria, assessing the significance of risk, and then proceeding with detailed assessment, effectiveness assessment, agreement among stakeholders, and prioritization of safety improvement measures. However, there is a case where judgment can be made only by deterministic consideration without performing a detailed evaluation of risk importance, so that it is not always necessary to take all steps.

In order to perform a detailed risk significance assessment for each plant in the above steps, quantitative input is preferable if possible, but in many cases, qualitative input must be used. In order to comprehensively examine all factors of qualitative input and qualitative input, it is necessary to understand and examine each boundary condition, assumption, uncertainty, and the like. Therefore, it is desirable to form a study team with many experts. These review processes must be logical, comprehensive, transparent, and documented, reproducible, and verifiable to allow for subsequent verification.

These examinations do not proceed in a uniform procedure, but will involve the coordination of stakeholders and the search for the optimal solution in a combination of various factors. Therefore, there is no specific procedure from an international perspective.

On the other hand, in Japan, there has been no backfit rule, and the improvement of safety measures has been examined by administrative guidance based on determinism such as defense-in-depth protection and safety margin. After Fukushima Daiichi accident, the backfit regulation was introduced by the revision of the Reactor Regulation Law in September 2012, but the current provisions stipulate the maintenance standards in Article 43-34 of the Act and establish Article 43 of the Act. It only stipulates suspension of use in the event of violating the rules of Article 2–33, and no specific rules or guidelines have been enacted. Then, in June 2014, the Nuclear Regulatory Commission deliberated and approved the concept of basic measures to be taken if new safety-related knowledge was obtained as part of the policy on "vulnerability in power system design." However, it is not clear how to reflect the new knowledge, the place for discussions to certify the new knowledge has not been clarified, and the concept, procedure, and criteria for determining whether interpretation needs to be revised are not shown. Also, the way of thinking when considering the grace period has not been clarified.

4.2.3.2 Initiatives for voluntary safety improvement of utilities and backfit requirements of the government

1. The situation of backfit in Japan

If new safety-related knowledge is gained in efforts to ensure the safety of nuclear power plants, it is clear that safety must be maintained as high as reasonably practicable in accordance with the as low as reasonably practical principles. However, there are various difficult aspects to grasp this in legal theory.

In September 2013, the Reactor Regulation Law was revised and the backfit provision was introduced in Japan, but before that, safety was confirmed by back-check. The purpose of the back-check is to reflect retroactively the contents of the technical standards and guidelines after they have been established or revised, such as ministerial ordinances. In response to a request from the government, the utility carried out reevaluation-confirmation work, etc., and the government received the report from the utility, and if there were any problems, consulted with the utility to determine countermeasures. For example, in the case of changes in seismic standards, safety has been confirmed by evaluating whether or not the change in standards satisfies structural strength standards. Basically, ensuring safety by regulatory standards is minimum security measure, and it is the same idea at the time that higher level of safety was ensured through the voluntary activities of utilities. However, the conventional method of back-checking has been a problem for both utilities and the government. In other words, for a utility, the safety level after taking safety improvement measures based on the reevaluation-confirmation is not legally determined, and therefore it is difficult to explain to the general public with the legal basis for the safety of its own plant. On the other hand, there was a problem for the government that safety could not be improved without the cooperation of the utilities.

In response to this, the Japanese regulatory authority issued the new regulatory rules, such as an order to conform to the technical standards and guidelines after enactment and revision in the term backfit and an order to take necessary measures to comply with them. In many cases, it means that existing nuclear power plants, etc., conform to the contents of technical standards and guidelines after they have been established and revised. The purpose of the backfit introduced in the revised Reactor Regulation Law is also to that effect. In this case, the government can strengthen safety measures regardless of the intentions of the utilities, but there are still difficult problems such as retrospective application of laws and regulations, security of property rights, and improvement of compensation system. For example, in Japan, the Building Standards Law has strengthened the seismic and fire resistance standards, but clearly defines the scope and conditions for retroactive application.

In connection with the backfit, the Supreme Court ruling in the Ikata Nuclear Power Station stipulates that the hearing in the action to revoke the license to set up a reactor will be conducted "in light of the current scientific standards." Some measures are needed to bridge the gap between the level of safety required to be complied with at the start of operation, and the level of safety required to be complied with when it becomes more severe due to the discovery of new knowledge.

4.2.3.3 The situation of backfit in overseas

In the United States and Europe, lessons learned from operating experience have been reflected in various laws and regulations, plant operating procedures and management plans (training, security, quality assurance and emergency planning). They are not only applied to new reactors, but are also steadily improving the safety of nuclear power plants by actively backfitting existing reactors.

While the term backfit means addressing new safety issues, different countries have different regulatory approaches. In the United States, backfit implies new or changed regulatory requirements after regulatory and utility discussions. They change the operating conditions of the plant, the plant's Structure, system and component (SSC), management plans or procedures for plant operation, the organization for plant operation, and qualifications or training. In Europe, on the other hand, regulations only set goals and existing rules will not be revised, but backfit will be settled through regulatory and business discussions.

The reasons for the need of a backfit by the government are: (1) maintaining plant safety at the required level; (2) requiring compliance with existing regulations; (3) taking action with new knowledge or analysis and requiring practical implementation in such a case that the necessity and the way to substantially improve safety is found. Of course, utilities often backfit themselves. Some utilities make continuous improvement a basic policy. In other words, the backfit is based on the independence of the utility to improve safety. This policy of continuous improvement revolves around periodic self-assessment of safety indicators, and involves corrective action management plans, assessments of operating experience, and comparisons with nuclear industry best practices.

4.2.3.4 Comparison of backfit and voluntary safety improvement activities

Based on the above, the meaning of backfit in Japan is understood as follows with reference to the US NRC definition.

[Backfit]

Backfitting means: (1) change or addition of SSC (system, structure, component: system, structure, equipment); (2) change or addition of license documents; Or add. In both cases, the rules and regulations of the government or the provisions of the by-laws and guides related to them are newly created or changed based on the consideration of risk significance.

However, the provisions of the by-laws and guides shall not be excluded if it is judged that the same or higher safety can be secured.

The backfit is stipulated in the revised Reactor Regulation Law, Article 43-3-14 (Maintenance of Facilities.

[Continuous voluntary safety improvement of utilities]

The voluntary safety improvement of the utility means (1) maintaining the safety of the plant at a higher level by setting voluntary numerical targets under the policy of the utility's continuous improvement, (2) the latest regulatory standards (3) When new findings or analysis indicate that it is necessary to take countermeasures, measures that can be implemented in practice and that substantially improve safety are implemented in both hardware and software.

In addition, the revised Reactor Standards Act: Article 43-3-29 is required to submit a safety improvement evaluation report, in which the effects of the voluntary safety improvement measures of the utility will be evaluated.

4.2.4 Summary

The purpose of establishing voluntary consensus codes is to contribute to the prosperity of society by applying technologies developed and pioneered based on disciplines to society, and to avoid the occurrence of accidents and problems in applying the technologies then safety is to be ensured. In other words, activities are underway to establish codes and standards to contribute to the prosperity and safety of society. The significance of establishing voluntary consensus codes is summarized in the following three points.

1. Consolidation of technologies: To systematize technologies developed by people in various fields of industry and academia so that anyone can use them without mistake.
2. Optimization of cost: To enable anyone to make products that meet demands with the same appropriate margin.
3. Ensuring safety: To ensure the safety of its users and those around them at any time.

In other words, this standardization is expected to further improve technical capabilities, simplify the industrial structure, streamline economic activities, and thereby secure international competitiveness. It is thought that it can contribute to securing public safety.

References

[1] Thermal and Nuclear Power Engineering Society of Japan, Thermal and Nuclear Power Journal, Relevant Regulations and Application for Thermal Power Station. April 2017: Site Selection to Operation Start.
[2] MHI Review, 38 (2) (2001-3); Super Large Modular Construction of 700MW Coal Fired Boiler.
[3] Turbomachinery Society of Japan, Journal March 2019; Masahiro OTA, Recent Trend of International Standard for Steam Turbine.

Issues in power generation and future prospects

5

Takao Nakagaki
Waseda University, Shinjuku, Tokyo, Japan

5.1 Global effort toward common issues

The 2030 Agenda for Sustainable Development and its 17 goals were adopted at the UN Summit in September 2015. Many of the 17 goals relate with energy and environment issues, and the government of Japan (GOJ) declared to contribute to implement the Agenda particularly for climate change, energy, and sustainable consumption and production. The Paris Agreement was adopted at the COP21 in December 2015 and quickly entered into force on November 2016 approved by more than 55 countries accounting in total for at least an estimated 55% of the total global greenhouse gas (GHG) emissions. It aims to strengthen the global cooperation to tackle with the climate change by keeping a global temperature rise "well below 2°C" above preindustrial levels and to pursue efforts to limit the temperature rise even further to 1.5°C. It requires all countries including Japan to put forward their best efforts through nationally determined contributions (NDCs), and also requires all countries to assess the collective progress toward and to inform further individual actions as a global stocktake every 5 years (Pledge and Review). Japan's NDC declared 26% reduction of GHG compared to total GHG emissions in 2013 by 2030.

5.2 Domestic policy statements and megatrends of energy system

In July 2018, the 5th Strategic Energy Plan endorsed by the GOJ's cabinet confirmed the basic viewpoint of the energy policy of "3E + S" that is to first and foremost ensure stable supply (Energy Security), and realize low-cost energy supply by enhancing its efficiency (Economic Efficiency) on the premise of safety. It is also important to make maximum efforts to pursue environment suitability (Environment) and to build a multilayered and diversified flexible energy supply and demand structure. On top of that, renewable energy was clearly positioned as a main power source that play a key role of Japan's energy supply. In particular, solar power using photovoltaic panels and wind turbines are expected to be installed more than 64 and 10 GW by 2030, respectively, and have been accelerated by feed in tariff policy after the power shortage caused by the great east Japan earthquake. The Agency for

Fundamentals of Thermal and Nuclear Power Generation. DOI: https://doi.org/10.1016/B978-0-12-820733-8.00005-4

Natural Resource and Energy showed long-term prospect of supply and demand of energy toward 2030, so-called "energy mix." Variable renewable energies (VREs) such as solar and wind power, which are not able to control power generation in accordance with varying demand by themselves, are expected to increase to 22%–24% of annual power generation in 2030 (Fig. 5.1).

A longer-term prospect of energy system toward and beyond 2050 is included in the 5th Strategic Energy Plan. As for the Paris agreement, "Japan will aim at the long-term goal of reducing greenhouse gas emissions by 80% by 2050 while reconciling global warming countermeasures and economic growth." However, "such a large-scale emissions reduction is difficult to do by merely continuing existing efforts," and therefore, "Japan will pursue to the maximum solutions through innovation including the development and adoption of innovative technology that makes thoroughgoing reduction of emissions possible, encourage domestic investment, strengthen international competitivity, and aim at large-scale emissions reduction through long-term, strategic initiatives and contribute to worldwide reduction."

Summarizing the aforementioned international and domestic policy statements, global megatrends of energy system prospecting toward and beyond 2050 are commonly abstracted in the following three terms.

Decarbonization: Primary energy sources should be decarbonized (no use of fossil fuel).

Decentralization: Energy supply system will shift from a centralized and large-scale system to a distributed and small-scale system, especially led by residential and commercial sectors.

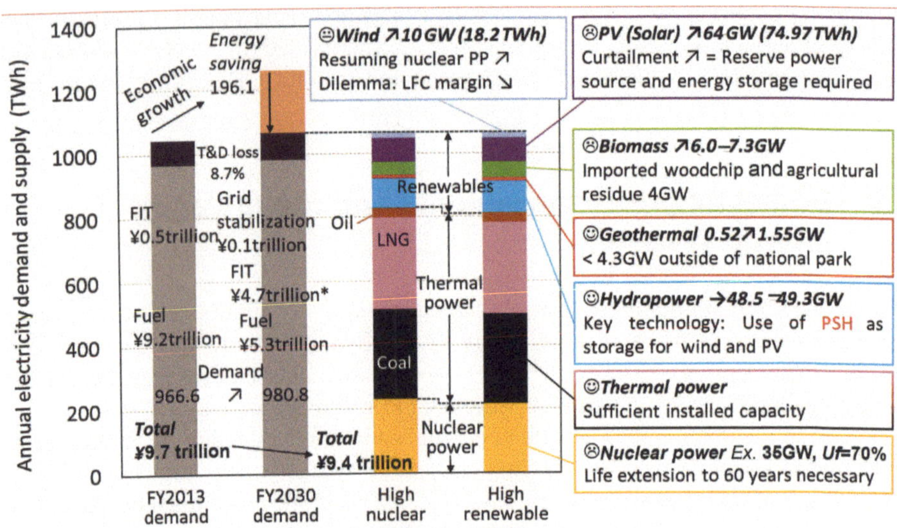

Submitted to 8th Long-term Prospect of Supply and Demand of Energy Subcommittee on 28 April 2015

*CRIEPI report Y16507 (2017)

Figure 5.1 Japan's "Energy mix" in 2030 [1].

Electrification: A large part of utilities, including transportation and industrial sectors, will be electrified concurrently with penetrating decarbonized electric power sources.

5.3 Domestic energy system outlook

Energy plan is a key pillar in the national strategy for all countries. The global megatrends are strategically localized according to specific situations, circumstances, and constrains in Japan. Energy and material flow must be uninterrupted to continue economic and social activities, allowing all fluctuation in various time ranges such as hourly, weekly, monthly, and temporal cataclysm in case of unexpected natural disasters. Universal service of energy should be ensured by distribution networks to far and wide of all islands for electric power and logistics of fuel. Utilities and amenities in contemporary civilization should be sustained with proper management of finite natural resources and waste considering intergenerational equity. Considering the above fundamentals, what to essentially ensure on the energy policy of "3E + S" ultimately result in "Sustainability," "Economical affordability," and "Stable accessibility." Japan is made up of many big and small islands in the distant ocean. On top of that, due to difficult geopolitical situations with east Asian countries, international power grid interconnection is hardly feasible and therefore, self-reliant, stable, and dispatchable power sources fueled by long-term stockpiles are indispensable for Japan's power system. More than 120 million people mostly live in many densely populated cities and use 1000 TWh annual electricity consumption to produce the third gross domestic product (GDP) in the world. Two-thirds of the annual electricity consumption in Japan is shared in the central area of Japan, specifically Tokyo, Chubu, and Kansai areas, and 40% in these areas generated electricity is supplied for the industry sector, most of which needs stable electric power (Fig. 5.2). Japan is long from north to south and inconveniently has a high VRE potential such as a preferable area of large-scale wind farm or mega solar power is unevenly distributed especially on Hokkaido and Kyushu islands, which are the north and south ends of the country far from the center of Japan. Pumped storage hydropower (PSH) stations, which can contribute to local production for local consumption (LPLC), are installed only 0.8 and 2.35 GW in Hokkaido and Kyushu area, where peak electric power loads reach 5.5 and 13 GW, respectively. Wide area transmission to the high-demand central area is restricted by capacities of the high voltage DC tie line interconnected between Hokkaido and Honshu islands (0.9 GW, overhead and submarine cable) and frequency converter stations between 50/60 Hz (3 GW by 2027), which is an unchangeable and inconvenient legacy in Japan (Fig. 5.3). The inherent problem for high VRE penetration is how to manage spatial and temporal mismatch of demand and supply and flexibility and energy storage are keywords of essential solutions.

Summarizing the aforementioned situations, each role and challenge of nuclear and thermal power generation is as follows.

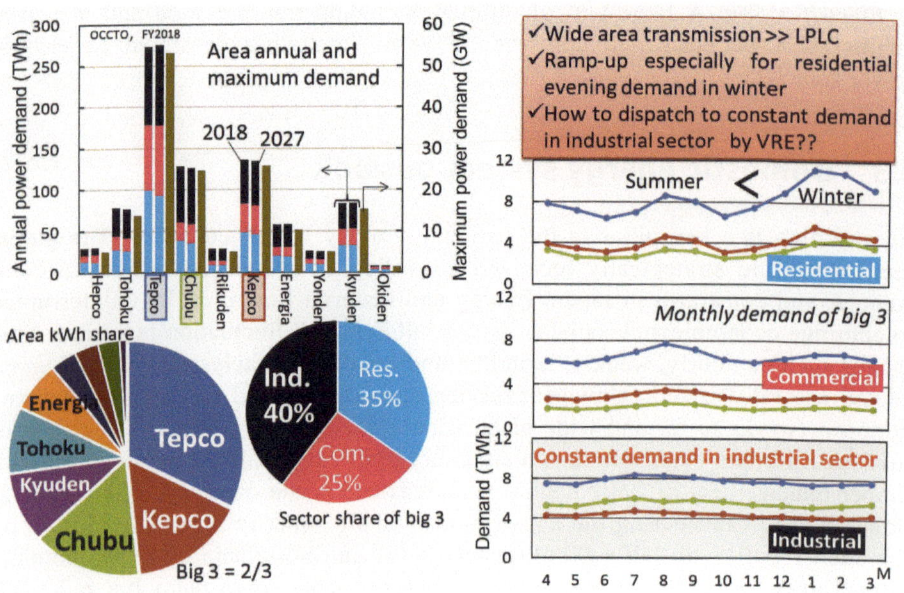

Figure 5.2 Concentrated population and industrial demand [1].

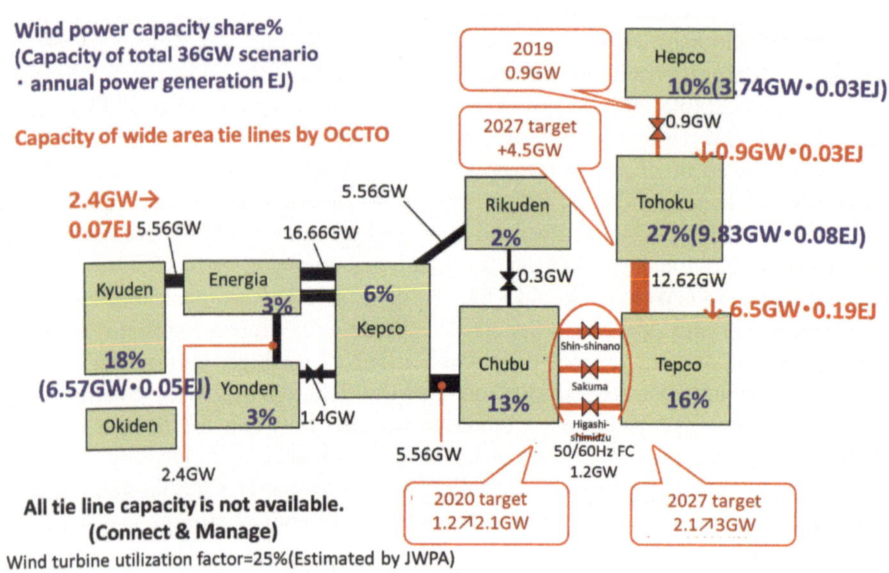

Figure 5.3 Wide area grid system topology: East 50 Hz, West 60 Hz [1].

5.3.1 Nuclear power

5.3.1.1 Role

Stable, reliable, low-carbon, and immediate (unknown term) baseload power source corresponding to constant demands of industry sector.

5.3.1.2 Challenge

Constant partial-load operation to ensure capacity of downward spinning reserve as well as offsetting regular inspection. Collaborative operation with PSH stations, large-scale battery stations, and on-grid power to gas stations in order to absorb and manage the fluctuations and excess generation of VRE with minimizing curtailment and demand generation in the whole power grid system.

5.3.2 Thermal power

5.3.2.1 Role

Flexible, dispatchable power source as primary, secondary, and ternary spinning reserves. Contributing to stabilize grid frequency provided by mainly inherent rotational inertia like governor free or load frequency control operations.

5.3.2.2 Challenge

Higher flexibility such as higher ramp rates, quick start and shutdown operation, and larger capacity of downward spinning reserve minimizing efficiency drop due to partial-load operation, decarbonization such as cofiring biomass, CO_2-free hydrogen, or carbon capture utilization and storage.

The trend in CO_2 emissions can be analyzed using a decomposition method called "Kaya identity." The Kaya identity is useful for separating the effects from changes in GDP, energy use per unit of GDP, and CO_2 emissions per unit of energy which depends on the share of fossil fuels in the total energy mix and also on the carbon intensity of the fossil fuel mix shared by coal, oil, and natural gas. CO_2 emissions per unit of GDP is defined by combination of energy use per unit of GDP and CO_2 emissions per unit of energy. Even in a phase of increasing GDP, a constant or downward trend of the CO_2 emissions per unit of GDP implies decoupling between economic growth and CO_2 emissions, which is a preferable tendency to the future energy society. According to published literatures and materials, the decoupling has not necessarily been a solid trend yet for a long time, while a few signs for the decoupling have been found in residential, commercial, and transportation sectors. Since electrification and decarbonization of electric energy sources concurrently proceed as the megatrends mentioned earlier, electricity demand should not be forcibly suppressed, while continuous energy saving in all sectors is necessary no matter how primary energy is supplied. The change in energy and societal systems by 2030 can be incrementally forecasted based on the Japan's NDC of the Paris Agreement, but inconveniently, the big picture of 80% GHG emissions reduction by 2050 cannot be envisioned by forecasting change. In other words, there is still a large discrepancy between forecasting to 2050 and backcasting from 2050, and to compensate the gap, "disruptive innovations" are uncertainly expected in future energy and societal systems. However, nobody knows what the disruptive innovations substantially are, and when and how they come out at this time.

Even if the future energy and societal systems head toward an ultimate target that is "net zero emissions" based on the long-term strategy with a disruptive innovation in some situations, "Just Transition" should be required on the way to the goal. In the context of the just transition and job security, sustaining manufacturing industries, which has been a great advantage to Japan in the world trading, is one of political options to keep not only economic growth with robust domestic demand but also a trading equilibrium with indispensable import of natural resources. In the manufacturing industries of which CO_2 emissions per unit of value-added is mostly higher than the other industries, made-in-Japan high value-added materials or intermediate materials remain strong in world trading, while assembly industries such as electric appliances and electronic devices have decrease international competitiveness over the past decades. Even under such a situation, the manufacturing industries mostly have lost GDP share by outward shift due to high labor costs, changes in the market, and an unpromising increase in self-sufficiency of natural resources. Energy demand in industry sectors may greatly be changed by what grand design Japan aims in future. Even in the mega trend of technological change toward the future energy and societal systems, intervening "Bridging Technologies" that maximumly use societal existing assets such as infrastructures of energy system maintained by appropriate remaining life assessment and renewal technologies is a reasonable pathway from the social and economic viewpoint. Zooming out to the international viewpoint, external pressure to GHG emission reduction is expected to increase more and more hereafter. Even if Japan's NDC by 2050 is accomplished, it contributes only 1 Gt/y GHG emission reduction for more than 50 Gt/y in the world. Widely spreading Japan's high-level energy saving technologies can effectively reduce more and more GHG emission around the world and to comply with it, continuous research and development of such technologies as well as human resource development are necessary, which should be financially supported by governmental agencies and private investors.

5.4 Technical options for high variable renewable energy penetration and their complemental, cooperative, or competitive relationships

5.4.1 Overview

Since actual energy system is a complex "system of systems," requirement for each subsystem cannot be singly determined. The complex future energy system should be designed carefully considering complemental, cooperative, or competitive relationships, or their constraints among subsystems. Extended available technical options on electric power grid are main countermeasures for 22%–24% annual power generation share of nondispatchable VRE in 2030. Improving prediction accuracy on electric power demand and supply by using applied informatics technologies, especially on VRE power generation which is strongly influenced by local and occasional climate condition, is generally welcomed by all transmission system operators (TSO) without any negative effects on other technical options. The technology for improving prediction accuracy on VRE power generation has been developed and demonstrated in many research projects. Many grid and power generation side technologies on hardware and software are available to stabilize electric power grid with mutually complemental, cooperative, or competitive relationships. For example, there are wide area electric power transmission using interconnection tie lines as a grid side technology, operability improvement of thermal, and hydropower generation as a power generation side technology, and active demand control technologies such as virtual power plant, negawatt power, and demand response. On top of that, on-grid energy storage technologies such as PSH, compressed air energy storage, batteries, and power to gas storage systems are applicable to the wide range of power rating and discharge time by selecting an appropriate technology (Fig. 5.4). A portfolio of energy storage technologies will be determined considering limitation of on-grid VRE capacity supported by competitive countermeasures at affordable costs.

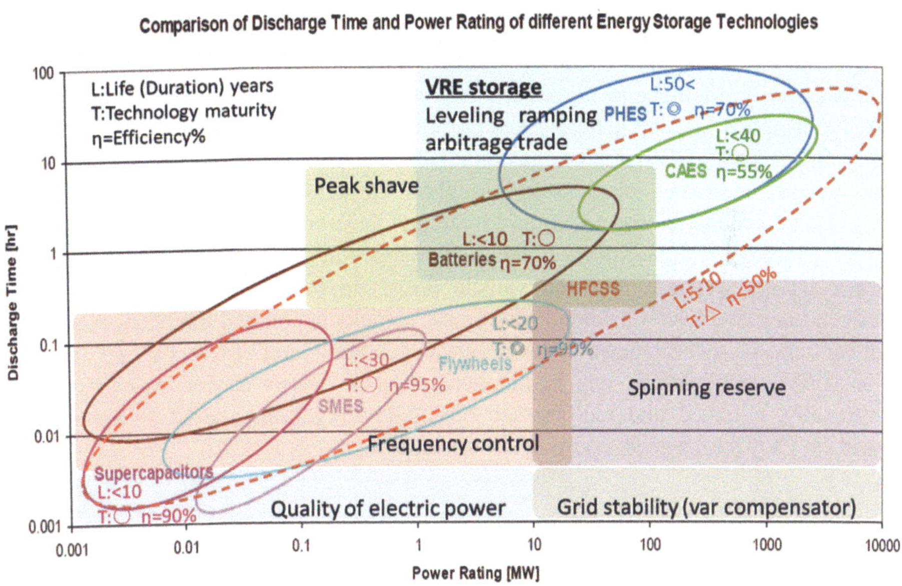

Figure 5.4 Applicability of energy storage technologies [2].
Source: *Modfied from* stoRED Project Report 2.1 Report summarizing the current Status, Role and Costs of Energy Storage Technologies.

5.4.2 Distributed energy storage technologies

Li-ion, sodium sulfide, and flow batteries have been already commercialized as a mainstream of distributed electricity storage and their costs have declined gradually. Plug-in vehicles such as PHEVs and battery electric vehicle can realize a bidirectional power flow between vehicle and home namely V2H (vehicle to home), which contribute not only ordinary use of demand response but also independent energy station in case of power outage possibly caused by natural disasters. CO_2 heat pumps with hot water tanks have been widely spread in Japan, which are compatible with solar power as an on-grid sensible heat storage devices, while they were formally introduced to raise the level of base demand during nighttime. Nevertheless, the curtailment necessity of excess power generation by VRE will increase with an increase in the installed capacity of VRE. Power to X, which is a general term including conversion from excess electric power to fuels such as hydrogen and its carrier compounds, methane, methanol, and so on, is one of the promising options to absorb excess power generation of VRE. The power to X is attractive as a relatively large-scale distributed energy storage due to its many advantageous features such as versatile usage as fuel for all sectors, transportable, quickly refuellable, high energy density, long-term and stable storage without degradation, and basically unlimited storage capacity in spite of its relatively low conversion efficiency comparing to batteries. The national target cost of commercial hydrogen in 2030 is 20 yen/Nm^3, which is corresponding to 12 yen/kWh-electricity. According to cost prediction in 2030 and 2050 by the literature [3], it is difficult to realize the target cost of imported hydrogen by promising hydrogen carriers such as liquification, ammonia, and Methyl cyclohexane. Domestic production of renewable hydrogen, which is thought to remain high cost ever, possibly reduces levelized cost of hydrogen soon through the increasing utilization of electrolysis cells combined with batteries at a suitable capacity. Methane or liquified hydrocarbon fuels at ambient temperature is still attractive in terms of easy handling and compatibility with existing infrastructures and supply chains such as 260,000 km gas pipelines in total, LNG terminals located near a large demand, and extant gas stations throughout the whole land of Japan. Methanation and methanol production from CO_2 and hydrogen have been already commercialized at a high technology readiness level. Olefins, Fundamental Aromatics (Benzene, Toluene, Xylene; BTXs), aviation fuel and vehicle fuels can be synthesized from CO_2 and hydrogen via these basic C1 chemicals. As such, methanation and methanol production from captured CO_2 and renewable hydrogen, which are a kind of CO_2 capture and utilization (CCU) technologies, can be another option of energy carrier for excess VRE. Additionally, essential and multifaceted discussions regarding the CCU technologies converting CO_2 into organic compounds are summarized as follows.

5.4.2.1 Unattractive cost inherently higher than fossil fuels and even than hydrogen

Actually, cost of power to X fuels is generally higher than that of fossil fuels at this time regardless whether conventional or unconventional energy resources and the situation will not change soon. Technical cost reduction can be incrementally expected according to frequently used experience curves. Even so, it mainly depends on the hydrogen cost. The role of financial key drivers becomes relatively more important to change the situation, which are introduction of carbon pricing policy, environmental, social, and governance investing, and divestment movement for the great CO_2 emitters.

5.4.2.2 Meaningless process converting CO_2 back into fuels in the context of fundamental energy conversion theory, or concern of net CO_2 emitter process from the viewpoint of life cycle assessment

Standard Gibbs free energy of formation of CO_2 is a quite low value of -394.4 kJ/mol. The change in Gibbs free energy (delta-G) in oxidation of hydrocarbon fuels is very large and the equilibrium of it remains on production side even at high temperatures of combustion, which is the principle of fossil-fueled thermal power generation. Inversely, converting CO_2 back into fuels requires ideally equal to the generable exergy, but actually more than that due to irreversible production processes such as electrolysis of water and temperature and pressure swing between operating condition of synthesis reactors and feedstock conditions. Furthermore, all utilities for not only hydrogen production and CO_2 capture, transportation, and temporal storage of them but also the above synthesis process are hopefully derived from CO_2-free energy sources because the whole process delimited with an appropriate boundary must substantially reduce CO_2 emission from the viewpoint of life cycle assessment (LCA) and avoiding carbon leakage. In other words, converting CO_2 back into fuels never exists unless it can contribute to reduce CO_2 emissions into atmosphere.

5.4.2.3 Comparison of direct combustion of hydrogen with combustion of fuel including hydrogen carrier such as ammonia

One of the ultimate pictures of future energy system is that all primary energy come from renewable energy and nuclear energy. It is expected that hydrogen is widely spread as the main energy carrier and net zero emission is achieved. However, it will take a long time. To connect to the ultimate picture of future energy system and present system with feasible technologies and affordable costs, intervening carbon recycling (CCU) as the bridging technology is a reasonable pathway depending on cost. In the short term, carbon recycling organic compounds replaces fossil fuels and thereby CO_2 emission can be reduced, while the substantial sequestration time of CO_2 before reemission is only several weeks at the longest. Net carbon neutral process with complete carbon recycling will possibly remain thereafter in some centralized CO_2 emission sources and even in the case of direct air capture technologies (i.e., capturing CO_2 from atmosphere) depending on monetizing situations. Direct use of renewable derived ammonia as fuel is another promising option especially for centralized power stations with tank truck or ship-based transportation system, while it is difficult to widely distribute via city gas pipelines.

5.4.3 Wide area transmission

The following two sections describe other options competing (cooperative in some cases) with the power to X. If VRE is directly used as on-grid electric power, operation service and equipment for grid stability are provided by a regional TSO in Japan. Increasing capacity of intraregional and interconnection power lines are necessary to mitigate constraints due to the specific topology of wide area power grid in Japan. Inconveniently, a huge amount of stock of electric power infrastructure have ever been piled up, and the maintenance and renewal of the stock will take for several decades. Levelized cost of electricity of VRE is expected to be cheaper hereafter, but cost for underlying support such as delta-kW spinning reserve; GW-class thermal power plants on standby in case of unexpected weather, which depends on costly dispatchable power sources as well as the transmission and distribution cost including reasonable maintenance and renewal, should be charged on the cost of VRE by an appropriate institutional design. It is important to install cost-effective overhead electric power cables considering power flow congestion with installation plan of VRE power stations and also to develop low-cost maintenance and diagnosis technologies using artificial intelligence and robotics.

5.4.4 Maximum utilization of existing pumped storage hydropower

There are 45 sites of PSH stations with 27.6 GW installed capacity in Japan, which account for 11% of total installed capacity of power generation, while European countries have less than 5% share. However, vintage stock over 30 years shares 57% of total PSH stations and the number of PSH stations that need upgrading or renewal of water turbines and generators are expected to increase hereafter. Since most of existing PSH stations were built for load leveling between day and night times especially during summer of peak demand season, variable speed reversible pump-turbines are far less than fixed speed pump-turbines. This legacy situation is hopefully overcome by sophisticated electricity storage for high VRE penetration with step-by-step countermeasures such as group control of fixed speed pump-turbines and downsizing and upgrading to variable speed pump-turbines at renewal timing of each station (Fig. 5.5).

5.4.5 Technical options of CO_2 emission reduction in thermal power generation

Thermal power generation undertakes an important role as dispatchable power sources and spinning reserve in grid stability, but have a big challenge to avoid CO_2 emissions into atmosphere. Triple basic countermeasures which are "less input," "less production," and "less emission" are simply required. Efficiency of thermal power generation theoretically increases with increase in working fluid temperature whether in Brayton cycle or Rankin cycle. Since higher thermal

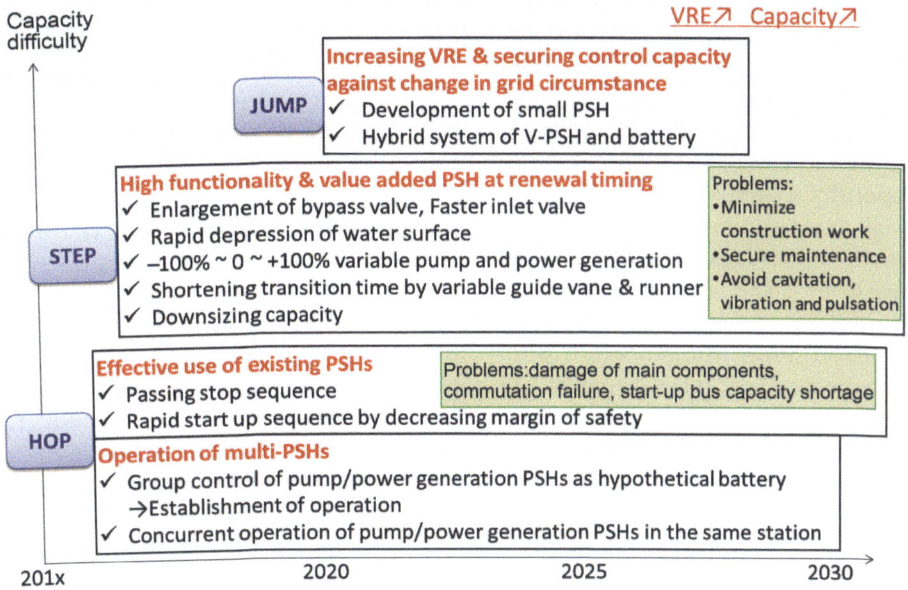

Figure 5.5 Upgrading strategy pumped storage hydropower stations [2].

efficiencies directly result in the "less production" of CO_2, technology development of thermal power generation has ever necessarily aimed for higher turbine inlet temperatures and higher main/reheat steam temperatures. Actually in several decades, thermal efficiency successfully improved from 1100°C to 1600°C in heavy duty gas turbines or from 538°C to over 600°C in ultra-supercritical steam turbines. In coal-fired power generation systems, integrated coal gasification combined cycle can boost up thermal efficiency comparing to conventional boiler and steam turbine system. While only fuel cells are not constrained by increasing operation temperature, incremental approaches ever done before such as overcoming the higher temperature by introducing materials, coatings, and film cooling technologies are reaching a technical plateau. Even if more improved, efficiency gain is relatively smaller than ever. Unfortunately, there is no answer to more than 80% CO_2 emissions reduction in fossil fuel-fired power plants by the "less production" because thermal efficiency is limited by combustion temperature.

Specific CO_2 emissions are different by fuel to fuel, which inherently depends on hydrogen per carbon ratio in the fuel. According to this hydrogen per carbon ratio index, the order is coal < natural gas < hydrogen (infinity) and leads the "less input" of carbon as the fuel. Biomass-firing or its cofiring in the boiler is also possible answer as the "less input" option because of its carbon neutral nature. From the same viewpoint, artificial carbon neutral fuels such as biofuels of which refinement or extraction process driven by CO_2-free electricity and carbon recycling fuels synthesized with captured CO_2 and

CO_2-free hydrogen can provide the same function of "less input." Various CO_2 emission sources from not only thermal power plants in terms of the last countermeasure of "less emission" but also iron and steel making, petrochemical, and cement industries are available to capture CO_2, while the amount of net-avoided CO_2 depends on additional CO_2 emissions due to capture and transportation processes. Although a completely closed carbon recycling system is hopefully established, the same amount of CO_2 equivalent to replaced fresh fossil fuels can be reduced even by single use of the synthesis fuel. There still remain big issues of affordable cost for the "less input" fuels and the "less emission," and the net CO_2 emissions reduction certificated by LCA. Ultimately, underground storage is a solution for the "less emission" giving a large CO_2 reduction potential, while profitable storage sites such as successful enhanced oil recovery projects are hardly found in Japan. Nevertheless, the "less input" and the "less emission" countermeasures will be of growing importance in thermal power generation in Japan because of retrofittable or minimum modification for existing facilities. Those will be only solutions for $>80\%$ reduction.

As mentioned earlier, the main technical challenge for thermal power plants is increasing flexibility to contribute grid stability on the condition of high VRE penetration. Manufactures intensify their thermal power generation products to adapt higher ramp rates, quick start and shutdown operation, and the lower minimum operation load (Table 5.1).

5.5 Future perspectives and risks

The 5th strategic energy plan notes "ambitious multiple track scenario approach," which covers omnidirectional possible decarbonizing technologies to hedge "uncertainty" considering Japan's specific circumstances and constrains. Energy transition toward the sustainable future has just started. Societal transformation will be caused by transition of energy technologies according to the megatrends. Concurrently, energy technologies required for society will change with societal change. Even deep decarbonization by higher VRE penetration and resuming nuclear power plants, net zero emissions cannot be achieved due to not only stockpiled fuels of dispatchable power plants for ternary reserve but also inevitable GHG emission derived from nonenergy such as waste, agriculture, lime calcination, and fresh iron and steel making. Negative emissions technologies (NETs) are a possible solution almighty to cancel any remaining emissions. However, some of NETs such as biomass energy with CCS and afforestation/reforestation additionally require geochemical essentials of a huge area of land use, a huge amount of freshwater and fertilizer, while all NETs should be operated by CO_2-free energy. Land fit for cultivation, fertilizer, and freshwater directly compete with increase in food demand and freshwater usage in the coming 10 billion population era, which namely "Energy, Water, and Land use

Table 5.1 Present status of flexibility of power generation [2].

		BTG			LNGCC	Hydropower (excluding run-of-river)				
Type		D: Drum (350 MW) K: Once-through (700 MW)			M: 1100°C (Single 150 MW) L: 1300°C ~ (Single 350 MW)	Retention basin	Reservoir	PSH		
Category		Oil	LNG	Coal	LNG	–		Output	Pump	V-pump
GF LFC		*** ***	*** ***	*** ***	*** ***	* *	** **	** **	Load dispatch control ×	Load dispatch control **
Control ability		D ** K ***	***	***	Single: M */L** Unit: ***	**	**	**		
Minimum output		D 30% K 15%	D 20% K 15%	30%	Single: M 80%/L 50% Unit: 20%	50%			–	
Ramp rate %/min		D 3% K 5%	D 3% K 5%	D 1% K 3%	M 7% L 10%	< 1 min for change in available range				
Start up time	Cold	D 20–30 h K 30–40 h			12 h	Start 3–5 min Stop 1–2 min			Start 5–10 min Stop 1–2 min	
	Hot	D 3–5 h K 5–10 h			1 h					

***Most suitable, **Suitable, *Applicable, × Poor.

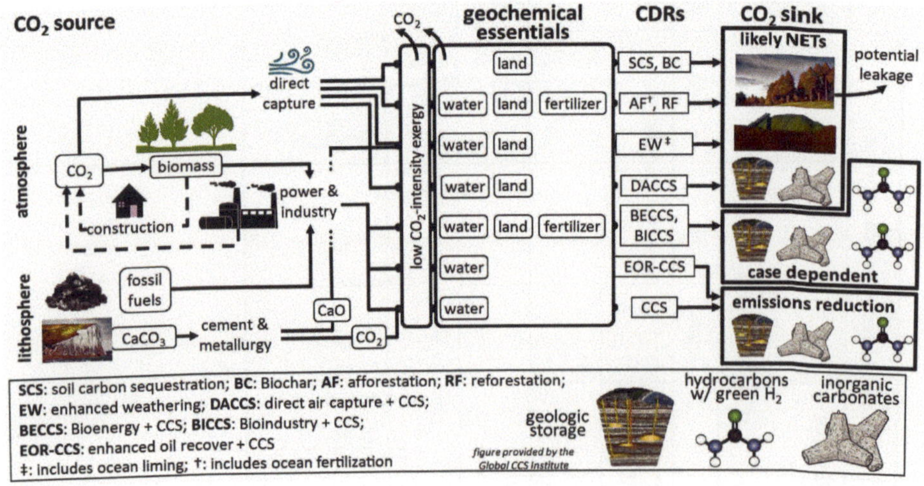

Figure 5.6 Energy, water, and land use nexus concerning negative emission technologies [4].

Nexus," may become another big issue in the near future (Fig. 5.6). Essential to do is controlling GHG emissions into atmosphere with no downgrading of modern civilization and growth in our "inclusive wealth." Although it is hard to find and to reach affordable consensus, we have to solve the problem.

References

[1] T. Nakagaki, Perspectives of low carbon fuels toward future energy systems, Therm. Nucl. Power J. Therm. Nucl. Power Eng. Soc. 70 (10) (2019).

[2] T. Nakagaki, Role of energy storage system for introducing massive variable renewable energy, J. Turbomach. Turbomach. Soc. Jpn. 46 (5) (2018).

[3] Y. Mizuno, Y. Ishimoto, S. Sakai, K. Sakata, Economic analysis on international hydrogen energy career supply chains, J Energy Res 38 (3) (2017) 11–17.

[4] T. Nakagaki, C.A. Myers, Review and technology perspectives for CCS/CCUS, JSME Mag. 122 (1203) (2019).

Index

Note: Page numbers followed by "*f*" and "*t*" refer to figures and tables, respectively.

O

P

Q

T

FSC
www.fsc.org
MIX
Papier | Fördert
gute Waldnutzung
FSC® C083411

Kolibri 360 GmbH
Ferdinand-Jühlke-Straße 7
99095 Erfurt, Deutschland
produktsicherheit@kolibri360.de